AF789585

Springer Proceedings in Physics 59

Springer Proceedings in Physics

Managing Editor: H. K. V. Lotsch

44 *Optical Fiber Sensors*
Editors: H. J. Arditty, J. P. Dakin, and R. Th. Kersten

45 *Computer Simulation Studies in Condensed Matter Physics II: New Directions*
Editors: D. P. Landau, K. K. Mon, and H.-B. Schüttler

46 *Cellular Automata and Modeling of Complex Physical Systems*
Editors: P. Manneville, N. Boccara, G. Y. Vichniac, and R. Bidaux

47 *Number Theory and Physics*
Editors: J.-M. Luck, P. Moussa, and M. Waldschmidt

48 *Many-Atom Interactions in Solids*
Editors: R .M. Nieminen, M. J. Puska, and M. J. Manninen

49 *Ultrafast Phenomena in Spectroscopy*
Editors: E. Klose and B. Wilhelmi

50 *Magnetic Properties of Low-Dimensional Systems II: New Developments*
Editors: L. M. Falicov, F. Mejía-Lira, and J. L. Morán-López

51 *The Physics and Chemistry of Organic Superconductors*
Editors: G. Saito and S. Kagoshima

52 *Dynamics and Patterns in Complex Fluids: New Aspects of the Physics-Chemistry Interface*
Editors: A. Onuki and K. Kawasaki

53 *Computer Simulation Studies in Condensed Matter Physics III*
Editors: D. P. Landau, K. K. Mon, and H.-B. Schüttler

54 *Polycrystalline Semiconductors II*
Editors: J. H. Werner and H. P. Strunk

55 *Nonlinear Dynamics and Quantum Phenomena in Optical Systems*
Editors: R. Vilaseca and R. Corbalán

56 *Amorphous and Crystalline Silicon Carbide III and Other Group IV – IV Materials*
Editors: G. L. Harris, M. G. Spencer, and C. Y.-W. Yang

57 *Evolutionary Trends in the Physical Sciences*
Editors: M. Suzuki and R. Kubo

58 *New Trends in Nuclear Collective Dynamics*
Editors: Y. Abe, H. Horiuchi, and K. Matsuyanagi

59 *Exotic Atoms in Condensed Matter*
Editors: G. Benedek and H. Schneuwly

60 *The Physics and Chemistry of Oxide Superconductors*
Editors: Y. Iye and H. Yasuoka

61 *Surface X-Ray and Neutron Scattering*
Editors: H. Zabel and I. K. Robinson

62 *Surface Science: Lectures on Basic Concepts and Applications*
Editors: F. A. Ponce and M. Cardona

63 *Coherent Raman Spectroscopy: Recent Advances*
Editors: G. Marowsky and V. V. Smirnov

64 *Superconducting Devices and Their Applications*
Editors: H. Koch and H. Lübbig

Volumes 1–43 are listed on the back inside cover

G. Benedek H. Schneuwly (Eds.)

Exotic Atoms in Condensed Matter

Proceedings of the Erice Workshop
at the Ettore Majorana Centre for Scientific Culture,
Erice, Italy, May 19 – 25, 1990

With 142 Figures

Springer-Verlag
Berlin Heidelberg New York
London Paris Tokyo
Hong Kong Barcelona
Budapest

Professor Dr. Giorgio Benedek
Dipartimento di Fisica dell' Università,
Via Celoria 16, I-20133 Milano, Italy

Professor Dr. Hubert Schneuwly
Université de Fribourg, Institut de Physique,
Pérolles, CH-1700 Fribourg, Switzerland

ISBN-13: 978-3-642-76372-4 e-ISNB-13: 978-3-642-76370-0
DOI: 10.1007/978-3-642-76370-0

Library of Congress Cataloging-in-Publication Data. Exotic atoms in condensed matter: proceedings of the Erice workshop at the Ettore Majorana Centre for Scientific Culture, Erice, Italy, May 19–25, 1990 / G. Benedek, H. Schneuwly (eds.). p. cm. – (Springer proceedings in physics; 59) Includes bibliographical references and index. 1. Condensed matter–Congresses.
2. Exotic atoms–Congresses. 3. Muons–Congresses. I. Benedek, G. (Giorgio) II. Schneuwly, H. (Hubert), III. Series: Springer proceedings in Physics; v. 59. QC173.4.C65E96 1992 530.4'16–dc20 91-45454 CIP

This work is subject to copyright. All rights are reserved, whether the whole or part of the material is concerned, specifically the rights of translation, reprinting, reuse of illustrations, recitation, broadcasting, reproduction on microfilm or in any other way, and storage in data banks. Duplication of this publication or parts thereof is permitted only under the provisions of the German Copyright Law of September 9, 1965, in its current version, and permission for use must always be obtained from Springer-Verlag. Violations are liable for prosecution under the German Copyright Law.

© Springer-Verlag Berlin Heidelberg 1992
Softcover reprint of the hardcover 1st edition 1992

The use of general descriptive names, registered names, trademarks, etc. in this publication does not imply, even in the absence of a specific statement, that such names are exempt from the relevant protective laws and regulations and therefore free for general use.

Typesetting: Camera ready by authors

54/3140 - 5 4 3 2 1 0 – Printed on acid-free paper

Preface

The topics of the fifth course of the International School of Physics of Exotic Atoms, which took place at the "Ettore Majorana" Centre for Scientific Culture, Erice, in Spring 1989, were the electromagnetic cascade and the chemistry of exotic atoms. The observed and reported influences of chemical structures on the formation and deexcitation mechanisms of exotic atoms suggested to G. Torelli, the Director of the School, the idea of an interdisciplinary workshop, where condensed matter physicists and exotic-atom specialists would have the opportunity to meet and to exchange views and information.

To bring people to an effective interdisciplinarity is not easy. Everyone feels generally more comfortable among specialists of his own domain. In this workshop, we had the luck that the specialists of either domain had a real interest in the other domain. Nobody expected to return to his home institution with a recipe for an interdisciplinary miracle. Everybody knows that it takes time, even if one starts early. This workshop intended to be such a start.

High-T_c superconductors and muon catalyzed fusion do not seem to have a lot in common. Through the talks of the various speakers, one gets, however, finally the impression that one can throw a bridge between these two fields and that the building of the bridge might become more important than the bridge itself.

The high intensity facilities for slow muons that seem likely to be constructed in the next decade have the potential to increase the maximum available flux by nearly three orders of magnitude. Such an increase should usher in a new era in cold muon studies of condensed matter. Their interaction with electrons promises new progress in core and deep level spectroscopies, hyperfine field analyses and spin-polarization spectroscopies of magnetic materials, polymers and high-T_c superconductors.

New impetus is expected for muon catalyzed fusion in solid hydrogen. Progress in this area, aiming at alternative fusion energy sources, has been envisaged as theoretically possible in high-density metallic deuterium. Fascinating alternatives in low-energy fusion, e.g., cluster fusion, are also discussed.

This book constitutes the proceedings of the Erice Workshop on Exotic Atoms in Condensed Matter, which was held at the "Ettore Majorana" Centre for Scientific Culture, between 19 and 25 May 1990. It reviews the state of the art in this domain of research, from meson factories and facilities to the basic interactions of muons and pions in condensed matter. It covers the applications and perspectives of muon- and pion-based analyses of solid state structural, magnetic and superconducting properties. Some basic aspects of condensed matter physics which may be enlightened by exotic atoms spectroscopies are reviewed, together with applications to new sophisticated methods of chemical analysis. A section is devoted to muon catalyzed fusion in metallic deuterium.

The Workshop was jointly organized by the International School of Physics of Exotic Atoms and the International School of Physics of Solids under the auspices of the European Physical Society and the Regional Office for Science and Technology in Europe of UNESCO. It was sponsored by the Italian Ministry of Education, the Italian Ministry of University, Scientific and Technological Research, the National Research Council (CNR), the National Agency for Alternative Energies (ENEA), the Physics Department of the University of Milano and the Sicilian Regional Government.

The Ettore Majorana Centre for Scientific Culture was a very efficient and pleasant host for this workshop, and our sincere thanks go to its Director, Prof. Antonio Zichichi, and to all the Centre staff.

Milan, Fribourg
March 1991

G. Benedek
H. Schneuwly

Contents

Part I

Exotic Atom Formation and Experimental Techniques

Solid State Effects in Atomic Capture of Negative Muons

H. Schneuwly

Institut de Physique de l'Université, CH-1700 Fribourg, Switzerland

Abstract. The formation mechanism of exotic atoms is not yet well known. The muonic x-ray intensity patterns are, however, found to be sensitive to solid state structures, showing that the muon capture mechanism is influenced by the spatial distribution of bonding electrons. In compounds, pion and muon capture ratios are both influenced by the bond structure. Hydrogen compounds are cases of particular interest.

A negative muon is decelerated in a gas essentially through inelastic collisions with the atoms or molecules of that gas. When it has a sufficiently low kinetic energy, it is captured into the Coulomb field of an atomic nucleus. This capture process is not well known. One imagines that the muon loses its energy by ejecting an electron, and takes, in a certain sense, its place. Therefore, the muon is, at first, in an excited state and deexcites through Auger electron ejection and radiative transitions in a cascade finally to the ground state. These radiative transitions of energies characteristic to each element can easily be measured and their intensities can give information about the capture and cascade processes.

The intensities of muonic x-rays have been measured in argon gas at various pressures [1-4]. As may be seen from Table 1, the relative intensities of the Lyman series change with pressure. This change is due to the slower refilling of the electron shells, depopulated during the muonic cascade, at lower pressure. Indeed, the measured intensities can be reproduced by a muon cascade calculation [5] in all three cases by assuming an initial statistical

Table 1: Measured muonic x-ray intensities of the Lyman series in argon and reproduction of the muonic cascade with a statistical initial angular momentum distribution at n = 14. P_L is the population of the electronic L-shell used in the cascade calculation.

	Expt. 1.4 bar	Calc. P_L=2%	Expt. 22 bar	Calc. P_L= 3.5%	Expt. 170 bar	Calc. P_L=15%
2p-1s	.914(26)	.909	.905(25)	.899	.875(24)	.870
3p-1s	.0400(22)	.0395	.0410(14)	.0427	.0537(16)	.0509
4p-1s	.0097(11)	.0112	.0116(8)	.0128	.0187(8)	.0185
5p-1s	.0083(8)	.0066	.0095(6)	.0085	.0154(6)	.0161
6p-1s	(.0080)	.0065	(.0094)	.0086	(.0133)	.0155
7p-1s	.0077(8)	.0075	.0093(6)	.0090	.0111(15)	.0119

Springer Proceedings in Physics, Vol. 59 Exotic Atoms in Condensed Matter
Editors: G. Benedek · H. Schneuwly
© Springer-Verlag Berlin, Heidelberg 1992

angular momentum distribution $P(l) \propto (2l + 1)$ in a level of main quantum number $n = 14$, and by adjusting the population of the electronic L-shell or, equivalently, the K-shell refilling width.

The observations made in argon gas is confirmed in neon. That the measured intensities can, in both elements, be reproduced by assuming a statistical angular momentum distribution, is of particular interest. Indeed, the muon capture process being insufficiently known, one can predict neither the orbit n of the capture level nor the angular momentum state l. One knows from quasi-classical approximations that, for muons cascading down from high n levels (n = 30-40), the angular momentum distribution remains practically unchanged, i.e. if the distribution is statistical in a high n level, it remains statistical down to levels of about n = 14. This means that in noble gases like neon and argon, the initial angular momentum distribution of captured muons is statistical.

At first sight, this statistical angular momentum distribution should reflect a spherical symmetry. If the muon capture proceeds through electron ejection, this distribution should be an image of the spatial symmetry of those electrons through which muons are captured.

In neon and argon, all electron shells are of spherical symmetry. However, in a one-to-one mixture of neon and argon, the ratio of muons captured in neon to those captured in argon has been measured to be A(Ne/Ar) = 1.01(3) [2], which indicates that the electrons do not all have the same weight in the capture process [6].

If the initial angular momentum distribution reflects the spherical symmetry of the electrons through which muons are captured, one expects that, when captured in Cl^- and K^+ ions, the muons have the same initial statistical angular momentum distribution. This is what is actually observed [3,4].

There is no reason to exclude muon capture through ejection of bonding electrons. In the case of silicon or diamond, where the spatial structure of the bonding electrons is essentially given by hybridization of the $3sp^3$ or $2sp^3$ orbitals, the symmetry is the same as that of full atomic (3s,3p) or (2s,2p) shells, respectively. One would, therefore, not be surprised to reproduce the muonic x-ray intensities measured in silicon and diamond by a cascade calculation assuming again a statistical initial angular momentum distribution. The measured muonic x-ray intensity structures in silicon and diamond are, indeed, reproduced by such a statistical distribution [4,7,8]. In addition, the muonic x-ray intensity pattern measured in nitrogen of cubic boron nitride, where the spatial electronic structure is comparable to diamond, can also be reproduced with the same distribution [8].

In graphite and in nitrogen of hexagonal boron nitride, the muonic x-ray intensity structures are different [8], showing a clear solid state effect. They can only be reproduced by assuming a greater population of lower angular momentum states. The difference in the angular momentum population may be attributed in both cases to a different symmmetry in the spatial distribution of the bonding electrons [9]. In the diamond structures one has four σ-bonds and in the hexagonal structures one of the four bonds is a delocalized π-bond.

In addition to the differences observed in the muonic x-ray intensity structures in nitrogen between cubic and hexagonal boron nitride, one has also found a significant difference in the atomic muon capture ratios A(B/N) between the two allotropies. This difference is explained in terms of differences in B-N bond dipole moments, expressed through bond ionicities [8], and its explanation is consistent with the one given for the muonic x-ray intensity structures.

These observations and their explanation stimulate our confidence in the understanding of the main mechanism of muon capture.

In the ions F^- and Na^+, which have the same electronic structure as neon, one expects that the captured muons have also a statistical initial angular momentum distribution. The muonic x-ray intensities have been measured in these elements in the ionic compounds KF and NaCl, where both K^+ and Cl^- showed a statistical muon distribution. The muonic fluorine x-ray intensities can, indeed, be reproduced by assuming a statistical initial angular momentum distribution, whereas those of sodium cannot. The pattern of the muonic sodium x-rays measured in NaCl is not a special case. In a whole series of compounds, where sodium can, in good approximation, be considered as an ion, the muonic sodium x-ray intensities are the same [10,11]. Even metallic sodium has the same pattern [11]. From this last observation one deduces that the electron refilling in a conductor does not be measurably different from the refilling in insulators.

The difference between the muonic sodium x-ray intensity pattern and those measured in the other isoelectronic elements of neon and argon may be related to different weights of the electronic shells and subshells in muon capture. Indeed, one might expect that the per atom capture ratio A(Na/Cl) measured in NaCl were equal to the capture ratio A(Ne/Ar). One does not expect that muon capture ratios like A(K/Cl), A(Na/F) and A(Na/Cl), measured in KCl, NaF and NaCl are exactly equal to unity, because the bonds are not ideally ionic. For the first two ratios, the mean values A(K/Cl) = 1.15(5) and A(Na/F) = 0.96(5) are near to unity, but A(Na/Cl) = 0.73(5) indicates that about 25% less muons are captured by sodium than by chlorine. By assuming that only electrons with binding energies less than E = 60 eV participate in muon capture, the measured A(Na/Cl) capture ratio can be reproduced [6].

The inital statistical angular momentum distribution of captured muons observed in the other elements is then explained by capture via the electrons of the outermost full s- and p-shells. The capture in sodium would then proceed only via the six electrons of the 2p shell [3,4,6]. This approach, even if it explains these and a lot of other muon capture data [12], does not account for the capture ratios A(K/F) and A(Na/F) measured in KF and NaF. The present model [6] would predict about 1.0, resp. 0.75, and the measured values are 1.89(18), resp. 0.96(5).

Solid state effects are not only observed in the muonic x-ray intensity structures of carbon and of nitrogen in boron nitride between hexagonal and cubic structures, but also in those of selenium between the metallic and amorphous allotropies [13]. In phosphorus, differences have been found between the white modification and the red and black ones, but no difference between the last two (Fig. 1) [13].

Pions are also very sensitive to the chemical bond as has been shown by Imanishi et al. [14,15]. They have measured the pion capture in boron compounds and corresponding mechanical mixtures. The differences in the per atom capture ratios between the compounds and the mechanical mixtures range from a factor of two to seven. Different chemical bonds in beryllium borides

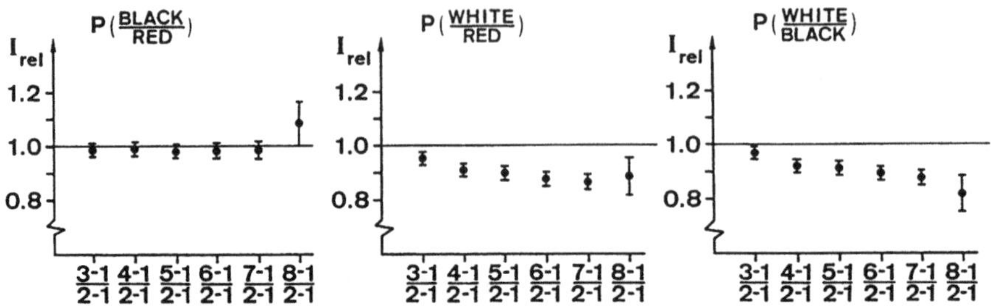

Figure 1: Relative muonic x-ray intensity ratios of the Lyman series for the three phosphorus modifications : white, red and black.

Table 2: Experimental atomic (per atom) Coulomb capture ratios of negative pions, A(Z/B), for compounds and mechanical mixtures of boron with beryllium and carbon.

Elements	x/y	xB + yZ	B_xZ_y	$(xB+yZ)/B_xZ_y$
B and Be	1/4	0.76(2)	0.40(2)	1.9
B and Be	1/2	0.75(3)	0.36(2)	2.1
B and Be	2/1	0.73(2)	0.25(2)	2.9
B and Be	6/1	0.75(5)	0.10(3)	7.1
B and C	4/1	0.78(12)	0.55(6)	1.4

change the per atom capture ratios A(Be/B) by up to a factor of four (Table 2). Measurements of pion capture in LiF, NaF and KF show that the pion capture ratios A(M/F) are identical to the corresponding muon capture ratios [16].

The muon capture in NaCl and $NaClO_2$, $NaClO_3$, $NaClO_4$ has been investigated in great details [17]. The muonic x-ray intensity structures of oxygen are identical in all three oxygen containing compounds. The same proves true for the chlorine x-rays. The only difference observed has been in chlorine between NaCl and the $NaClO_x$ compounds, where the relative intensities of the Lyman series change by up to 50%. The muonic x-ray pattern in sodium is the same in all four compounds, which presents no surprise, the sodium bond being in all four compounds highly ionic.

Surprising is, however, that the atomic muon capture ratio A(Na/Cl) is practically the same in NaCl and in the $NaClO_x$ compounds and that the A(Cl/O) capture ratio (per atom) decreases with increasing oxygen content (Fig. 2). The measured muon capture ratios A(Na/Cl) and A(ClO_x/Na) agree nicely with model predictions (Table 3). The A(Cl/O) and A(Na/O) capture ratios can only be reproduced by model calculations if one assumes that the Cl-O bond dipole moments or, equivalently, the ionicity of the Cl-O bonds are different in the three compounds.

In empirical intensity correlations in muonic x-ray spectra of oxides, Daniel et al. [18] found a weak correlation, e.g., between the muonic K_α intensity in oxygen and the Pauling ionicity of the oxygen bond. They found a strong correlation between the atomic radii and the ionicity.

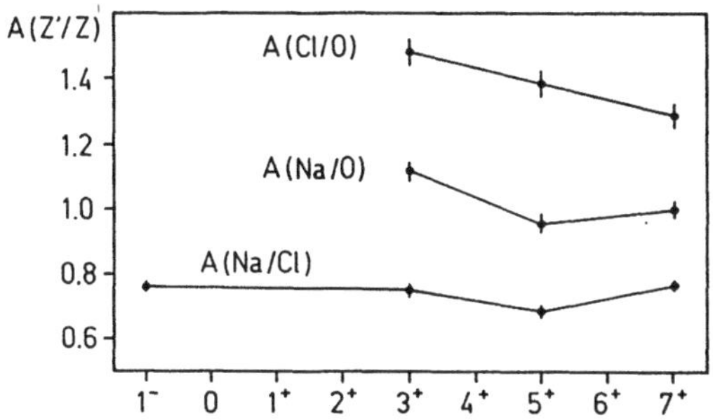

Figure 2: Per atom muon capture ratios A(Cl/O), A(Na/O) and A(Na/Cl) measured in $NaClO_2$, $NaClO_3$, $NaClO_4$ and NaCl as a function of the formal chlorine valency.

Table 3: Comparison of measured muon capture ratios $A(ClO_x/Na)$ in NaCl, $NaClO_2$, $NaClO_3$ and $NaClO_4$ with model predictions [6].

	NaCl	$NaClO_2$	$NaClO_3$	$NaClO_4$
Experiment [17]	1.32(3)	3.13(7)	4.56(11)	5.32(12)
Model [6]	1.32	3.33	4.33	5.33

Figure 3 displays the muonic K_β/K_α intensity ratios of oxygen measured in oxides of the third row elements [19] in dependence of the Pauling ionicities defined through the electronegativities of the elements. At first glance, the correlation is not obvious unless one were guided by the broken line, drawn to suggest the correlation. However, the oxygen bond in Na_2O_2 has not the ionicity of a Na-O bond (σ=82%) because, naively speaking, one oxygen atom is also bound to the other such that the mean ionicity of the oxygen bond is about 40 to 50%. Then the correlation is plausible.

Following our model [6] in its simplest version, the capture into oxygen should proceed through four core L-electrons on one side and partly through the two pairs of binding electrons on the other side, the part being determined through the bond ionicity and the nuclear charges of the bond partners. In a highly ionic compound like MgO (σ=73%), 43% of the muons reaching the 1s state in oxygen are assumed to be captured via the binding electron pairs, compared to 28% in P_2O_5 (σ=39%). In oxygen of the oxides of the third row elements, this percentage increases with increasing ionicity, whereas the muonic K_β/K_α intensity ratio decreases.

A similar but much less pronounced trend is observed in chlorine of corresponding compounds of elements of the same row (Fig. 4). The K_β/K_α intensity ratios seem, however, much less sensitive to the change in ionicity, whereas the higher transitions show a clear decrease in intensity [20]. The explanation for the observed trend is analogous to that for the oxides [21].

The proposed explanation for the decrease of the muonic K_β/K_α intensity ratios in oxygen and chlorine of third row compounds with increasing ionicity cannot be simply extended to other compounds. Indeed, in boron nitride in cubic and hexagonal structures, one had to assume different bond

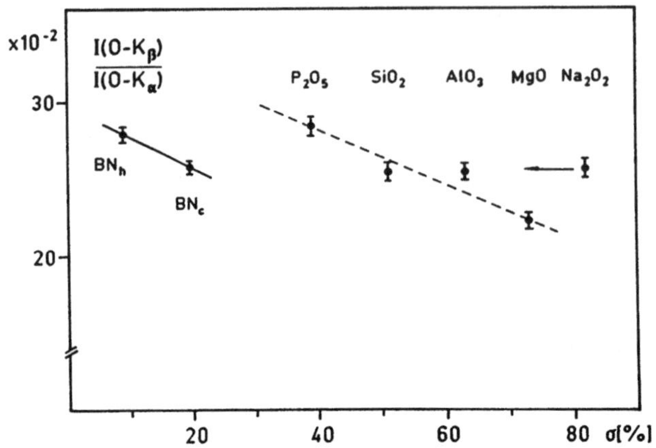

Figure 3: Muonic K_β/K_α intensity ratios in oxygen of oxides of third row elements as a function of bond ionicities. Muonic K_β/K_α intensity ratios in nitrogen of boron nitride in hexagonal (BN_h) and cubic (BN_c) structures are also shown. The lines are only drawn to guide the eye.

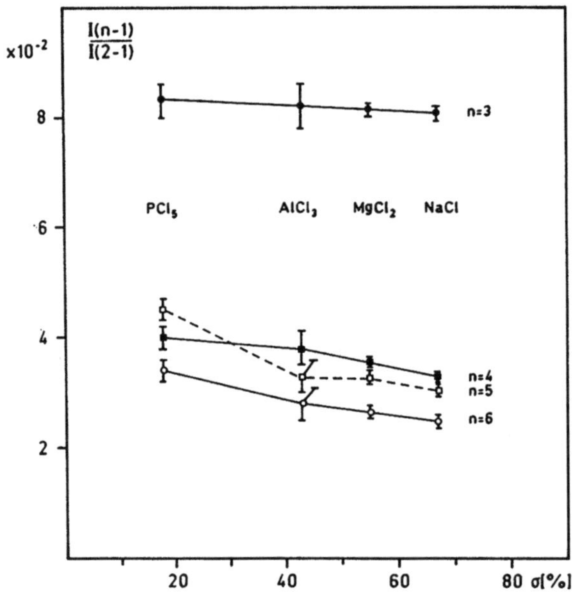

Figure 4: Muonic I(n-1)/I(2-1) intensity ratios (n=3,4,5,6) in chlorine of chlorides of third row elements as a function of the bond ionicities. The lines are drawn to guide the eye.

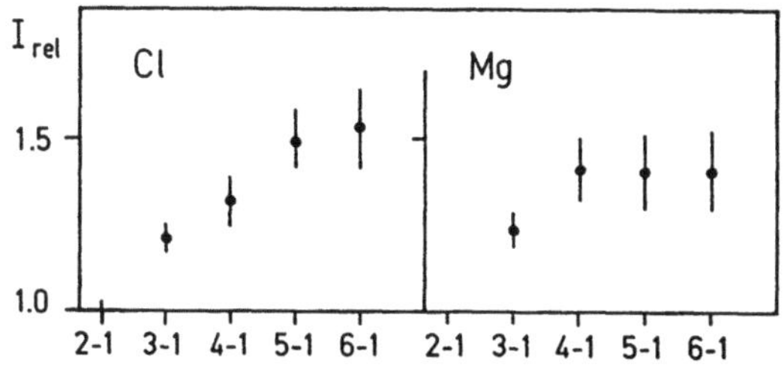

Figure 5: Muonic Lyman series intensities of chlorine and magnesium, normalized to the respective K_α intensities, in 4M relative to those of anhydrous $MgCl_2$.

ionicities in order to reproduce with our model the measured A(B/N) muon capture ratios. The muonic nitrogen K_β/K_α intensitiy ratios in the two structures of boron nitride agree with the trend observed in oxygen (Fig. 3).
However, these K_β/K_α intensity ratios in the two structures can be explained independently from the ionicity like in carbon,in diamond and graphitic structures [9]. Another example are the muonic K_β/K_α intensity ratios in oxygen of Cr_2O_3 and CrO_3 [22]. The ionicity of the Cr-O bond is greater in Cr_2O_3 than in CrO_3. The oxygen intensity ratio increases with increasing ionicity, which is opposite to the trend observed in the oxides of third row elements. However, in chromium compounds like in iron compounds d-electrons are involved in the bonds, whereas third row elements have only s- and p-electrons. The observed trends represented in Figs. 3 and 4 might well be related to the fact that the bond partners contribute only through s and p electrons to the bonds to oxygen and chlorine.
The observed correlations to the bond ionicities confirm the dominant role played by the outer electrons of the bond partners in atomic muon capture.

As already shown, the initial angular momentum distribution of muons captured by Cl^- is statistical as for argon. For muons which are transferred to argon from muonic hydrogen atoms, this angular momentum distribution is strongly peaked to low angular momenta [23]. The resulting muonic x-ray intensity pattern is, consequently, also different. In particular, the intensities of the higher members of the Lymans series are strongly enhanced.
In an experiment, Daniel et al. [24] have compared the muonic x-ray intensity patterns in magnesium and chlorine in crystalline $MgCl_2$ to those measured in a 4M solution. (Fig. 5). From the strong enhancements of the intensities of the higher members of the Lyman series in both elements, they concluded that muon transfer is responsible for the observed hydrogen effects while mechanisms involving muon-molecular orbitals are ruled out.

In a similar experiment, we have investigated the muonic x-ray intensities in calcium and chlorine in $CaCl_2$ anhydride and in $CaCl_2 \cdot 2H_2O$ and $CaCl_2 \cdot 4H_2O$. In chlorine, the observed differences in the intensities are correlated to the quantity of water of crystallization. In calcium, the enhancement of the intensities due to the presence of hydrogen is hardly visible (Fig. 6). If the enhancement of the muonic x-ray intensities were essentially due to muon transfer from the ground state of muonic hydrogen atoms, one would expect that the enhancements were approximately the same in both elements as was the case in $MgCl_2$. If the enhancement were due to transfer from excited μp^*

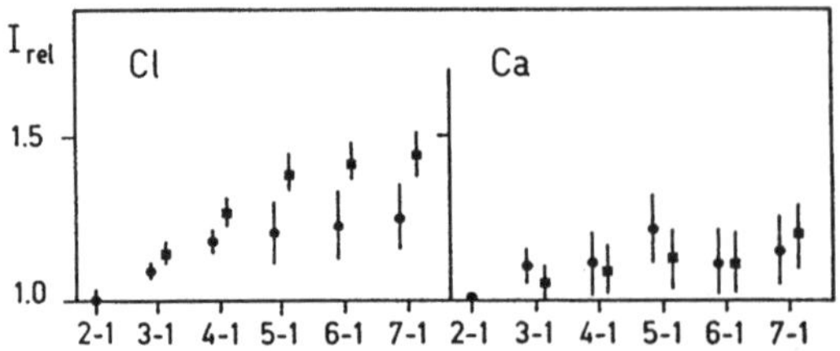

Figure 6: Muonic Lyman series intensities in chlorine and calcium, normalized to the respective K_α intensities, in $CaCl_2{\cdot}2H_2O$ (circles) and in $CaCl_2{\cdot}4H_2O$ (squares) relative to anhydrous $CaCl_2$.

states, then the cations should even be favoured [25]. By comparing the per atom capture ratios A(Ca/Cl) in $CaCl_2$ anhydride to those in the water containing samples:

$$\frac{\text{A(Ca/Cl) anhydride}}{\text{A(Ca/Cl) in } CaCl_2{\cdot}2H_2O} = 1.036(18)$$

$$\frac{\text{A(Ca/Cl) anhydride}}{\text{A(Ca/Cl) in } CaCl_2{\cdot}4H_2O} = 1.079(18)$$

one observes that capture in chlorine is increased with increasing water content as suggested by the muonic x-ray patterns.

The observed intensity enhancements can, therefore, not be explained through muon transfer alone. Actually, comparable enhancements have been observed in elements bound to hydrogen compared to the pure element or other compounds, i.e., the hydrogen bond has a particularly strong influence on the initial angular momentum distribution of captured muons.

There are already a lot of experimental data on muon capture available. Some systematic investigations have been performed. Some understanding of the capture machanism has been achieved. No model seems, however, exempt from contradictions with experimental results. Further effort has to be made to increase our knowledge on the atomic capture of muons and pions in mixtures of elements and compounds. Combined analyses of muonic x-ray structures and muon capture ratios seem to be a promising way. New ideas seem, however, necessary to make decisive progress in the understanding of the atomic capture mechanism.

References

1. J.D. Knight, C.J. Orth, M.E. Schillaci, R.A. Naumann, F.J. Hartmann and H. Schneuwly, Phys. Rev. A **27**, 2936 (1983).
2. P. Ehrhart, F.J. Hartmann, E. Köhler and H. Daniel, Z. Physik **A311**, 259 (1983).
3. R. Jacot-Guillarmod,F. Bienz, M. Boschung, C. Piller, L.A. Schaller, L. Schellenberg, H. Schneuwly and D. Siradovic, Phys. Rev. A **37**, 3795 (1988).

4. R. Jacot-Guillarmod, Muon Catalyzed Fusion **4**, 113 (1989).
5. V.R. Akylas and P. Vogel, Comput. Phys. Commun. **15**, 291 (1978).
6. H. Schneuwly, V.N. Pokrovsky and L.I. Ponomarev, Nucl. Phys. **A312**, 419 (1978).
7. R. Jacot-Guillarmod, F. Bienz, M. Boschung, G. Piller, L.A. Schaller, L. Schellenberg and H. Schneuwly, Phys. Rev. A **38**, 3106 (1988).
8. H. Schneuwly, M. Boschung, K. Kaeser, G. Piller, A. Rüetschi, L.A. Schaller and L. Schellenberg, Phys. Rev. A **27**, 950 (1983).
9. H. Schneuwly, Lett. Nuovo Cimento **44**, 209 (1985).
10. L.F. Mausner, R.A. Naumann, J.A. Monard and S.N. Kaplan, Phys. Rev. A **15**, 479 (1977).
11. K. Kaeser, B. Robert-Tissot, L.A. Schaller, L. Schellenberg and H. Schneuwly, Helv. Phys. Acta **52**, 304 (1979).
12. D. Horváth and F. Entezami, Nucl Phys. **A407**, 297 (1983).
13. K. Kaeser, T. Dubler, B. Robert-Tissot, L.A. Schaller, L. Schellenberg and H. Schneuwly, Helv. Phys. Acta **52**, 238 (1979).
14. N. Imanishi, T. Furuya, I. Fujiwara, A. Shinohara, H. Kaji and S. Iwata, Phys. Rev. A **32**, 2584 (1985).
15. N. Imanishi, S. Miyamoto, S. Okashi and A. Shinohara, Phys. Rev. A **35**, 5244 (1987).
16. N. Imanishi, S. Miyamoto, Y. Takeuchi, A. Shinohara, H. Kaji and Y. Yoshimura, Phys. Rev. A **37**, 43 (1988).
17. G. Piller, Thesis (No. 881), Faculty of Science, University of Fribourg, 1985, unpublished.
18. H. Daniel, W. Denk, F.J. Hartmann and W. Wilhelm, Phys. Rev. Lett.. **41**, 853 (1978).
19. T. von Egidy, W. Denk, R. Bergmann, H. Daniel, F.J. Hartmann, J.J. Reidy and W. Wilhelm, Phys. Rev. A **23**, 427 (1981).
20. H. Daniel, R. Bergmann, G. Fottner, F.J. Hartmann, and W. Wilhelm, Z.Physik **A300**, 253 (1981).
21. H. Schneuwly, *in* Nuclear Techniques and Structural Chemistry, edited by A. Vértes and B. Lévay (in press).
22. M. Hild, K. Kaeser, G. Piller and H. Schneuwly, J. Phys. B: At. Mol. Opt. Phys. **18**, 2093 (1985).
23. R. Jacot-Guillarmod, F. Bienz, M. Boschung,C. Piller, L.A. Schaller, L. Schellenberg, H. Schneuwly, W. Reichart and G. Torelli, Phys. Rev. A **38**, 6151 (1989).
24. H. Daniel, F.J. Hartmann, R.A. Naumann and J.J. Reidy, Phys. Rev. Lett. **56**, 448 (1986).
25. L. Bracci and G. Fiorentini, Nuovo Cimento **A50**, 373 (1979).

Atomic Capture of Exotic Particles in Elements, Mixtures, and Compounds

F.J. Hartmann

Physik-Department, E18, Technische Universität München, W-8046 Garching, Fed. Rep. of Germany

Abstract. Slowing-down, atomic capture and atomic cascade of exotic particles are outlined. The influence of chemical and solid-state properties of the target on these processes is described.

1. Introduction

Coulomb capture of exotic particles (μ^-, π^-, $\bar{p}$): To cover this subject really is a task. The bibliography on exotic atoms fills a book [1] of more than 600 pages. And this for the time up to 1983. So I can only try to make clear how the electronic structure of an element, mixture or compound may influence the processes of slowing-down, atomic capture and cascade of exotic particles. But first I would like to describe these processes in more detail.

A few facts about the influence of the electronic structure immediately come to mind:

- Slowing-down of exotic particles proceeds, even at very low energies, predominantly by collisions with quasi-free electrons and therefore the availability of these electrons plays an important role.
- In the atomic capture of the exotic particle an electron is emitted and the particle falls to a bound state in the screened Coulomb potential of the nucleus. Where this electron comes from, what binding energy it has, all this governs the capture process.
- The atomic cascade of the exotic particle down the ladder of bound states is determined by a competition between radiative and Auger transitions. The exotic atom is heavily ionized during the cascade. Whether the ejected electrons are refilled or not depends on the structure of the target.

As you see the state of the target is important during all stages of the exotic particle's lifetime. From the observation of all these processes we may learn something about the structure of matter.

Among the exotic particles the muon is, without doubt, best suited for experiments on the Coulomb capture and atomic cascade, because it interacts only

 © Springer-Verlag Berlin, Heidelberg 1992

weakly with the nucleus: nuclear capture of the muon takes place only after it has reached the 1s level. As the cascade is fast and only radiative transitions populate the 1s level (see below) one characteristic x ray from the Lyman series is emitted for each muon which is Coulomb-captured. This is the reason why most of the experimental work was done with muons and I will restrict myself nearly all of the time to these particles.

2. Slowing-down of exotic particles

Pions are generated in the meson factories by interaction of high-energy protons with light nuclei, as e.g. from carbon or beryllium. The pion and its decay product, the muon, usually is fast when it enters the experimental area and has to be degraded from a kinetic energy T of at least a few MeV. The energy loss is well described down to T $\approx$ 100 keV by the familiar Bethe formula [2], with some corrections included at lower energies (see, e.g. [3]). At velocities v $\approx$ α*c (with α the fine-structure constant and c the velocity of light) Bethe theory breaks down [3]. Fermi and Teller [4] were the first to point out that the v^{-2} behaviour in the stopping power S= -dT/dx goes over into a linear v dependence. A rigorous treatment of the slowing-down process based on linear response theory was developed by Lindhard et al. [5].

At very low energies the spatial distribution of loosely bound electrons becomes important. To take the influence of the solid state into account Latta and Scanlon [6] renormalized the total electron density $\rho(r)$ under the assumption that in a solid the electrons of the target atom are confined within a sphere of finite radius. They found a pronounced periodicity of S with atomic number Z, with maxima for the alkali metals and minima roughly halfway in between. At still lower energies the deflection of the exotic-particle path in the screened nuclear potential has to be taken into account. Especially the track of the muon (with $m_\mu = 206 * m_e$) is easily bent. In the atomic potential the particle gains an energy which is large compared with its kinetic energy outside the atom. The stopping power becomes energy-independent. As found by Wilhelm [7] S is still periodic in Z.

Finally, the particle is Coulomb-captured.

3.Atomic capture

3.1 Outline of the problem

During Coulomb capture the exotic particle goes over from an unbound to a bound state while moving in the screened field of the atomic nucleus. The problem of Coulomb capture would be solved if one knew two quantities:

- The first one is the probability $B^{capt}(\hat{s},E)$ for capture at total energy E into a bound state characterized by a set $\hat{s}$ of quantum numbers (in the case of a spherically symmetric central potential these could be the principal quantum number n, the angular-momentum quantum number ℓ, and the magnetic quantum number m).

- The second quantity is the probability F(E) that an exotic particle reaches energy E without being Coulomb-captured. F(E) is determined by the competition between capture and slowing down of the particle. To perform proper Coulomb-capture calculations it is *essential* to solve the transport equation properly.

In reality $B^{capt}(\hat{s},E)$ cannot be determined directly. One has to be content with measuring more global quantities. One of them is the per-atom capture ratio A(Z,Z'). If we shoot, let us say, muons into a compound $Z_k Z'_\ell$ they will be captured into atoms of species Z with probability w_Z and into those of species Z' with probability $w_{Z'}$. A(Z,Z') is the ratio of probabilities divided by the stoichiometric ratio k/ℓ,

$$A(Z,Z') = \frac{w_Z}{w_{Z'}} * \frac{\ell}{k} = P(Z)/P(Z'). \tag{1}$$

The per-atom capture probability $P(Z)= w_Z/k$, $B^{capt}= \sum_{\hat{s}} B^{capt}(\hat{s},E)$ and F(E) are connected by the relation

$$P(Z) = \int B^{capt}(E) * F(E) * dE. \tag{2}$$

In principle quantum mechanics has to be applied to solve the problem of Coulomb capture. Such calculations are available only for low Z elements, mostly hydrogen and helium (for an overview, see Ref. [8]). The complexity of the problem has prevented a solution for heavier elements. From the calculations, however, it could be deduced that the kinetic energy of the exotic particle at capture should be roughly equal to the first ionization potential of the atom. For hydrogen and helium this is the binding energy of electrons in the K shell.

3.2 Semiclassical models for Coulomb capture

The number of "semiclassical" calculations performed in the seventies and eighties is large [8]. A number of assumptions enter these calculations:

- The exotic particle follows a *classical* trajectory in a spherically symmetric atomic potential. This is justified as at not too low energies the deBroglie wave length of the particle is small.

- The electrons are viewed as a degenerate electron gas in the screened nuclear potential.

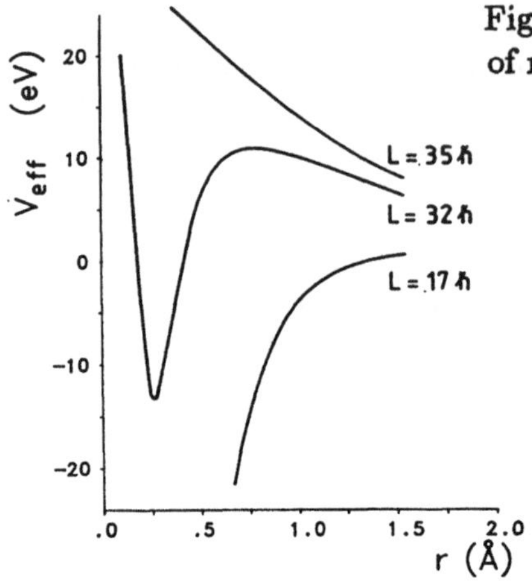

Figure 1: Effective potential for the movement of muons in Zr.

- The exotic particle looses energy by raising bound electrons to unoccupied states via shielded Coulomb interaction.
- Capture is assumed to have taken place as soon as the total energy of the exotic particle drops below zero while the particle passes through the electron cloud. The centrifugal barrier (see Fig. 1) may very well trap the particle already at positive energies.

It would lead too far to describe all the conclusions that can be drawn from semiclassical calculations. Table 1 only shows selected results for P(Z).

Table 1: Results for the per-atom capture probability P(Z) from different semiclassical calculations

$P(Z) \propto Z$	Ref. [4]
$P(Z) \propto Z^{7/6}$	Ref. [9]
$P(Z) \propto Z^{1/3} * \ell n(0.57*Z)$	Ref. [10]
$P(Z) \propto Z^{1/3} * \ell n(0.57*Z)/ R(Z)$ R(Z) = atomic radius	Ref. [11]

3.3 The "SPP" model

Schneuwly, Pokrovsky and Ponomarev (SPP) [12] were quite successful in deriving per-atom capture probabilities by calculations based on the model of large mesic molecules [13]. According to their ideas only loosely bound electrons can take part in the capture process, because only for these electrons the energy transfer from the exotic particle to the electron is sufficient to lift the electron to an unoccupied state. If we, for the moment, consider only binary compounds

$Z_kZ'_\ell$, an exotic particle can be either captured directly into the constituents or it can be trapped in molecular orbits around the atoms of the compound. In the binary compound there are n loosely bound core electrons per atom which cause direct capture into element Z, n' loosely bound core electrons cause direct capture into component Z' and $\alpha_m = (k * \nu + \ell * \nu')$ valence electrons lead to capture into mesomolecular orbits ($k*\nu = \ell*\nu'$). From these molecular orbits the exotic particle goes over into levels in atom Z and Z' with probabilities ω and ω', respectively. An expression for the per-atom capture ratio

$$A(Z/Z') = \frac{\ell}{k} * \frac{k * n + \alpha_m * \omega}{\ell * n' + \alpha_m * \omega'} \tag{3}$$

results. The authors now make the following assumptions:

- Only loosely bound core electrons take part in the direct capture process. Which electrons are involved is determined by adjustable parameters [12].
- Due to the ionicity σ of the chemical bond a fraction $p = (1-\sigma)/2$ of the valence electrons is localized near element Z, the fraction $p' = 1-p$ near element Z'. The localization of the exotic particle in the mesomolecular orbit follows this electron distribution.
- The exotic particles are redistributed from their original location with probabilities q and q', respectively. The redistribution probability itself is assumed to be proportional to Z^2: $q = Z^2/(Z^2+Z'^2)$, $q' = 1-q$. ω is then given by $\omega = p*q$, ω' by $\omega' = p'*q'$.

3.4 A semiempirical model based on quantum-mechanical calculations

A few years ago a semiempirical expression for the per-atom capture probability P(Z) was derived [14]. Exotic-particle capture is proposed to be effected by electrons from different shells with different probability $B_i(E_i, n_i, \ell_i, Z, k_{ex}, k_e)$. P(Z) is then given by

$$P(Z) \;\propto\; \sum_i N_i * B_i(E_i, n_i, \ell_i, Z, k_{ex}, k_e) \tag{4}$$

where N_i electrons with binding energy E_i are in shell i (characterized by quantum numbers n_i, ℓ_i). B_i also depends on the momentum k_{ex} of the exotic particle and on the momentum k_e of the emitted Auger electron. If we use first order perturbation theory, an expression for $B_i(k_{ex}, k_e)$

$$B_i(k_{ex}, k_e) \;=\; c_0(k_{ex}) * k_e * \frac{n_i}{Z} \tag{5}$$

with

$$k_e \;=\; \sqrt{2(k_{ex}^2/2m_{ex} + E_{ex} - E_e)} \tag{6}$$

follows. A parameterization

$$B_i = c * \sqrt{1 - E_i/E_0} * Z^a * n_i^b, \quad E_i < E_0, \tag{7}$$

and

$$B_i = 0, \qquad E_i \geq E_0 \tag{8}$$

was suggested. Parameters a, b, c and E_0 were adjusted to yield best agreement with the experiment.

4. Exotic-atom cascade

As can be deduced from the principle of geometric similarity [15], the exotic particle is captured into a highly excited bound state. The exotic atom then deexcites by radiation and also by Auger transitions [16]. Other ways of deexcitation are relevant only for exotic hydrogen atoms and shall not be considered here.

Radiative transitions tend to have as high an energy as possible: The change Δn in principal quantum number is maximized [16]. Radiative transitions therefore preferentially populate circular orbits, levels with maximum possible angular momentum ℓ = n-1. Auger transitions, on the contrary, prefer minimum Δn. They dominate at high n, whereas radiation prevails at principal quantum numbers below n $\approx$ 8. In hadronic atoms the exotic particle will react with the nucleus as soon as it has reached states with low angular momentum.

An isolated exotic atom would be heavily ionized during the cascade by Auger transitions [17]. In reality electrons are refilled from outside, either from the conduction or valence bands (for solids) or by charge transfer (for gases). It is the competition between radiative transitions, Auger transitions and refilling which determines the intensity pattern of exotic-atom x-ray transitions [18].

It is not too difficult to simulate the exotic-atom cascade. Those cascade calculations start at an intermediate level (n= 20 or n= 14 for muonic atoms). The initial distribution of exotic particles over the angular momentum levels and parameters relevant for refilling are adjusted to obtain optimal agreement with the experiment [18].

5. Basic facts about Coulomb capture: Experimental results

5.1 Experimental procedures

Let me now switch over to experimental findings which helped to establish our picture of the Coulomb-capture process.

The experimental techniques have already been described in the talk by L. Simons. I only want to emphasize that more than 90% of all measurements on *muonic atoms* were performed using solid state detectors to measure the characteristic muonic x-rays, which typically range in energy from 10 keV to about 10 MeV. Only in one experiment a spectrum of low energy (10-50 keV) Auger electrons emitted during the cascade was taken [19]. Therefore the Ge detector is *the* detection device of our choice. Please remember that for muonic atoms the sum of the Lyman series intensities equals the number of muons captured by the relevant element. For pions such a normalization does not exist. Therefore, to the best of my knowledge, only one experiment [20] attempted to determine absolute pionic x-ray intensities.

Per-atom capture ratios for muons can also be determined by observing the electrons emitted in muon decay

$$\mu^- \to e^- + \bar{\nu}_e + \nu_\mu$$

The decay electrons have energies up to roughly 50 MeV and are easily detected with scintillation counters. The nuclear capture proceeds only from the muonic 1s level with a rate which is characteristic for the element. Therefore the time spectrum of the decay electrons contains one exponential for each element in the sample. These exponentials can be disentangled to give the probability w_Z for atomic muon capture into component Z [21]. In the following I want to present a few results from work done at the modern meson factories.

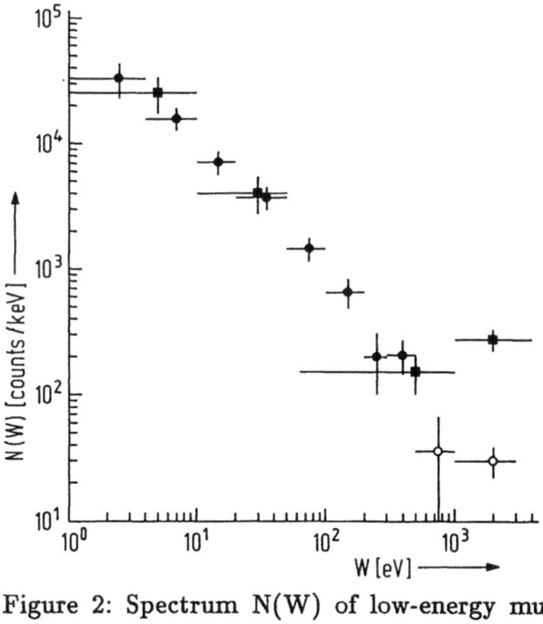

Figure 2: Spectrum N(W) of low-energy muons versus energy W of muons leaving a polyethylene moderator. W: total energy. Circles: $Au_{80}Pd_{20}$ catcher, squares: Si catcher. Open circles: the muons were too energetic to stop in the catcher.

5.2 Detection of muons of very low energy

In a series of measurements the fraction of very low energy muons leaving a moderator could be measured using a time-of-flight technique [22]. The energy spectrum of muons leaving a CH_2 moderator is shown in Fig. 2. The number of particles per energy interval leaving the moderator increases with decreasing energy down to very low energies. This has an interesting consequence: *No* muons would be left at a certain energy E if they all had been captured already at higher energies. This is obviously *not* the case down to a few eV of energy. At these low energies muons can transfer only a few eV to the electrons in a solid. Only therefore one may hope that changes in the energy distribution of electrons by chemical or solid-state effects may affect the atomic capture.

5.3 Per-atom capture probabilities

During the nearly twenty years which have passed since the meson factories at SIN (now PSI)/Villigen, at TRIUMF/Vancouver, and at LAMPF/Los Alamos went into operation a large data set has been collected on per-atom capture ratios in mixtures, solutions (solid and liquid), alloys, and compounds. Compilations of experimental results by Horvath and Entezami [23] (321 entries) and by von Egidy and Hartmann [24] (146 compounds) give a good basis for the study of the Coulomb capture of *muons*. Von Egidy et al. derived from their data set per-atom capture probabilities P(Z) for 65 elements, averaging over different valence states and disregarding the influence of the bond partner(s). Fig. 3 shows the result. A pronounced periodicity of P(Z) with Z leaps to the

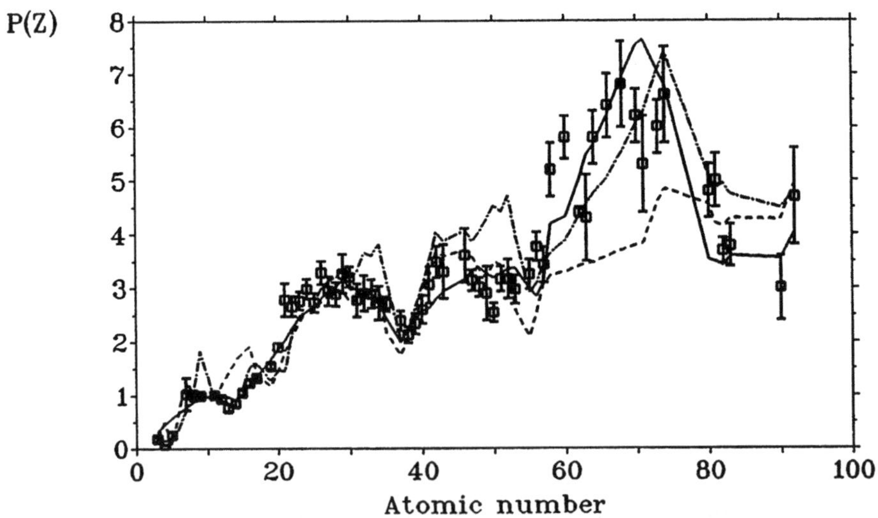

Figure 3: Per-atom capture probabilities P(Z) as function of Z. Solid line: calculations Ref. [14], dashed line: calculations Ref. [11], Dot-dashed line: calculations Ref. [12].

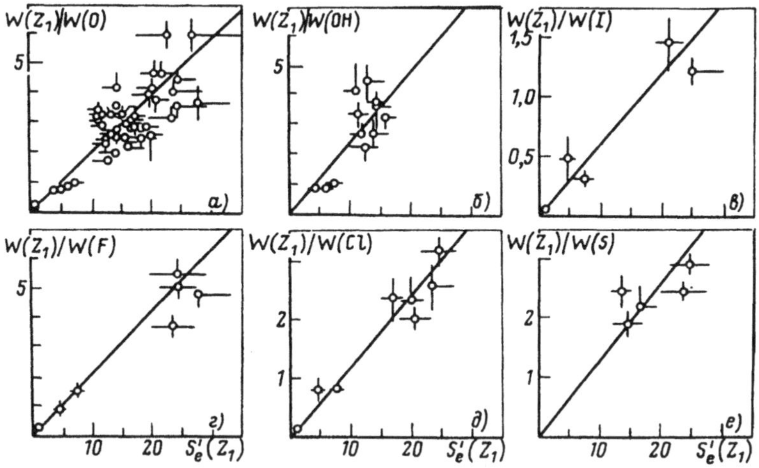

Figure 4: Correlation of per-atom capture ratios A(Z,Z') = W(Z)/W(Z') with the stopping power $S_e(Z)$ for different partners Z'.

eye. P(Z) is small for alkali metals and reaches local maxima roughly halfway in between. Also shown are predictions for P(Z) from a semiclassical calculation [11], from the SPP model [12] and from the semiempirical model [14] (here a number of parameters were adjusted to optimize the result). The periodicity is reproduced by all models. Differences, however, still exist between experiment and calculation.

A series of solid state properties has been correlated with per-atom capture probabilities:

- The density of oxides [25] follows the P(Z) curve. This is in accordance with the semiclassical calculation [11] which demands an anticorrelation with the atomic radius and hence a correlation with density.
- Positron lifetime in annealed metals [26] and P(Z) are anticorrelated: The larger the electron density in the periphery of the atom the larger the interaction (and capture) probability and the smaller the positron life time.
- The correlation between P(Z) and the stopping power S is not conclusive. Whereas a strong *anticorrelation* exists between the results for S [6] and P(Z), Evseev et al. [27] claim to see a *correlation* (see Fig. 4) when correcting the S data for charge transfer from cation to anion with a factor characteristic for the preferred valence number of the element.
- A very interesting correlation was discovered between per-atom capture ratios A(Z,F) for fluorides of the rare-earth elements [28] and the number of possible 4f-to-4f electronic rearrangement excitations from the ground state valence shell electron configuration of the rare-earth metal ion [29].

Only two such states arising from the ground state exist for Ce^{3+} and Yb^{3+}, respectively; the capture ratios are low: A(Ce,F) = 5.40±0.34, A(Yb,F) = 5.05±0.55. For Gd^{3+}, on the other hand, 99 such states exist and the capture ratio is high: A(Gd,F) = 7.29±0.70. It is the high density of low-lying electronic states which leads to a large energy loss of the exotic particle inside the ion and hence to a large capture probability.

6. Exotic-atom x-ray intensities and per-atom capture ratios: The influence of physical and chemical properties

To get a global picture on Coulomb capture I, in the last section, deliberately neglected influences of specific target properties on capture ratios and x-ray intensity patterns. Now is the time to give details of such influences.

6.1 Atomic concentration and atomic capture

As stated earlier the probability for capture into an atom of a mixture or compound may very well be changed by the influence of the partner in the target. If the arrival probability F(E) approaches zero at low energies as all exotic particles have been captured already at higher energies by one partner and the other atomic species can capture only at these low energies,the capture probability for just this species would tend to zero. Indications for such an effect were observed for muons in mixtures of noble gases [30]. The per-atom capture ratio A(Ar,Ne) increases by (35±7) % when changing the atomic ratio Ar/Ne from 0.25 to 4.0 [please remember: the atomic concentration has been divided out already to derive A(Z,Z')]. Obviously the atoms of the heavier element (Ar) capture μ^- at higher energies than those of the light one (Ne): Comparatively few muons are slowed down to low energies without being captured. The effect is even more pronounced for mixtures Kr/Ar, but disappears for Xe/Kr. Apparently at lower Z (up to Kr) the limit of the capture region increases with Z. At high atomic number it tends, however, to be independent of Z.

No concentration dependence was found, again with muons, for A(Nb/V) in a series of solid Nb/V solutions with strongly varying atomic composition [31]. Table 2 shows the results. A possible explanation lies in the solid state prop-

Table 2: Per-atom capture ratios in Nb/V solid solutions as function of the stoichiometric ratio Nb/V.

Stoichiometric ratio Nb/V	0.046(3)	0.182(5)	0.97(2)	4.05(9)	18.5(4)
A(Z,Z')	1.16(9)	1.26(6)	1.17(5)	1.23(6)	1.11(10)

erties of the solid solution. In the picture of the semiclassical model the muon may be trapped behind the centrifugal barrier already at positive energies if a "pocket" exists (as in the potential for L= $32*\hbar$ shown in Fig. 1). This seems to be the case for noble gases with their screened 1/r potential but not in solids with a potential which goes like $1/r^2$.

6.2 Mixtures and compounds

Very nice examples of the influence of the solid state on muonic atom data are the differences which were found between mixtures and compounds of the same elements [32,33,34]. Fig. 5 shows the result of such a measurement [34]. The reason for the puzzling observation could lie in the grain structure of the mixture [35]. Electron transfer from Be to B taking place when forming the compound is another explanation. This view is corroborated by an experiment by Knight et al. [36], who found the per-atom capture ratio A(N/O) to increase by (15±4)% when going from a gas mixture N_2/O_2 to the gaseous compound NO. The authors were even able to attribute this increase to the bond structure of the molecule.

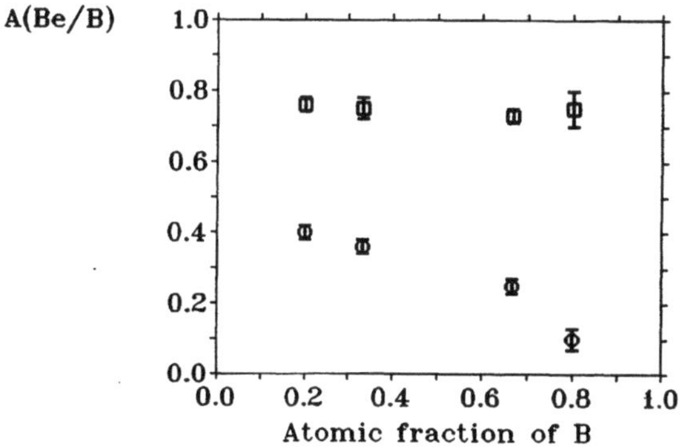

Figure 5: Per-atom capture ratio for pions in mixtures and compounds of beryllium and boron [34]. Squares: mixtures, circles: compounds.

6.3 Valence number and capture

When extracting averaged per-atom capture probabilities for muons, von Egidy et al. averaged over capture ratios for the same elements and different valence numbers. Is this procedure justified? Let us look at capture ratios for oxides in different oxidation states [21,37,38,39] (cf. Table 3). The influence of the oxidation states rarely exceeds 15%, the lower valence number yielding the lower capture ratio. One could be tempted to say: The attractivity of the cation increases with increasing ion charge. One has to be careful with such a statement. Comparing capture ratios of highly ionic compounds, e.g. fluorides, it was found

Table 3: Ratios of per-atom capture ratios in oxides

Oxides	Ratio	Authors	Oxides	Ratio	Authors
V_2O_4/V_2O_5	0.85±0.10	Ref. [37]	SiO/SiO_2	1.09±0.04	Ref. [21]
TiO/TiO_2	0.98±0.02	Ref. [38]	TiO/TiO_2	0.98±0.08	
V_2O_4/V_2O_5	0.94±0.02		MnO/MnO_2	0.88±0.06	
Co_3O_4/Co_2O_3	0.91±0.13		SnO/SnO_2	0.81±0.05	
PbO/PbO_2	0.97±0.11		PbO/PbO_2	0.92±0.07	
UO_2/U_3O_8	0.93±0.10		Cr_2O_3/CrO_3	0.82±0.08	Ref. [39]

Table 4: Experimental results for the per-atom capture ratio in fluorides

Cation	Na^+	Mg^{2+}	Al^{3+}	K^+	Ca^{2+}
A(Z,F)	0.97(5)	0.92(3)	0.95(17)	1.89(19)	1.47(26)

Cation	Rb^+	Sr^{2+}	Y^{3+}
A(Z,F)	2.41(18)	1.83(28)	2.54(19)

[40] that A(Z,F) does *not* change when going from singly charged ions (NaF) to triply charged ones (AlF_3) (see Table 4). In this special case the cation has a noble gas configuration. Please note that if we extrapolate the results from noble gas mixtures Kr/Ar to low Kr concentration (thus minimizing the influence of Kr on the energy spectrum of muons, see Sect. 6.1), A(Kr,Ar) = 1.49±0.10 becomes compatible with the ratio A(Rb,F)/A(KF) = 1.28±0.16.

A conclusive explanation for oxidation-state effects has been given in Ref. [39] for the experimental results found for chromium oxides. With ionicity values of the Cr-O bond of $\sigma = 0.59$ for Cr_2O_3 and of $\sigma = 0.35$ for CrO_3 [41], values for A(Cr,O) of 2.54 and 3.32, respectively, were derived with the help of the SPP model. They agree nicely with the experimental findings [A(Cr,O) = 2.65±0.18 for Cr_2O_3, A(Cr,O) = 3.23±0.22 for CrO_3].

Extensive studies on the influence of the valence state on the capture probability have been performed in the seventies with muons [42,43,44]. A good example comes from Schneuwly et al. [44], who found that the probability of capture into N, S and Se, repectively, in oxidic compounds with sodium (nitrites, nitrates etc.) decreases with increasing valence number. Oxygen is more electronegative for the higher oxidation state of the cation, attracting more electrons which are no longer available in N, S, and Se to cause capture.

In a dedicated experiment on the influence of ion charge and ionicity on atomic muon capture the capture ratios in Cu(SCN) and $Cu(SCN)_2$ were compared [45]. *No* difference was found at an error of a few percent for the probability of capture into S and Cu, although the ionicity of the Cu-S bond decreases by 25% when going from Cu(SCN) to $Cu(SCN)_2$.

6.4 Modification and capture

A striking example for the influence of the modification on atomic muon capture and cascade [37,46] is shown in Table 5. Other measurements also exist [47]. Significant differences show up for different modifications of boron nitride BN. By applying the SPP model and attributing changes in the capture ratio to changes in the ionicity of the bond, the authors derived a ionicity $\sigma = 0.20$ for the cubic case, and $\sigma = 0.10$ for the hexagonal structure. The SPP model hence would allow the determination of ionicities from per-atom capture ratios. Such a drastic influence in σ, however, seems unlikely [41].

Table 5: Effects of the modification in boron nitride

	Ref. [46]	Ref. [37]
A(B/N) cubic	0.233±0.11	0.235±0.20
A(B/N) hexagonal	0.275±0.12	0.258±0.20
A(hexagonal)/A(cubic)	1.18±0.03	

Lyman series intensity ratios			
Cubic		*Hexagonal*	
B(3-1)/B(2-1)	0.276±0.008	B(3-1)/B(2-1)	0.283±0.008
B(4-1)/B(2-1)	0.107±0.004	B(4-1)/B(2-1)	0.109±0.004
B(5-1)/B(2-1)	0.020±0.003	B(5-1)/B(2-1)	0.021±0.003
N(3-1)/N(2-1)	0.262±0.005	N(3-1)/N(2-1)	0.282±0.005
N(4-1)/N(2-1)	0.155±0.003	N(4-1)/N(2-1)	0.177±0.003
N(5-1)/N(2-1)	0.053±0.002	N(5-1)/N(2-1)	0.062±0.002
N(6-1)/N(2-1)	0.013±0.002	N(6-1)/N(2-1)	0.017±0.002

6.5 Conductivity and capture

If we take the semiclassical calculations seriously, there should be an influence of the conductivity on atomic capture as slowing-down is attributed to the energy-loss of the exotic particle to free or nearly free electrons and these electrons certainly should be more abundant in conductors or superconductors than in insulators. This was explicitly shown by comparing the energy spectrum of muons emerging from an insulator with a gap between the valence and conductivity bands (MgF_2 with $E_g \approx 11$ eV) and those leaving a good conductor, in this case Au [48]. The measurement showed that the stopping power in MgF_2 is strongly influenced by the lack of loosely bound electrons. In a collision with a free electron a muon of 5 keV looses only 50 eV at the maximum. As the band gap in MgF_2 is of the same order of magnitude the influence becomes plausible.

6.6 Electronic structure and x-ray intensities

Significant differences in the Lyman-series intensity pattern were found between simple iron compounds like Fe_2O_3 or $FeCl_3$ on the one hand and $K_3Fe(CN)_6$ (solid and in solution) on the other hand [40]. The results for the initial distribution of dedicated cascade calculations are shown in Fig. 6. The simple compounds show steeper initial ℓ distributions than the ferricyanides. This may be attributed to d-to-d electron transitions from the ground state. In the ferricyanides the ground state has spin S=1/2 and the muon has to supply angular momentum in order to enable transitions to higher states, thus loosing angular momentum itself. This is not the case for the simple iron compounds with ground state spin S=5/2. Therefore low-angular-momentum states are populated much more likely in the S=1/2 compounds.

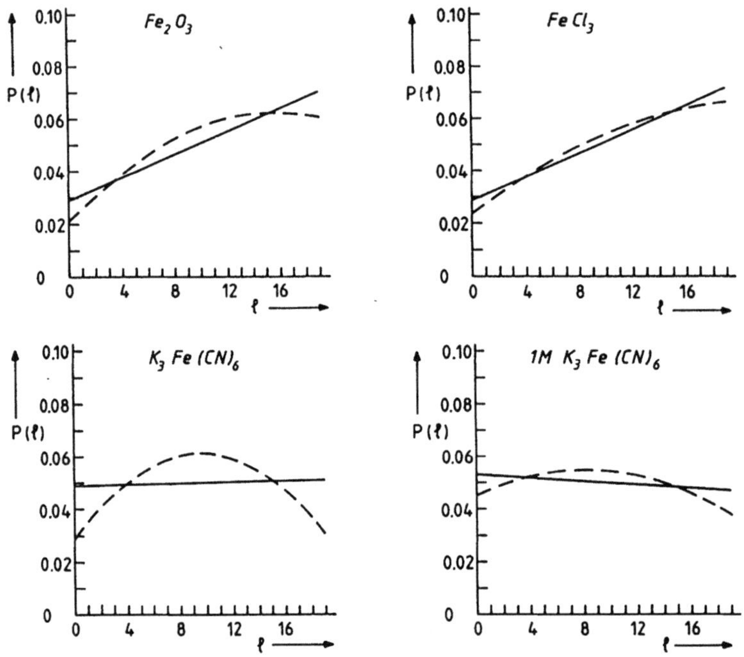

Figure 6: Angular momentum distribution of muons in Fe at n= 20 obtained from a fit of cascade-calculation results to Lyman-series intensities (solid line: linear distribution, dashed line: quadratic distribution).

7. The negative muon as a probe for the composition of matter

Rather few attempts have been made to apply the knowledge on atomic muon capture the analysis of matter. This is not astonishing as the effort for a measurement at one of the meson factories is great . On the other hand materials research with muons has some advantages compared with other methods:

- The analysis can be performed without destroying the object (a minute activation, however, exists).
- The analysis is sensitive to all elements except hydrogen at roughly the same level (atomic concentrations down to $\approx 10^{-5}$).
- The interior even of larger objects can be easily accessed by simply increasing the muon energy. The muonic x rays generated usually have sufficiently high energies to penetrate to the outside of the body.
- Bulk analysis allows one to average over an extended region of the sample, thus avoiding accidental fluctuations in the composition to influence the results.

Scanning analysis with muons was shown to be possible [49]. Spot analysis seems feasible by combining a technique described already long time ago [50] and applied for the visualization of pion stop distributions [51,52] and for the localization of μ^+ stops in μSR experiments [53] with x-ray detection. Surface analysis (even with monolayers) was shown to be feasible as very-low energy muons could be observed [22]. It should become easily possible as soon as improved accelerators and dedicated low-energy muon beams come into operation within the next few years.

Table 6: Scanning analysis of Roman coins. Composition in at. %.

Object	Part	Cu	Zn	Sn	Pb
Coin	interior	88.6(15)	11.4(15)		
	surface	88.6(14)	11.4(14)		
Coin 'Lucilla'	interior	80.7(20)		4.4(8)	14.9(17)
	surface	81.7(19)		4.2(6)	14.1(16)
Coin 'Roma'	interior	83.0(20)			17.0(20)
	surface	84.3(18)			15.7(18)

Biological and archeological [49] objects were analyzed with muons. Let me show two examples. The first one is a determination of the composition of Roman sesterces from different periods of the Roman empire. Just by changing the energy of the muons hitting the coins the body of the coins could be scanned. The results are shown in Table 6. Interior and surface of the same coin have the same composition. The composition, however, changes from coin to coin. The early coin (line 1) was stamped from brass. Later on this alloy, which was expensive because the zinc had to be brought to Rome from far away, was replaced by a Cu/Pb/Sn bronze (coin 'Lucilla') or just a Cu/Pb bronze (coin 'Roma'). The coins became cheaper and cheaper, an ancient example of devaluation.

Scanning analysis was also performed with ancient ceramics from the islamic world [49]. Considerable differences existed in the use of fluxes (sodium carbonate and lead oxide, for instance) and of stabilizers (natural calcium and

Table 7: Scanning analysis of Islamic tiles and pottery. Composition in at.%.

Element	Bowl, Cairo fourteenth century		Tile, Cairo fourteenth century		Bowl, Syria fourteenth century	
	Glaze	Body	Glaze	Body	Glaze	Body
O	59.3(11)	57.2(7)	55.0(7)	59.2(6)	60.5(7)	64.0(6)
Na	0.08(8)	0.21(4)	6.1(2)	2.44(9)	6.42(16)	3.26(8)
Al	8.7(11)	11.7(7)	1.9(6)	1.4(2)	0.7(5)	0.19(15)
Si	22.8(8)	24.8(5)	32.0(6)	34.6(6)	28.4(5)	29.2(5)
K	1.18(16)	1.61(9)	0.84(6)	0.67(6)	1.29(10)	0.75(6)
Ca	2.5(2)	2.25(11)	2.31(10)	1.26(6)	2.53(10)	2.37(9)
Fe	0.91(15)	0.67(8)	0.30(9)	0.25(5)	0.20(3)	0.24(2)
Sn	.	.	0.55(7)	0.06(3)	0.04(2)	0.003(3)
Pb	4.6(4)	1.6(2)	1.06(14)	0.16(3)	0.07(2)	0.009(6)

aluminium compounds). The results for three ceramic objects, manufactured in the fourteenth century, are shown in Table 7. The Egyptian bowl is characterized by a lead glaze with an Al stabilizer, whereas for the tile a soda glaze and no stabilizers were used. The Syrian bowl has a very clean soda glaze (nearly no lead and tin) and again no stabilizers. In the islamic countries glazes containing higher amounts of potassium and tin, as used in Europe, were unusual.

References

[1] D. Horvath, R.M. Lambrecht, Exotic Atoms - a bibliography 1939 - 1982, Elsevier, 1984.

[2] H.A. Bethe, Ann. Phys. **5**, 325 (1930).
H.A. Bethe, Z. Physik **76**, 293 (1932).

[3] J.F. Janni, At. Data Nucl. Data Tables **27**, 147 (1982).

[4] E. Fermi, E. Teller, Phys. Rev. **72**, 399 (1947).

[5] J. Lindhard, Dan. Mat. Fys. Medd. **28**, No. 8 (1954).

[6] B.M. Latta, P.J. Scanlon, Phys. Rev. **A13**, 1370 (1976).

[7] W. Wilhelm, Ph. D. thesis, Technische Universität München, 1980.

[8] F.J. Hartmann, Talk given at the "International Workshop on Electromagnetic Cascade and Chemistry of Exotic Atoms", Erice, May 14-21, 1989.

[9] P. Vogel, P.K. Haff, V. Akylas, A. Winther, Nucl. Phys. **A254**, 445 (1975).

[10] H. Daniel, Phys. Rev. Lett. **35**, 1649 (1975).

[11] H. Daniel, Z. Phys. **A291**, 29 (1979).

[12] H. Schneuwly, V.I. Pokrovsky, L.I. Ponomarev, Nucl. Phys. **A312**, 419 (1978).

[13] L.I. Ponomarev, Yad. Fiz. **2**, 223 (1965).

[14] T. von Egidy, D.H. Jakubassa-Amundsen, F.J. Hartmann, Phys. Rev. **A29**, 455 (1984).

[15] S.S. Gerstein, V.I. Petrukhin, L.I. Ponomarev, Yu.D. Prokoshkin, Usp. Fiz. Nauk **97**, 1 (1969).

[16] G.R. Burbidge, A.H. deBorde, Phys. Rev. **89**, 189 (1953). Y. Eisenberg, D. Kessler, Nuovo Cimento **19**, 1195 (1961).

[17] R. Bacher, P. Blüm, D. Gotta, K. Heitlinger, M. Schneider, J. Missimer, L.M. Simons, Phys. Rev. **A39**, 1610 (1989).

[18] F.J. Hartmann, Talk given at the "International Workshop on Electromagnetic Cascade and Chemistry of Exotic Atoms", Erice, May 14-21, 1989.

[19] R. Callies, H. Daniel, F.J. Hartmann, W. Neumann, Phys. Lett. **91A**, 441 (1982).

[20] F.J. Hartmann et al.,to be published.

[21] S. Stanislaus, F. Entezami, A. Bagheri, D.F. Measday, D. Garner, Nucl. Phys. **A475**, 630 (1987).

[22] H. Daniel, G. Fottner, H. Hagn, F.J. Hartmann, P. Stoeckel, W. Wilhelm, Phys. Rev. Lett. **46**, 720 (1981). G. Fottner, H. Daniel, P. Ehrhart, H. Hagn, F.J. Hartmann, E. Köhler, W. Neumann, Z. Phys. **A304**, 333 (1982).

[23] D. Horvath, F. Entezami, Nucl. Phys. **A407**, 297 (1983).

[24] T. von Egidy, F.J. Hartmann, Phys. Rev. **A26**, 2355 (1982).

[25] S. Stanislaus, F. Entezami, D.F. Measday, Nucl. Phys. **A475**, 642 (1987).

[26] R. Kunselman, J. Law, M. Leon, J. Miller, Phys. Rev. Letters **36**, 446 (1976).

[27] V.S. Evseev, T.N. Mamedov, V.S. Roganov, N.I. Kholodov, Yad. Fiz. **35**, 513 (1982).

[28] F.J. Hartmann, R. Bergmann, H. Daniel, T. von Egidy, G. Fottner, R.A. Naumann, J.J. Reidy, W. Wilhelm, Z. Phys. **A308**, 103 (1982).

[29] H. Daniel, R.A. Naumann, Muon Catalyzed Fusion **4**, 121 (1989).

[30] P. Ehrhart, F.J. Hartmann, E. Köhler, H. Daniel, Phys. Rev. **A27**, 575 (1983).

[31] R. Bergmann, H. Daniel, T. von Egidy, F.J. Hartmann, J.J. Reidy, W. Wilhelm, Phys. Rev. **A20**, 633 (1979).

[32] V.M. Abazov, V.S. Butsev, D. Chultem, D. Knotek, A. Mastalka, V. Presperin, Nucl. Instr. Methods **169**, 423 (1980).

[33] G.A. Grin, R. Kunselman, Physics Letters **31B**, 116 (1970).

[34] N. Imanishi, T. Furuya, I. Fujiwara, A. Shinohara, H. Kaji, S. Iwata, Phys. Rev. **A32**, 2584 (1985).

[35] H. Daniel, Nucl. Instr. and Methods **150**, 609 (1978).

[36] J.D. Knight, C.J. Orth, M.E. Schillaci, R.A. Naumann, F.J. Hartmann, J.J. Reidy, H. Schneuwly, Phys. Letters **79A**, 377 (1980).

[37] J.D. Knight, C.J. Orth, M.E. Schillaci, R.A. Naumann, H. Daniel, K. Springer, H.B. Knowles, Phys. Rev. **A13**, 43 (1976).

[38] T.von Egidy, W. Denk, R. Bergmann, H. Daniel, F.J. Hartmann, J.J. Reidy, W. Wilhelm, Phys. Rev. **A23**, 427 (1981).

[39] M. Hild, K. Kaeser, G. Piller, H. Schneuwly, J. Physics **B18**, 2093 (1985).

[40] R.A. Naumann, H. Daniel, P. Ehrhart, F.J. Hartmann, T. von Egidy, Phys. Rev. **A31**, 727 (1985).

[41] L. Pauling, The nature of the chemical bond and the structure of molecules and of crystals: an introduction to modern structural chemistry, 3rd edition. Ithaca, NY, 1960.

[42] K. Andert, V.S. Evseev, H.-G. Ortlepp, V.S. Roganov, B.M. Sabirov, H. Haupt, H. Schneuwly, R. Engfer, Sixth International Conference on High-Energy Physics and Nuclear Structure, Santa Fe, NM, 9-13 June 1975.

[43] T. Dubler, K. Kaeser, B. Robert-Tissot, L.A. Schaller, L. Schellenberg, H. Schneuwly, Report IPF.ME-75-02, Fribourg University, 1975.

[44] H. Schneuwly, T. Dubler, K. Kaeser, B. Robert-Tissot, L.A. Schaller, L. Schellenberg, Nucl. Phys. **A312**, 419 (1978).

[45] H. Daniel, R. Bergmann, V. Dornow, F.J. Hartmann, J.J. Reidy, W. Wilhelm, Hyperfine Interactions **5**, 215 (1978).

[46] H. Schneuwly, M. Boschung, K. Kaeser, G. Piller, A Ruetschi, L.A. Schaller, L. Schellenberg, Phys. Rev. **A27**, 950 (1983).

[47] L. Tauscher, G. Backenstoss, S. Charalambus, H. Daniel, H. Koch, G. Poelz, H. Schmitt, Phys. Lett. **27A**, 581 (1968).

[48] W. Schott, H. Daniel, F.J. Hartmann, W. Neumann, Talk given at the "International Workshop on Electromagnetic Cascade and Chemistry of Exotic Atoms", Erice, May 14-21, 1989.

[49] H. Daniel, F.J. Hartmann, E. Köhler, U. Beitat, J. Riederer, Archaeometry **29**, 110 (1987).

[50] H. Daniel, Nuklearmedizin **8**, 311 (1969).

[51] H. Daniel, J. Reidy, R. Hutson, J. Bradbury, J. Helland, Radiat. Research **68**, 171 (1976).

[52] G. Eaton, H. Kaspar, F. Pozar, K. Shrager, P. Shrager, Nucl. Instr. Methods **135**, 211 (1976).

[53] R. Tedeschi, P. Podini, R. de Renzi, A.P. Jeavons, R. Magnanini, L.O. Norlin, Nucl. Instr. Methods **214**, 509 (1983).

On the Production of Exotic Atoms: From Basic Facts to Advanced Techniques

L.M. Simons[1], *E. Morenzoni*[1], *and F. Kottmann*[2]

[1]Paul Scherrer Institut, CH-5232 Villigen, PSI, Switzerland
[2]Institut für Hochenergiephysik, ETH-Hönggerberg, CH-8093 Zürich, Switzerland

Abstract. After an introduction to different techniques of producing the conventional particle beams for the production of exotic atoms, the difficulties inherent in improving them are worked out. Several alternative approaches which have been used in the past are presented. New considerations and experimental investigations of the production of ultraslow particle beams are finally described.

1. Introduction and Problems

The usual method of producing negatively charged particles needed for the formation of exotic atoms is the scattering of high energy protons on a production target, where among the reaction products the desired particles can be created. As one of the more sophisticated examples, the antiproton production at LEAR/CERN can be mentioned (Fig. 1). Protons with an energy of 26 GeV are extracted from the proton synchrotron (PS) and lead to an antiproton production target. Here the produced antiprotons are focussed with an electromagnetic lens to a beam guiding system, which transports the antiprotons to an accumulator storage ring(AAC) where the particles are stochastically cooled, accumulated and stored for days.

According to the need of the low energy antiproton ring (LEAR), they can be extracted from the accumulator storage ring and brought to the proton synchrotron where they can be decelerated to 600 MeV/c. They are then guided to the low energy ring LEAR where they are cooled again and either accelerated or decelerated. Presently the minimum momentum which can be reached is 72 MeV/c. Antiprotons with this momentum could be extracted from LEAR and brought to a physics experiment.

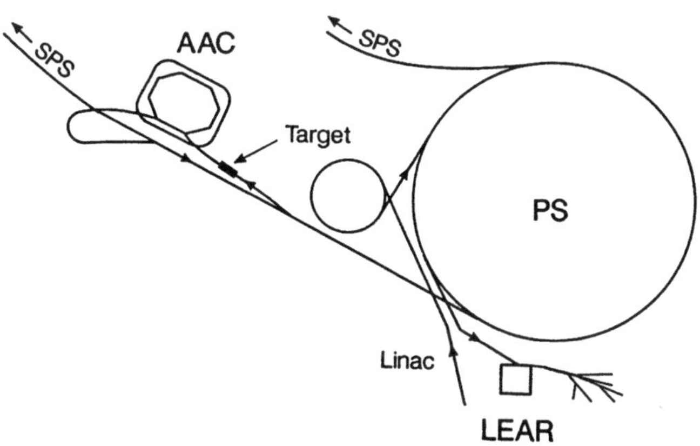

Fig. 1: The antiproton complex at CERN/LEAR

Exotic Atoms in Condensed Matter
© Springer-Verlag Berlin, Heidelberg 1992

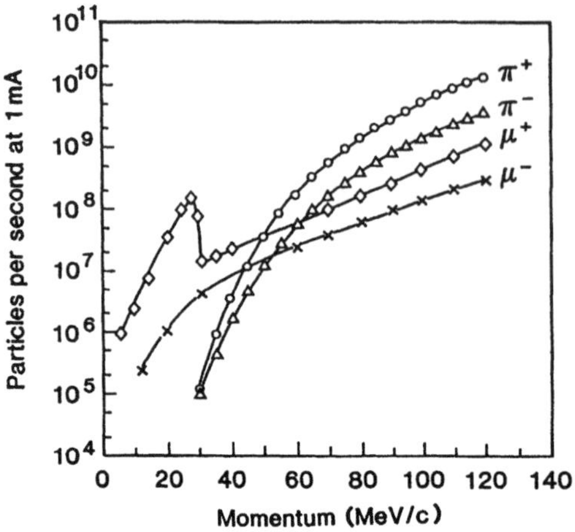

Fig. 2: Pion and muon fluxes at the newly designed π E5 channel at PSI as a function of the particle momenta. A proton current of 1 mA is assumed.

The foci achieved with LEAR beams are of the order of 1 mm^2 with intensities between 10^4 particles/s for the lowest momenta achieved and 10^6 for 105 MeV/c. A comparison with low energy proton beams used, for example, in PIXE studies, where space localisations of the order of μm are routinely reached, shows the orders of magnitude still to be gained.

Considering other particles used for exotic atoms studies shows an even worse situation. For pions and muons the finite lifetime which corresponds to decay lengths of

$$\lambda_\pi = 5.393 \text{ cm / MeV/c for pions}$$
$$\lambda_\mu = 623.7 \text{ cm / MeV/c for muons}$$

does not permit common cooling techniques. This is most regrettable for muons, since they are not absorbed by strong interaction and exhibit the full de-exitation cascade possible for exotic atoms. In addition, they can have a high polarisation whose developement can be followed in the ground state via the weak decay of the muon. These features make them useful probes for the studies discussed during this workshop.

However, the control over the beam parameters is much worse than for antiprotons. Focal spots of the order of $\approx$ 100 cm^2 are reached at the very best with a typical number of 10^7 muons in this focus. The number of muons stopped per gram of target matter and second [$\hat{=}$ stop density] has a maximum value of $2 \cdot 10^5 \mu^-$/gs at the μ-E1 channel at PSI/SIN. For μ^+ this value should be multiplied by a factor of 4. The polarisation of the usual muon beams reaches values around 80 % for beam transport channels optimized to high fluxes.

Much higher fluxes than for muons can be reached for pions, at least at higher energies. As an illustrative example, the expected intensities [1] of the newly designed pion channel πE5 at PSI are shown (Fig. 2).

This channel uses the production of pions at extreme backward angles, which combines high fluxes with good optical properties. The number of particles at the focus as a function of momentum exhibits some interesting features. First, a substantial number of muons is transported to the user. They stem from pion decay in the region between the target and the first bending magnet. Very intriguingly, a peak of μ^+ shows up at about 30 MeV/c. These muons originate from pion decay at rest very near the target surface. Below momentum values of 40 MeV/c, the number of pions rapidly decreases permitting a muon beam with low pion contamination. Such muon beams, which may be called parasitic, are used only occasionally. For most muon experiments, dedicated muon channels are used.

2. Muon Channels

The need for very pure muon beams of high intensities led to the invention of dedicated muon channels making use of special features of the pion decay kinematics [2]. With the mass ratio of the muon and pion mass $\alpha = m_\mu/m_\pi = .7571$, the kinematic parameters of the muon in the rest system of the pion (CM-system) are given by

$$\begin{aligned}
\beta^* &= \tfrac{1-\alpha^2}{1+\alpha^2} &&= .2712 \\
p^* &= \tfrac{1-\alpha^2}{2\alpha}\, m_\mu \, c &&= 29.78 \text{ MeV/c} \\
E^* &= \tfrac{1+\alpha^2}{2\alpha}\, m_\mu c^2 &&= 109.82 \text{ MeV} = m_\mu c^2 + 4.17 \text{ MeV}
\end{aligned}$$

E means the total energy of the particle

In the laboratory system the muons are found in a momentum region given by the limits

$$p_\mu{}^{max}_{min} = \frac{\beta_\pi \pm \beta^*}{\beta_\pi(1+\beta^*)} p_\pi \ .$$

The decay angles with respect to the pion direction are given by

$$\cos\vartheta = \frac{E_\pi E_\mu - m_\pi c^2}{p_\pi p_\mu c^2} \ .$$

The longitudinal polarisation is given by

$$Pol_{long} = \frac{e}{|\, e \,|} \quad \frac{(1-\beta^*)E_\pi - E_\mu}{\beta^* p_\mu c}$$

with e being the charge of an electron.

The distribution of the energies of the muons is flat in between the maximum and minimum energies. The distribution of the momenta, however, is given by

$$N(p_\mu)dp_\mu = \frac{(1+\beta^*)}{2\beta^* p_\pi} \beta_\mu dp_\mu \ .$$

For the sake of illustration, different interesting kinematical points or regions are discussed with the help of Fig. 3 [3]. For pions of momentum $p_\pi = 0$, the laboratory system coincides with the CM-system (Point I). The kinetic energy of the muon is 4.1 MeV and the particles are fully polarized in their direction of flight. For pion momenta up to about 40 MeV/c (Region II), muons decay in the forward and backward direction with respect to the pion flight direction in the laboratory system. The minimum energy of the muons decreases from 4.1 MeV down to 0 energy and the maximum energy increases up to 16.8 MeV. For an illustrative intermediate value, at $p_\pi = 20$ MeV/c the kinetic energy ranges from $\sim$ 500 keV to 9 MeV and the polarization in the direction of flight exceeds the minimum value of 85 % for $\vartheta = 90°$. For point III, the zero kinetic energy of the muons represents the transition to only forward decay in the laboratory system i.e. the muons of this point are standing still in the laboratory system. They are polarized in the flight axis of the pions. The decay angle is 90° in the laboratory system. Pions with momenta above 40 MeV/c represent the normal case used for muon channels. In order to achieve a high purity of the muon beams, the momenta of the muons are chosen near the minimum momentum of the momentum range. The normal procedure is to inject pions into a decay channel (quadrupole or solenoid channel). It is designed long enough to let most of the pions decay with a sufficient acceptance of a

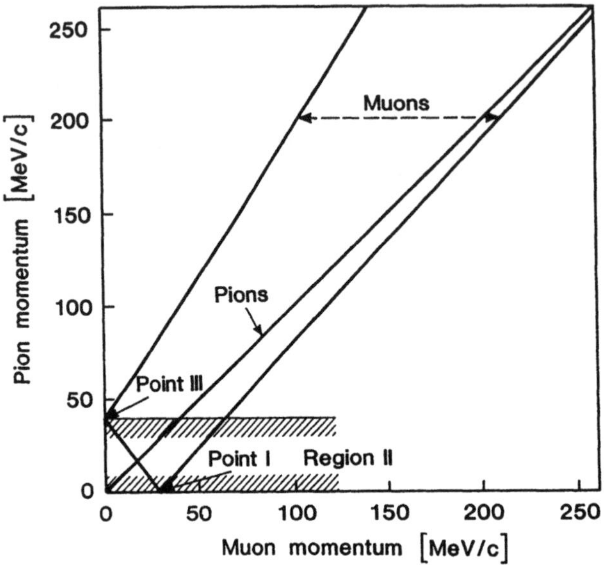

Fig. 3: Pion-muon decay kinematics

sizeable fraction of the muon phase space. High fluxes can be achieved only for higher pion momenta because of the angular acceptance of the muon channels. Thus only the lowest and highest part of the decay muon spectrum is accepted with high efficiency. For these muons also the magnitude of the polarization is highest. As indicated above, the stop density at the presently best muon channel (μE1 at PSI/SIN) is nominally about 10^5 stopped muons per gram of matter and per second for a proton beam current of 100 μA. At the beam momenta typically used, about 100 g of matter are necessary to fully stop the muon beams. For most of the experiments planned for the future, such an amount of target material is prohibitive. As shown below the amount of target material can be reduced for beams with low energies. The difficulties of accepting and transporting muons of lowest energies in present channels are a severe obstacle. Especially solenoidal channels followed by a momentum analysis are unable to transport very low momentum particles because such particles will adiabatically follow the solenoidal field lines and be lost.

3. Problems and Present Solutions

The inherent problems in reaching higher stop densities with the presently available beams are of two different natures. First, the beams currently used are not ideal for muons in the sense that the original beam properties, such as momentum resolution and high emittances, determine the focal structures of the stopped beam. Secondly, even initially ideal beams of infinitely small phase space volume will suffer from scattering and straggling processes while decelerating in matter. Such a situation occurs with the antiproton beams at LEAR with a momentum resolution of less than 10^{-3} and emittances less than 10 πmm mrad in the horizontal plane and 20 πmm mrad in the vertical plane. Even with such exceedingly good beam parameters the longitudinal and transversal straggling, which is only of the order of 0.5 % to 1 % of the range of 100 MeV/c antiprotons in hydrogen, determines the stop distribution at the end. A remedy to this situation would be the decrease of the initial momentum while maintaining the other parameters like emittance and relative momentum resolution unchanged. Another example for muon beams is provided by the μ^+-beams of 30 MeV/c stemming from pion decay at rest. A good approximation to the range-momentum relation is

$$R = ap^{3.5}$$

with the constant a being material dependent. Therefore

$$\Delta R = 3.5ap^{2.5}\Delta p = 3.5(\Delta p/p)R\ .$$

The straggling contributes almost 10 % to the range so that

$$\Delta R_{total} \sim [(.1)^2 + (3.5(\Delta p/p))^2]^{1/2} \cdot R = a[(.1)^2 + (3.5(\Delta p/p)^2]^{1/2} p^{3.5}\ .$$

From this it is evident that a decrease of the initial momentum resolution to values below 3 % is not of much use, but that a decrease of the initial momentum would be a dramatic improvement. All the techniques discussed in the following can be subdivided into two classes although some approaches contains elements of both. The first is the use of high energy beams and the deceleration of them to the desired momentum. The other class is the production of particles almost at rest and the direct use of these systems or the acceleration to the desired momentum. In the following, three techniques are presented which made it possible to produce and stop low energy particles in a controlled way.

3.1 Wedge Technique

A group from the Technical University München used an idea to correct for the momentum straggling produced in a first big deceleration step with a wedge moderator [4,5]. The beam is guided through a magnetic imaging field converting the momentum distribution into a position distribution at a focal plane. Therefore it is possible to correct for the dispersion. In this way a factor of 10 improvement of the number of particles per momentum bin could be achieved. A set up of the experiment is shown in Fig. 4. In front of the wedge moderator,

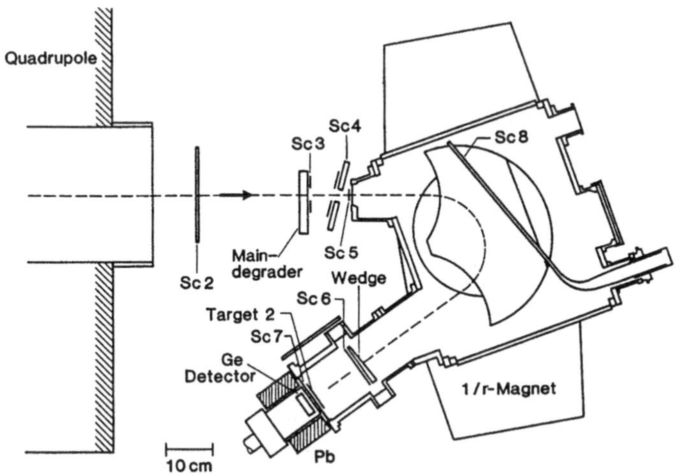

Fig. 4: Experimental set-up for the production of low energy muons

the muons had kinetic energies of about 400 keV. The time-of-flight between scintillator "6" and the target served to measure the energy of the muons leaving the scintillator "6" into the direction of the experimental target 2. The arrival at the target signalled by the X- rays emitted from the muonic atoms in the target was measured by a Ge-detector. Experiments based on this technique were the first to measure the atomic capture probability as a function of energy; they found that exotic atoms are formed at energies between 1 and several tens of eV.

3.2 Use of the Transfer Process

Negative muons stopping in hydrogen almost always reach the 1s-ground state, where they will have a typical life time of 2.2 μs during which the muonic hydrogen atom travels distances of the order of 10 mm. Collision with an atom of higher Z will led with high probability to a transfer process. This mechanism can be used to increase the number of exotic atoms for a specific material by orders of magnitude. One recent example is the investigation of nuclear shape parameters in Xenon isotopes [6]. Working with a high pressure hydrogen target (200 bar) made it possible to fully stop the beam in hydrogen. A contamination of only 600 mg of Xe isotopes leads to a 85 % transfer. This may be interpreted as having reached a 10-fold higher stop efficiency in Xe.

More in the spirit of this workshop is a proposal by L. Bracci and G. Fiorentini to use meson transfer reactions to study effects of surface structure [7]. The idea is to observe the intensity distribution of the X-rays after transfer. An experimental arrangement in order to detect transfer to solid samples is described in a recent proposal to PSI. The proposal did not foresee looking for solid state effects [8]. During the experiment, however, it turned out that the method was heavily dependent on such effects. As an example, the intensity distribution of X-rays after transfer to a gold foil showed an enhanced population of lower angular momentum states. This observation conflicts with the observation of Bertin et al. [9] , who obtained a higher population of higher angular momentum states after transfer to a high-Z gas contamination. Methods discussed below will make it possible to use the transfer process in a wide pressure range from liquid hydrogen down to subatmospheric pressure. Current experiments will establish a knowledge of the velocity distribution of mesonic hydrogen and deuterium atoms in their ground state as a function of pressure. Thus, a means will be available to change the initial conditions before transfer. Transfer from πp-atoms changes the initial conditions to higher main quantum numbers than for transfer from μp-atoms.

3.3 The Muon Bottle

To stop a reasonably high number of negative muons in gases at pressures much lower than one atmosphere, a particular experimental technique known as the "muon bottle" was developed some years ago at SIN/PSI [10]. It is designed to trap low energetic muons in a magnetic bottle, a configuration of a magnetic field produced by two similar coils placed at an axial distance about equal to their diameter. The field at the center of each coil is then about 1.5 times stronger than at the central point between the two coils. This field can be regarded as two magnetic mirrors in which the helical path of a charged particle is reflected back and forth. In the presence of a gas, muons are slowed down and finally stopped even at very low densities - the lower limit being given only by the initial muon energy and the condition that the muon has to be stopped before decaying, i.e. within 2.2 μs.

The injection of a muon in the magnetic bottle is a special problem: In order to be trapped, either the momentum has to be changed or the muon has to be generated within the field of the magnetic bottle. The solution is provided by the in-flight decay of pions with momenta around 40 MeV/c traversing the bottle parallel to its axis. A fraction of all pions

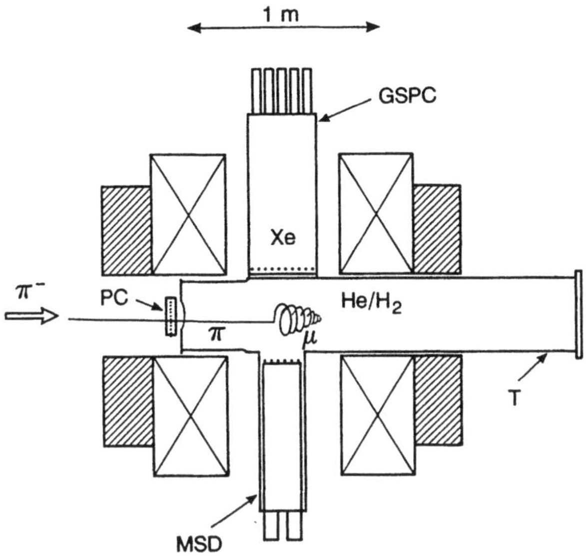

Fig. 5: Experimental set-up: PC = multiwire proportional chamber, MSD = muon stop detector (wavelength shifter, lucite light guide and RCA 8854 photomultipliers), GSPC = xenon gas scintillation proportional chamber (wire planes, mirror light guide and 19 photomultipliers), T = stainless steel tank.

decays within the bottle, and muons emerging from backward decays (in the CMS-system) have very low kinetic energies in the lab system (see point III of Fig. 3) and, simultaneously, decay angles around 90° relative to the pion momentum - a necessary condition for being trapped in the bottle. The dimension of the bottle must be of the same order as the decay length of 40 MeV/c pions, which is 2.2 m.

In the only case where this principle has been realized so far (shown in Fig. 5), the axial distance between the coils was 1 m, and the field on the axis varied between 0.9 and 1.4 T. Muons with an angle bigger than 55° with respect to the field axis were captured inside the mirror field and bounced back and forth with a frequency of $\approx 10^7$/s. In this way, it was possible to stop 500 μ^-/s in hydrogen gas at a pressure as low as 1 mbar (100 N/m^2, T=300 K), corresponding to a density of 10^{-7} g/cm^3, which is by orders of magnitude lower than any pressure yet achieved with other techniques for μ^-.

Originally designed with the aim to measure the 2S-2P energy difference in muonic hydrogen with laser spectroscopy, the development of the muon bottle resulted in much useful experimental information about the stopping process and the electromagnetic cascade in muonic hydrogen and helium [11]. To study these two lightest muonic atoms at low densities, it was necessary to develop two detectors adapted to the peculiarities of the bottle: the muon stop detector, which distinguishes muons stopping in the gas from the numerous pions and muons with higher energies that leave the sensitive volume of the bottle without considerable interaction; and the xenon gas scintillation proportional chamber (GSPC) which yields timing information by detecting the soft X-rays from the K- and L-transitions of the μp and μHe atoms (Balmer and Lyman lines, around 2 and 8 keV). The principle of the stop detector is simple: it is sensitive to the ultraviolet scintillation light that is emitted during the slowing down process of a muon that spirals in the bottle, thereby loosing its initial kinetic energy of typically a few hundred keV by the ionization and excitation of H_2 molecules (or He atoms). In this way, the moderation time in H_2 and He-gases could be

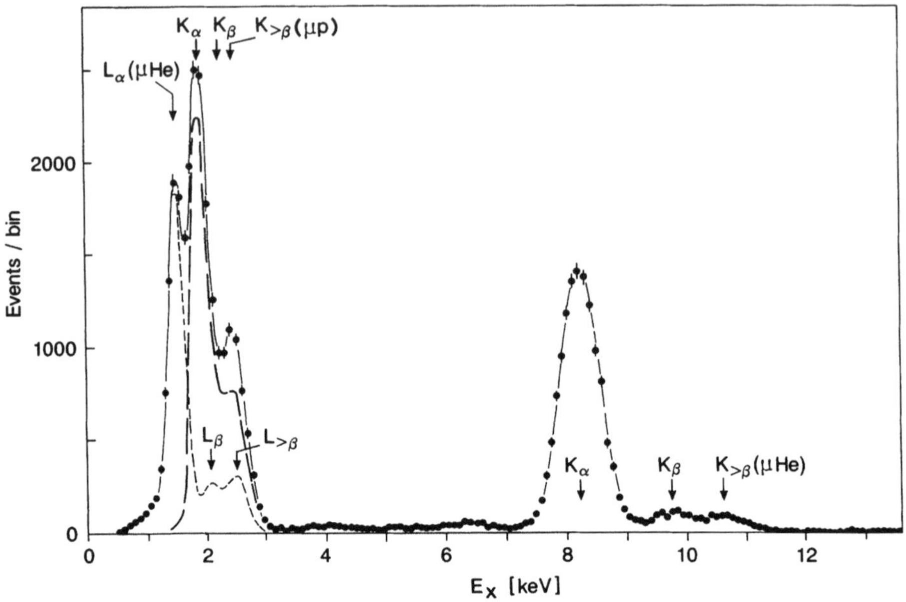

Fig. 6: Typical X-ray energy spectrum measured with μ^- stopped in a 4 mbar H_2 + 4 mbar He gas mixture.

measured for muon velocities around αc and below, a time which could not be measured for negative particles using conventional techniques [10],[12]. This information is necessary for the interpretation of experiments in which the deceleration time of negatively charged particles is important, such as those involving the chain of muon-catalyzed-fusion or the formation of pionic, kaonic and antiprotonic atoms.

The capture process of a free negative muon into a bound atomic (or molecular) orbit is the last step of the slowing down process and, simultaneously, the first of the muonic cascade. For more detailed quantitative understanding of the whole chain, it is crucial to know precisely the distribution of kinetic energies at which the muon is captured, and the quantum numbers of the initial bound state. Ongoing experiments, such as precision spectroscopy of pionic hydrogen or muon-catalyzed-fusion, rely on such knowledge, but experimental information on capture is scarce. A series of model calculations has been performed, which also make predictions of the ratio of probabilities that a muon is captured by helium or hydrogen atoms in a mixture of both gases. This ratio can be measured, e.g. by comparing the number of K-transition X-rays of μp and μHe atoms. There is, however, an additional effect - known for a long time from a series of experiments by Petrukhin et al. [13] - with the consequence that not all μp atoms end their cascade in the μp ground state, but are rather transferred to a helium atom in one of the numerous collisions during the cascade: $\mu p^* + He \rightarrow \mu He^* + p$. (This effect should be distinguished from the well-known transfer from ground state μp atoms to heavier nuclei, which has a high cross-section for elements other than helium or lithium.) The transfer during the cascade depends not only on the relative gas concentrations, but also on the total density, as do all non-radiative cascade processes. By varying the gas pressure by orders of magnitude it is possible to disentangle the pressure-independent ratio of initial capture from subsequent transfer during the cascade. Such measurements have been performed with He+H_2 gas mixtures at total pressures of 8 mbar and 5 bar using the muon bottle technique [12]. Additional information at 40 bar comes from an experiment at Dubna [14]. A typical energy spectrum of the measured X-rays is shown in Fig. 6. The good energy resolution of the GSPC made it possible to determine the number of K-transitions from μp-atoms above the background from the μHe-L-lines.

The main results of an analysis of these data are given here, as a simple example, for a gas mixture with equal partial pressures of He and H_2:

- The pressure-independent probability that the muon is initially captured by an H atom is 57 ± 5 %.
- At the lowest pressure (8 mbar), 49 ± 4 % of all muons reach the ground state of the μp atom. A non-linear dependence on the He concentration was observed, which implies that the difference between 57 % and 49 % is due to a residual transfer effect that necessarily occurs in high atomic levels n>10, since the muonic cascade is purely radiative below n=10 at this low density.
- At 5 bar, only 27 ± 3 % of all muons reach the μp_{1s} state, and at 40 bar this fraction decreases to 21 ± 1 %. This means that one-half to two-thirds of all muons initially captured by hydrogen transfer to helium before reaching the ground state. Cascade calculations as well as additional information from pionic atoms show that this part of the transfer occurs at rather low n-levels around n=4.

These experimental results, in combination with more refined theoretical models and complementary measurements with antiprotonic atoms, give a much better picture of the various processes during slowing down, capture and cascade of muons and other heavy negative particles.

4. Cyclotron Trap

Whereas the muon bottle is especially suited for stopping muons at pressures in the region of mbar and below, the device to be described here is able to stop antiprotons, muons and pions in a more direct way in a wider pressure region [15]. The idea is to decelerate the particles further in matter (normally gases) which fills the decelerating volume defined by a weak focussing cyclotron field. The focussing properties will guide the particles on roughly spiral orbits to the center. The apparatus is shown in Fig. 7.

The method works as follows: A particle beam of momentum p_o is injected into a cyclotron field which accepts a maximum momentum p_{max} for particles bound into orbits in this field. This corresponds to a radius r_{max}. In most cases $p_o \approx 1.7\ p_{max}$. The mean injection point is chosen at a slightly smaller radius $r_{IN} \sim .9 r_{max}$. The azimuthal position of r_{IN} is chosen such that an average radial momentum $p_r \approx 0$ is established. A moderator of suitable thickness (Be or CH_2) then serves to decelerate the particles from p_o to a momentum p_{IN} slightly smaller than $p(r_{IN}) = \frac{e}{c} B r_{IN}$. In this way, betatron oscillations are excited such that the number of particles executing the first revolutions in the cyclotron field without hitting the moderator is maximized. Denoting the range of the particles of momentum p_{IN} injected into the cyclotron field at radius r_{IN} by R and a measure for the longitudinal range straggling by δR the radial spread δr_{stop} is in a first approximation given by

$$\delta r_{stop} = r_{IN} \cdot \frac{\delta R}{R} \, .$$

This should be compared with a normal linear set-up to stop particles which results in a longitudinal extension δR of the stop distribution. The transversal extension is, as a rule of thumb, almost twice as big so that in a normal linear set-up a stop volume of

$$V_{norm}^{stop} = 4(\delta R)^3$$

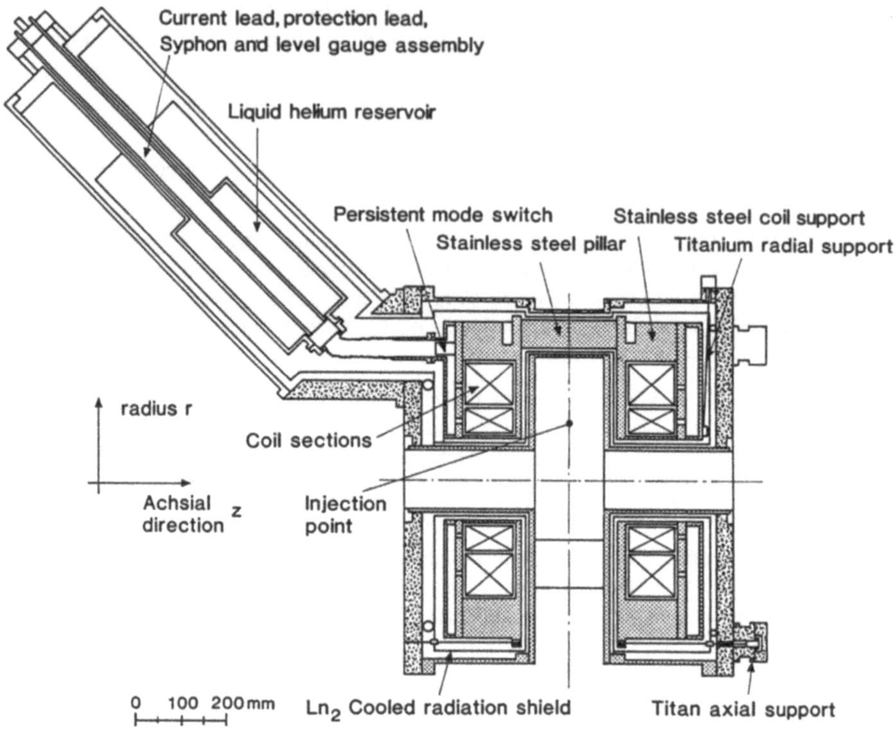

Fig. 7: The cyclotron trap

results. With the cyclotron trap, a stop volume of

$$\begin{aligned} V_{cyc}^{stop} &= 2\pi(\delta r_{stop})^3 \\ &= 2\pi(r_{IN}\frac{\delta R}{R})^3 \end{aligned}$$

ideally results. Thus, a gain factor of $2/\pi(\frac{R}{r_{IN}})^3$ could be reached in principle. Normally, the gain factors are orders of magnitude less. As an example, the deceleration of antiprotons at 100 MeV/c in hydrogen gas with a pressure of 30 mbar requires a range of R = 60 m. For a radius of injection of 0.12 m as used in a realistic set-up, a gain in stop density of almost 10^8 could be expected. During injection into the cyclotron trap of even ideal beams, however, betratron oscillation have to be excited in the way mentioned above. In addition oscillations coming from the momentum straggling in the moderator occur. All together the initial spatial extension of the beam inside the existing cyclotron is an order of magnitude bigger than estimated above. Thus, gain factors of 10^4 to 10^5 are measured. From the argumentation above, it should be clear that with special injection schemes possible for high quality beams such as the present antiproton beams of low momentum, the ideal case could be approached. This can be achieved by using a smaller ratio p_o/p_{max} with correspondingly smaller straggling in the injection moderator and an electric injection kicker to counteract the betatron oscillation. The limitations of the method are different for the different types of particles used. For muon and pion beams, the first limitation is given by the lifetime of the particles, which limits the deceleration time. The second is given by the quality of the beams. Taking the dimensions of the beam spots as a measure of the beam quality, antiproton beams with spot dimensions of 1 mm^2 are the best. The quality of pion beams is much less with typical dimensions of 100 mm^2. The situation is even worse for muon beams produced with pion decay channels where beam spots are still an order of magnitude bigger. Similar comparisons hold for the momenta of the beams and their momentum resolution. Antiproton beams of 72 MeV/c with a momentum resolution of 2 $^o/_{oo}$ are available now. The small momentum resolution together with the small beam spot makes possible a small beam spot at r_{IN}, which is necessary to obtain a small stop distribution at the end. The momentum of 72 MeV/c corresponds to a particle velocity of $\beta = 0.0744c$. Pion and muon beams of comparably low velocity, which is the quantity dominating the energy loss would have momenta of 10.4 MeV/c for pions and 7.88 MeV/c for muons. These are values not realizable for pions and available for muon beams only at a high cost of intensity. In this sense, antiproton beams are in many respects superior to muon and pion beams for exotic atoms studies. Before coming back to antiprotons, the results reached with present pion and muon beams are briefly discussed.

- Pions: Despite the short decay length, a two orders of magnitude gain of stop density could be reached at the $\pi E3$ pion channel at PSI. A beam momentum of 85 MeV/c ($\delta p/p \leq \pm$ 4.5 %) was used. As a target, gaseous hydrogen in a 40 mm diameter Mylar container was used. A specially designed foil structure for the fast deceleration of the pions allowed 3 % of the incoming beam at a gas pressure of 3 atm, to be stopped.

- Muons: A gain factor for the stop density only comparable to that of pions could be reached. The reason for this are the poorer phase-space parameters of the muons from decay channels, which destroys the gain to be expected from the longer lifetime. Progress can only be expected if the beam muons of low momentum (see muons in Fig. 2) could be used directly. In this case an injection efficiency of $\sim$ 50 % can be reached and an increase of 3 orders of magnitude compared to a linear set-up. As an example, 10^5 negative muons per second could be stopped in 10^{-3} g of hydrogen gas (100 mbar in 100 cm^3). This number represents the limit for muons and cannot be increased much for the assumed proton intensity of 100 μA.

- Antiprotons: The results for the stop efficiency as a function of the pressure of hydrogen filling are shown in Fig. 8 for different momenta of the antiproton beams.

  The peculiar feature of the sharp decrease below a certain value of pressure is most evident in the curve representing a beam momentum of 105 MeV/c. It reflects the fact that the energy loss per revolution in the target gas has to reach a well-defined value. Only then is the radial shrinking of the beam particle orbit sufficient to keep the antiprotons away from the region of the moderator. The maximum of stop density reached at 30 mbar was about 10^9/gs with 10^5 antiprotons stopping in a volume of 20 cm^3. In a normal linear set-up, without using a magnetic focussing device, a stop volume of about $4 \cdot 10^5 cm^3$ can be expected for 30 mbar, thus representing the above mentioned gain in stop density of more than 10^4.

As an example of a novel kind of experiment made feasible by the cyclotron trap, the detection of X-rays from exotic noble gas atoms is mentioned here [16]. In Fig. 9 a, b, a part of the X-ray spectrum emitted from antiprotonic neon and krypton at target gas pressures of 20 mbar and 25 mbar is shown. The energy region displayed is 0-23 keV. As an X-ray detector, a Si-detector of 30 mm^2 surface was mounted on the axis of the cyclotron trap at a distance of 75 mm from the center.

At the low pressures used, self-absorption in the gas can be neglected even for very low energies. In the spectrum of antiprotonic neon, a series of transitions corresponding to initial

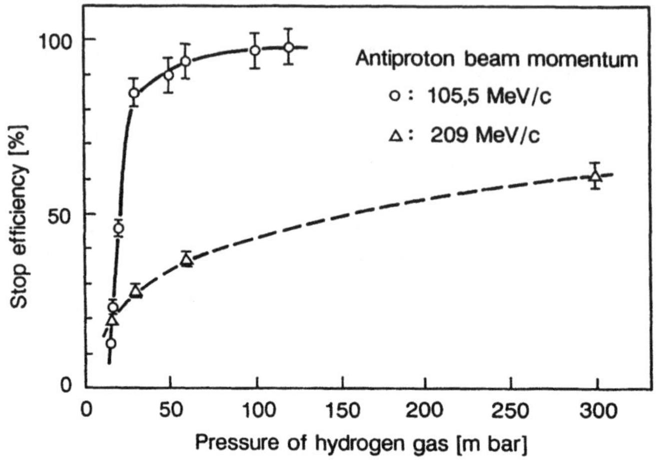

Fig. 8: Stop efficiency for antiprotons as a function of pressure for H_2-filling

main quantum numbers n = 15 to n = 7 can be seen. First it should be noted, that only Δ n = 1 transitions occur. This indicates that we are in a region of the radiative cascade where only the states n, l = n-1 are populated. This needs some explanation: In the beginning of the electromagnetic cascade of an exotic atom, the quantum mechanical states are almost statistically populated with a weight given by 2l+1. They are assumed for simplicity to be Bohr states characterized by the main quantum numbers n and angular momentum quantum numbers l ranging from 0 to n-1 for a specific n. Since the electron shell of the atom is assumed to be filled at the beginning of the cascade the first transition will be so-called Auger transitions with a de-excitation characterized by the smallest steps possible in Δn. If the transition energy is not sufficient to eject one of the remaining electrons after some steps, higher Δn transitions are also possible. This leads to a higher population of high (= outer) l-levels. If the electrons are too tightly bound after the depletion of a shell or if all electrons are ionized, a radiative transition has to occur which prefers to make the highest possible step in Δn. In this case mostly the n, l=n-1 - states - the circular orbits - are populated. The occurence of only circular orbits in the case of neon shows that no electron is left. A more direct proof is the calculation of the probability of radiative transitions below 5 keV. With only one electron present, the intensity of the radiative transition would be suppressed by orders of magnitude. The decrease in intensity at low energies below 3 keV is a result of the decreasing efficiency of the detector. Corrected for this efficiency, the yield of the low energy transitions visible in the spectrum would equal that of the transition of higher energy. The complete ionization of the electron shell is a common feature for antiprotonic noble gases at low pressures for which no refilling during the cascade occures. This is shown in the antiprotonic krypton spectrum shown in Fig. 9 b. This spectrum displays the same energy region as neon but because of the Z^2 dependence of the energies, it corresponds to a much higher part of the cascade. In contrast to the neon spectrum, a smooth dependence of intensity on energy corresponding to the efficiency curve is not visible. Instead, a depression of intensity for the 15-14, 16-15 and also for transitions above n = 25 shows up, whereas the Δn = 1 transitions between 24 and 16 are not attenuated. The natural interpretation of the phenomena is the depletion of the electronic L-shell above n=24 and the ionization of the remaining K-electrons at main quantum numbers n = 15 and 14. This corroborates the picture given above of an initially high depletion of the outer electronic shell together with an early development of a circular cascade. Thus, a strong correlation between the degree of

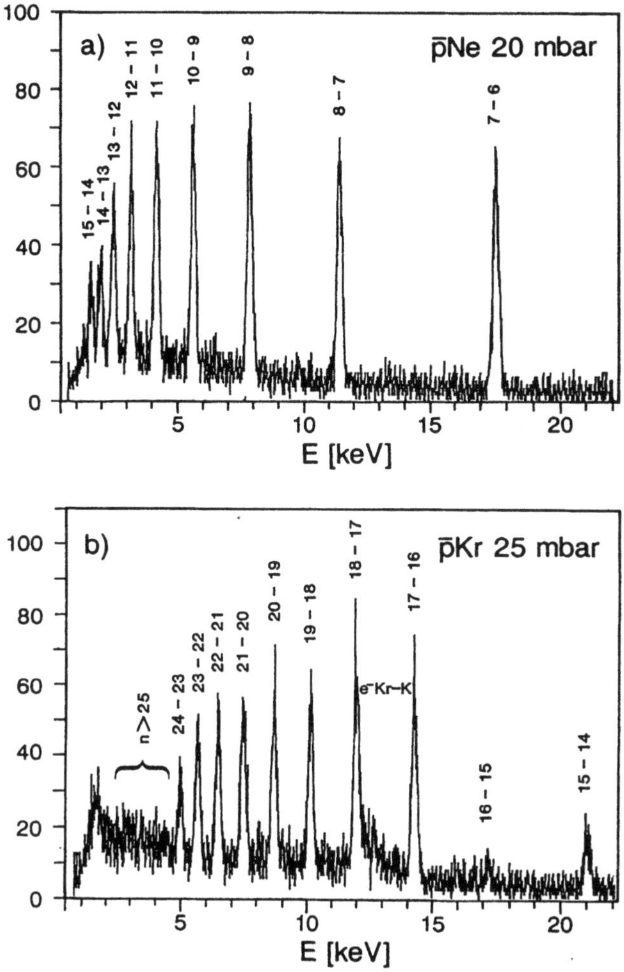

Fig. 9: Antiprotonic X-ray spectra taken with a 30 mm^2 Si(Li)-detector for an energy region below 23 keV
a) antiprotonic neon
b) antiprotonic krypton

ionization of the electronic shell and the main quantum number of the state of the antiprotons occurs very soon. Similar spectra have been measured also for nitrogen, oxygen, argon and xenon. They support the conclusion that the cascade peels off the electrons one-by-one and ionizes the antiproton atom completely.

These facts open a means to produce a well-defined state of the antiprotonic atom whose Q.E.D. binding energy can be calculated accurately. With this uncertainty eliminated, future high precision measurements of the binding energy will be able to determine fundamental quantities of interest like the polarizability of the antiproton or strong interaction parameters without the need of corrections for parallel or Auger transitions. It can also be considered to observe the change of transition intensities as a function of pressure. In this way, electron-refilling cross-sections of interest in plasma physics can be obtained.

A recent development is an attempt to use the cyclotron trap to produce low energy antiproton beams with energies in the region of 10 keV [17]. The idea is to inject antiprotons using beams with the lowest momenta presently feasible at LEAR, which is around 60 MeV/c. With the normal passive injection schemes, a pressure inside the target chamber below 1 mbar for hydrogen gas can be achieved. Much lower gas pressures can be reached by an active injection scheme which relies on the bunch structure of the beam ejected from LEAR. At the moment the first antiprotons stop, the rest of the bunch can be ejected from the trap by applying an electric field in axial direction. The energy gain will be about 10 keV, which is necessary in any case to transport the particles through a thin ($< 10\mu g/cm^2$) carbon window which separates the deceleration gas medium from the high vacuum in the extraction region. The energy gain is also necessary to transport the particles over the axial potential well (mirror field of the cyclotron trap). The purpose of this procedure is the production of low energy antiproton beams, which can be used directly or be injected into electrodynamic traps like a Penning trap. There particles can be cooled further, which would serve as a perfect ion source yielding low energy beams with very small phase space. Another project could be to take the keV particles directly and inject them into a phase space compressing device to be discussed below.

For muons an equivalent procedure is conceivable at least for the passive injection scheme. Beam muons of momenta below 40 MeV survive the injection procedure with probabilities above 70 %. They can then be decelerated in a foil structure to momenta between 1 and 3 MeV/c. An acceleration into the axial direction with an energy gain of 10 keV will transport them to an extraction channel outside the cyclotron trap. Again, such beams can be used directly or tailored by phase space compression. An alternative way to produce low energy muons in the axial region inside the magnetic mirror has been mentioned above in the discussion of the muon bottle technique. Using 40 MeV/c pions injected axially will produce low energy muons which can be guided by the combined action of decelerating foils and accelerating fields. Both schemes are presently being investigated.

5. Use of Stopped Particles

In the following, several methods will be discussed by which particles are brought to rest. They can then be used either in the form of atomic systems or as free particles which can be accelerated to the desired energy.

5.1 Slow Muonic Deuterium in Vacuum

A group at TRIUMF stopped negatively charged muons in natural solid hydrogen containing $\approx$ 150 ppm deuterium deposited on a cold finger [18]. Some muons are transferred from μp to μd before mesomolecular formation or muon decay takes place. During the transfer, an energy of 45 eV is gained which is then lost by the scattering of the μd atoms with the H_2-molecules.

The decisive effect is the minimal size of the scattering cross-section for the process $\mu d + p$ at energies of some eV. Thus the hydrogen layer becomes transparent to the atoms, which consequently leave the layer with high probability. The existence of μd-atoms in vacuum was signalled by measuring the direction of the decay electron of the muons. The set-up is shown in Fig. 10. In this way, a maximum yield of .007 of μd atoms formed in vacuum per incident negative muon could be measured with a hydrogen layer of thickness more than 5 mg/cm^2. By increasing the amount of deuterium in the layer, the yield could be enhanced up to values of 1 %. Such a technique, which could be applied under high vacuum conditions, would be useful in the study of surface effects by using the transfer process.

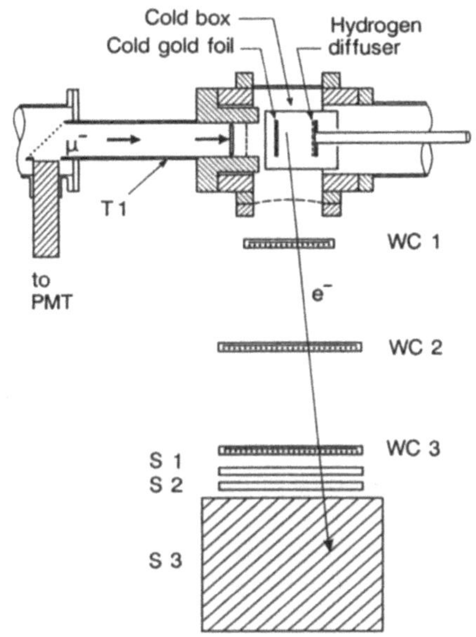

Fig. 10: Experimental set-up used at TRIUMF to detect μd - atoms recoiling into vacuum

5.2 Free Negative Muons after Fusion

A group from the University of Tokyo proposes to use the muon d-t fusion process of deuterium and tritium to ^{5}He in order to produce free negative muons [19]. The idea is that after fusion most of the negative muons do not stick to the α-particle in the dissociation of ^{5}He into an α and neutron, but are left free with kinetic energies of several keV (Fig. 11). The group proposes to deposit muons in a condensed D_2/T_2 layer on a thin cold plate.

Some muons may leave the layer and can be focussed with a magnetic lens system on a target. The evaluation of the number of available muons at small energies is still the subject of an experimental program.

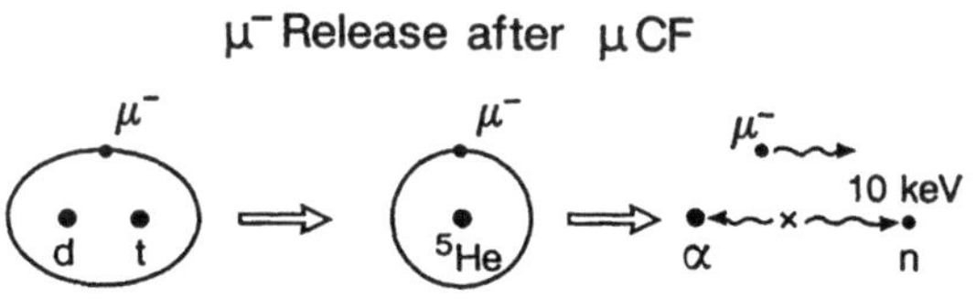

Fig. 11: Schematics of the production of free-negative muons after muon catalyzed fusion.

6. Production of Slow Positive Muons

Positive muons do not form exotic atoms with nuclei. They can be bound only in systems with electrons whose binding energies are in the eV region, and can be ionized again with suitable lasers. Two methods have been discussed in the past to get free positive muons in the eV to keV regions.

6.1 Slow Positive Muons from Noble Gases

It was found by a group at TRIUMF that in analogy to positrons, slow μ^+ beams could be produced by moderation [20]. The group took a beam of 30 MeV/c muons from pion decay at rest and stopped it in a layer of solid noble gases on a thin cold finger. Some of the stopping muons are decelerated to energies below 20 eV where electronic processes like ionization are strongly suppressed thereby strongly reducing the energy loss per unit path length. Eventually they may leave the noble gas environment. The yield per incident muon is on the level of several 10^{-5}, which represents a very small number for present μ^+ beam intensities. With future high intensity beam lines like the $\pi E5/\mu E5$ beam (Fig. 2) however, intensities of more than 10^3 muons could be expected.

6.2 Slow Positive Muons from Muonium

The japanese group working at KEK and TRIUMF pioneered the production of slow muonium from hot tungsten foils in spite of the positive work function of the metal [19]. They propose to ionize the muonium after it has left the metal surface with a suitable high-power pulsed laser and then to focus and form a beam of positive muons. By using the tungsten sheet as a production target for the proton beam they hope to produce a high number of muonium atoms and therefore obtain a reasonable number of μ^+.

As the proton beam is pulsed, the pulsed laser is ideally suited to maximize the number of free muons. A schematic picture of this technique is shown in Fig. 12.

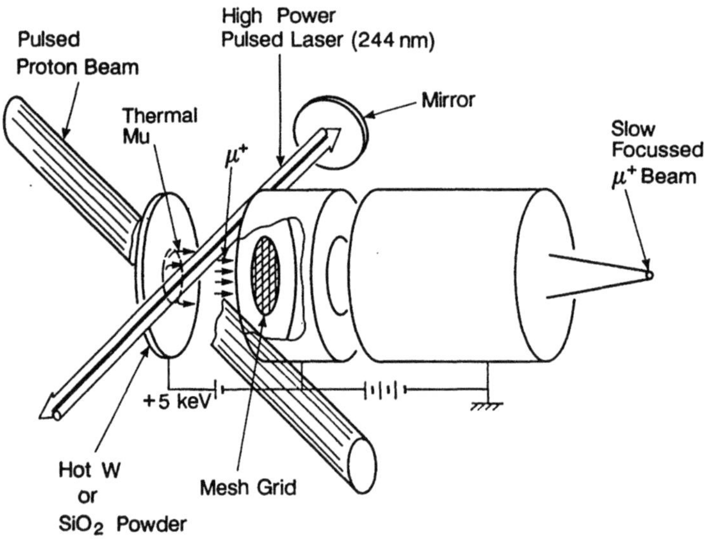

Fig. 12: Schematic set-up for the production of slow μ^+-beams

7. Phase Space Compression of Muon Beams

Phase space compression (PSC) has been successfully applied to beams of protons and heavy ions to improve the beam quality, and it is now implemented and routinely used at various accelerators (as for instance at the antiproton complex at CERN [21]. The methods applied, stochastic and electron cooling, are too slow for unstable particles like the muons, where rapid action is required within the lifetime. To circumvent this limitation, D. Taqqu [22] at PSI proposed a method which can be considered as the limiting case of stochastic cooling and defined as single particle cooling. In stochastic cooling, appropriate pick up elements provide information about the beam parameters (for instance lateral dispersion) from a sample of particles. In a successive step, according to the measured values, a correction is applied to reduce the spread. One can show that by repeating this process many times, the beam is cooled as a whole and a phase space compression is obtained.

In the PSC scheme for muons, each particle is treated separately: its position in the phase space volume is measured with position-sensitive particle detectors and it is steered to a desired position with pulsed electromagnetic deflection systems. Such a system is reminiscent of the Maxwell demon, which is supposed to reduce the entropy of a gas (thus violating the second law of thermodynamics) by a similar procedure.

The sequence, measurement-steering, has to be repeated many times and the process has to be divided into various consecutive stages, each one responsible for cooling action and for the optimal preparation of the particle for the next stage. A specific design has been proposed [22],[23], consisting of a many-stage linear beam line connected to an existing superconducting muon channel at PSI, where a degraded pion beam is allowed to decay at the magic momentum of 40 MeV/c. To reduce the required pulsed steering power and increase the response time, the accepted beam has to have a low longitudinal velocity. The first element after the muon channel is a S-solenoid which discards the muons with a total momentum larger than $\sim$ 12 MeV/c. After that they enter the first two stages of effective phase space compression. Each stage is a solenoid, producing an axial field of 5 Tesla. At the entrance of the first solenoid, the particles moving along spirals of radius up to 2 cm and with a spiral axis distribution of $\sim$ 2 cm (FWHM) with respect to the magnet axis are accepted. The first PSC stage reduces the spread of the radii, thus achieving transverse momentum compression. Parallel to this, a longitudinal phase space compression is affected by debunchers. Further stages making use of the arrival time of the particle are foreseen to extract the beam from the magnetic field and to further enhance the beam brightness with special lenses. The total processing time is few μs, thus limiting the input intensity to 10^5 - 10^6 muons per second.

It is evident that such a machine represents a formidable technological challenge. The accepted longitudinal phase space ($E_\mu < 20\ keV$) and the length of a PSC stage ($\sim$ 0.5 - 1 m) requires that within typically 100 ns the parameters of the muon trajectory are measured with a precision of a few percent or better without affecting the trajectory, that these parameters are processed to trigger pulsed electromagnetic elements and that the elements act on the muon phase space position to move it to the desired position. In order to assess the technological feasibility a prototype of the proposed second stage of the PSC beam has been built at PSI and tested [23]. The schematic layout is shown in Fig. 13. The solutions investigated with the prototype for performing the various operations are representative for the whole beam. The phase space accepted by the prototype consists of muons moving along spirals with radius smaller than 2 cm (corresponding to transverse momenta less than 12 MeV/c in the 5 T field of the solenoid) but centered on the magnet axis. The longitudinal energy ranges from 0 to 20 keV. The arrival time of the muon and its coordinates are determined by a detector system consisting of a thin foil at the beginning of the 5 T region of the solenoid and a MCP detector located on the axis at the exit side of the solenoid (at 1.2 T). Muons traversing a thin foil of $\sim$ 15 μg/cm^2 C+MgO emit electrons due to secondary electron emission. The electrons emitted in the forward direction are accelerated toward the detector since the foil is kept at a negative potential of 4 kV with respect to the detector front side. Due to the magnetic rigidity of the electrons they follow the field lines very tightly and hit the detector. In this way, it is possible with a position sensitive detector to obtain a magnified image of the foil plane and to determine the crossing point of the muons at the foil. Since

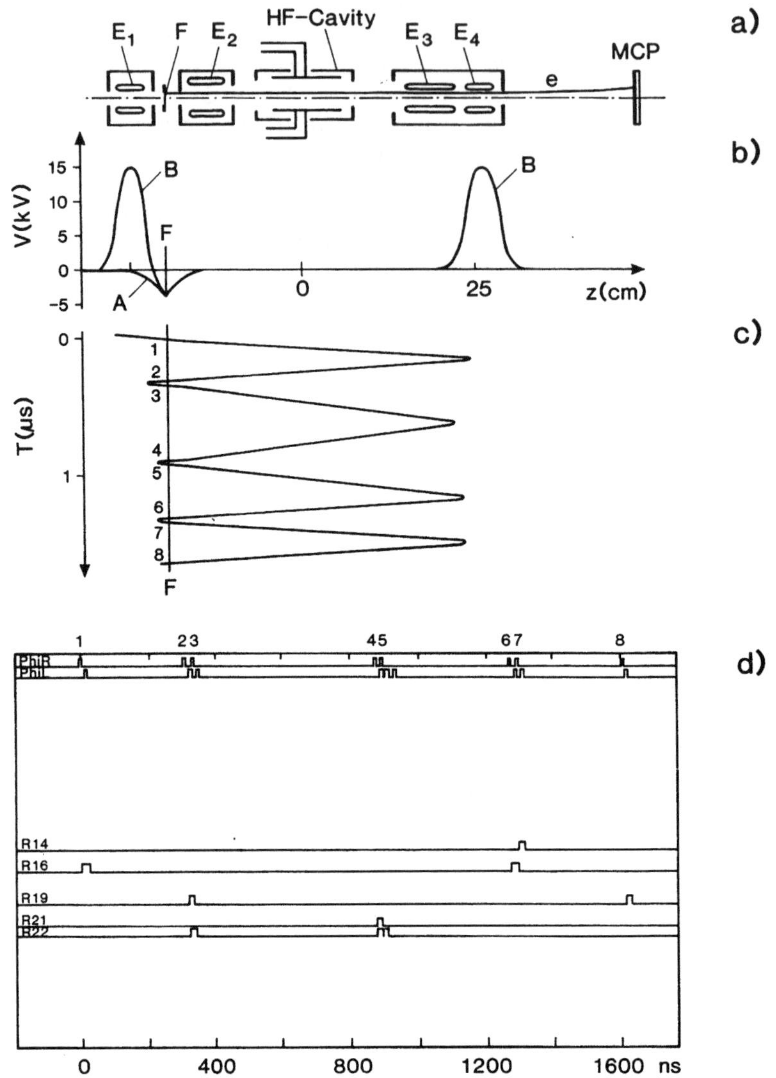

Fig. 13: a) Schematic lay-out of the prototype of the second stage PSC.
E_1 to E_4: pulsed cylindrical electrodes, F: C+MgO foil.
MCP: Microchannelplate
b) Potential configuration to trap a positive muon between E_1 and E_4
(A: High voltage at foil F,B: High voltage at E_1, E_4)
c) Pulse sequence of one muon traversing the foil
d) Measured pulse sequence of a trapped muon with a pipeline TPC.
PhiR, PhiL: angular coordinates, Ri radial coordinates.

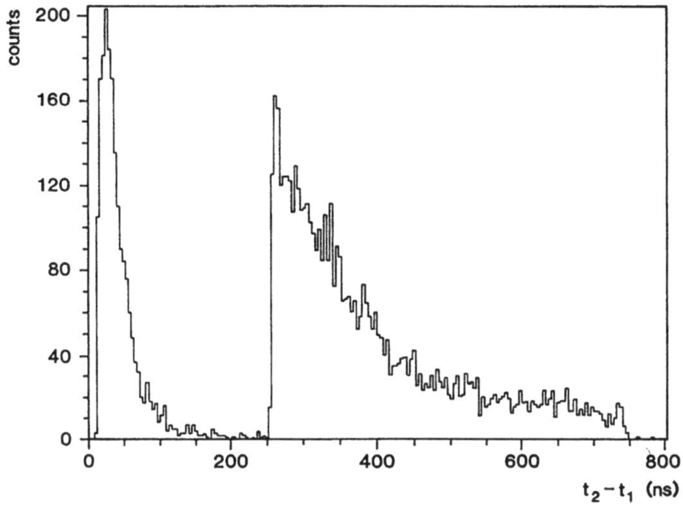

Fig. 14: Time distribution of trapped muons ($t_2 - t_1 > 250$ ns)
The broad peak at ≈ 30 ns is due to detector afterpulsing.

muons entering this part of the PSC beam move along centered spirals (i.e. spirals with the same axis as the magnetic field), the crossing point can be expressed in radial coordinates: radius from the magnetic axis and phase with respect to an arbitrary transversal axis. The radius is directly related to the transverse momentum of the muons. This information plus the crossing time information is used to trigger electrostatic steering elements (cylindrical electrodes and a high-frequency cavity) that trap the muons in the solenoid and act on their momentum, thus obtaining a phase space compression.

The position information must be available within a very short time in order to be able to act on the muons before they leave the solenoid (typically 100 ns). A special readout gives the position information within ~ 40 ns. The phase obtained directly from the propagation time of a delay line is used to trigger the high frequency. A central hole of 9 mm diameter is left in the detector to allow for the extraction of the muons out of the solenoid after their processing.

The processing of the muon proceeds as follows. First, to increase the available processing time, the muon is trapped between two cylindrical electrodes which are pulsed to 15 kV by using a fast timing signal from the detector. The trapped muon moves between E1 and E4 (Fig. 13 b) traversing the foil many times and producing each time a detectable signal with an efficiency of ~ 50 % (Fig. 13 c). Fig. 13 d shows the measured pulse sequence produced by a muon trapped during 1.6 μs. The time sequence is characteristic of a trapped muon with a first pulse followed by a series of double pulses. From the time difference between first and second pulse the longitudinal velocity distribution of the trapped muons can be measured (Fig. 14).

The sharp edge at ~ 250 ns corresponds to muons with a longitudinal energy of 8 keV, the fastest muons which can be trapped in the prototype. With a pulsed cylindric electrode acting as a debuncher (E3 in Fig. 13 a), the spread of the longitudinal energy is then reduced to prepare for the action of an E-field rotating in phase with the muon cyclotron frequency (HF-cavity), which compresses the transverse component. Fig. 15 shows the results of the debuncher action demonstrating for the first time active phase space compression of muons. A full PSC beam line giving an unprecedented increase of beam brightness and muon stopping rate [22] requires an additional intensive research and development program. Nevertheless,

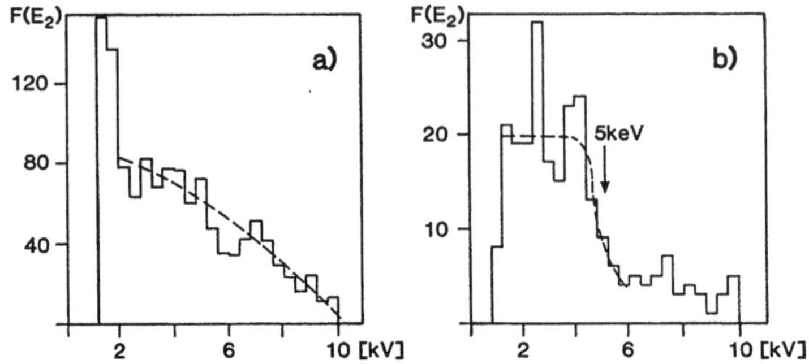

Fig. 15: Integral longitudinal energy distribution of muons
a) before debuncher action
b) after debuncher action.
The sharp edge at 5 keV indicates that the longitudinal energy of the muon has been stabilized at 5 keV.

the prototype study shows that techniques to cool single particles can be realized. Elements of such a technique implemented in a beam line especially designed for low energy muons could provide a sizeable improvement of the beam quality necessary for future generations of muon experiments.

References

[1] R. Abela, D. Renker, PSI Internal Report (1990).

[2] A. Citron et al. CERN Yellow Report, CERN 63-35 (1963)

[3] SIN-USERS HANDBOOK (1981)

[4] G. Fottner et al., Z. Phys. A 304, 333 (1982)

[5] H. Daniel et al, P.R.L. 46, 720 (1981)

[6] Th. Hennemann, diploma work KPH 2/84, University of Mainz

[7] L. Bracci, G. Fiorentini, Phys. Lett. 78A, 437 (1980)

[8] J.B. Kraiman, Thesis, College of William and Mary, WMH EG-89-2

[9] A. Bertin et al., N.C. A76, 35 (1983)

[10] H. Anderhub et al., Phys. Lett. 101B, 151 (1981)

[11] H.P. von Arb et al., Phys. Lett. 136B, 232 (1984)
H. Anderhub et al., Phys. Lett. 143B, 65 (1984)

[12] F. Kottmann, Proc. of II Intern. Symposium on Muon and Pion Interactions with Matter, Dubna 1987, p. 268

[13] V.I. Petrukhin and V.M. Suvorov, Zh. Eksp.Teor.Fiz. 70, 1145 (1976)/
Sov.Phys. JETP 43, 595 (1976)

[14] V.M. Bystritskii et al., Zh.Eksp.Teor.Fiz. 84, 1257 (1983) /
Sov.Phys.JETP 57, 728 (1983)

[15] L.M. Simons, Physica Scripta T22, 90 (1988)

[16] R.Bacher et al, Phys. Rev. A38, 4395 (1988)

[17] J. Eades, L.M. Simons, NIM A278, 368 (1989)

[18] G.M. Marshall, private communication and TRIUMF-proposal #453

[19] K. Nagamine, Contribution to this workshop

[20] D.R. Harshman et al. Phys. Rev. B36, 8850 (1987)

[21] S. van der Meer, Rev. Mod. Phys. 57, 689 (1985)

[22] D. Taqqu, NIM A247, 268 (1986)

[23] A. Fuchs et al, PSI/SIN Annual Reports 1987, 1988, 1989 and to be published

Slow μ^+ and μ^- Beam Production and the Next Generation of Muon Science

K. Nagamine

Meson Science Laboratory, University of Tokyo (UT-MSL),
Hongo, Bunkyo-ku, Tokyo, Japan and
Institute of Physical and Chemical Research (RIKEN), Wako, Saitama, Japan

Abstract. New techniques of intense keV beam production for both positive and negative muons are now under development at UT-MSL based upon the idea of re-emission of the muons from the surface of the stopping material. Descriptions are given on the principle, status of facility construction, future development and new possible muon science research with these slow muons.

1. INTRODUCTION

Muons, μ^+ or μ^-, are unstable particles with a free-decay life time (τ_μ) of 2.2 μs which have many applications such as μ^+SR or μ^-SR, μ^- catalyzed fusion, μ^- X-ray element analysis, etc. So far, in most of the accelerator laboratories, the production of muons is carried out through either pion (π) to μ decay in flight or π^+ to μ^+ decay at rest at the surface-skin of the production target (surface μ^+). There, conventional beams are characterized by a beam size of more than (1 cm)^2 and a range width of more than 100 mg/cm^2.

It is a dream of all muon scientists to have an intense muon beam with advanced features such as a ultra-low energy (keV or lower) and a small spot size (μm diameter or smaller) in order to realize stopping muons in the target within a volume of atomic distance. By using these beams, it is possible to perform all the muon experiments including either μSR or element analysis for the targets with "atomic scale" sizes. Thus the goal of the advanced beam is to obtain a high quality slow and cold muon beam, namely to convert a "lousy" MeV beam to a "delicate" keV beam with a high conversion efficiency within microseconds. Technically, the present subject may have a strong relevance to that discussed at the ISOL(Isotope Separator on Line) facility by heavy ion physicists or chemists.

There have been two proposals up to now of how to produce such a slow/cold beam : a) the re-emission method by using the re-emission phenomena of the thermal muonium (Mu) or muons from materials surface after stopping conventional muons inside the specially selected materials [1-3]; b) the beam cooling method by using electro-magnetic acceleration and deceleration after detecting the phase space of each of the injected conventional muons [4].

In this report, we would like to concentrate on the re-emission method which was mainly proposed and developed by the present author and co-workers at UT-MSL/KEK (Meson Science Laboratory of the University of Tokyo located at National Laboratory for High Energy Physics). The beam cooling method proposed by D. Taqqu is now extensively under development at PSI (Paul Scherrer Institute) [4].

2. SLOW μ^+ BEAM

The re-emission of the slow beam of the μ^+ can be relatively easily achieved compared to that of the μ^-, since μ^+ thermalization and diffusion can be realized in various condensed matter systems. The requirement is therefore to find special materials in which the MeV μ^+ can be converted into the keV μ^+ with a high efficiency.

 © Springer-Verlag Berlin, Heidelberg 1992

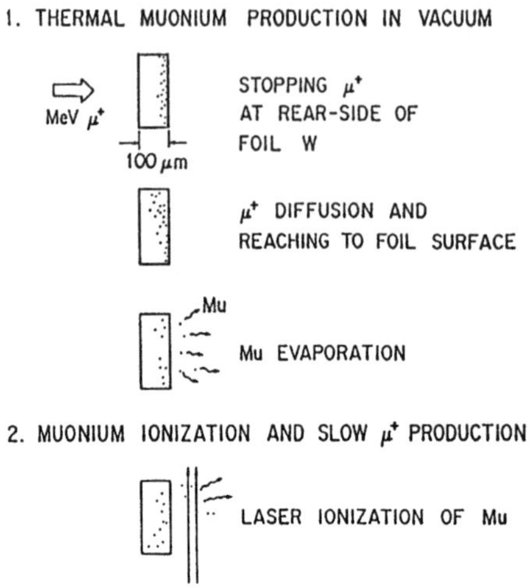

Fig. 1. Schematic picture of the concept of slow μ^+ beam production from hot tungsten.

At UT-MSL/KEK, we have been considering the following three steps towards the realization of the intense slow μ^+ source : 1) generation of thermal muonium (Mu) in vacuum after stopping μ^+ in special materials; 2) production of slow μ^+ by ionization of thermal Mu; and 3) further increase in slow μ^+ production by placing these two steps at the primary proton beam. A schematic view of these steps can be seen in Fig. 1.

The actual first step of the project was initiated by the first successful experiment done at UT-MSL/KEK on the production of thermal Mu in vacuum from a hot tungsten foil [5]. There, in the W foil, the following successive processes take place (see also Fig. 1): the MeV μ^+ stop near to the rear surface of the foil; then a part of the μ^+ diffuses towards the front surface of the foil, and there, after pickking up unpaired electrons, Mu evaporates as a thermionic emission process. The production rate, which is roughly equal to the ratio of the μ^+ diffusion length in W versus the μ^+ stopping length was found to become close to 4 %.

Following our hot W experiment, the efficient thermal Mu production phenomena was also found from SiO_2 powder [6]. Recently, we found that efficient thermal Mu production might be realized in SiO_2 coated (30 A thickness) Si crystal along the (100) direction [7].

As the next step, an experiment on the ionization of thermal Mu was carried out at UT-MSL/KEK by using a laser three-photon ionization method (1S→2S → unbound) [8]. There, although the main concern of the experiment was not intending to achieve an efficient Mu ionization, the laser ionized μ^+ was successfully detected from the thermal Mu in vacuum.

There are several other possible methods for ionizing thermal Mu in vacuum other than the 3 photon ionization: laser 2 photon ionization (1S → 2P → unbound); electron impact ionization; ionization in D^+e^- plasma; laser blasting ionization on Mu on the surface. On the other hand, the direct production of thermal or epithermal slow μ^+ was found from the surface of solid Ar or alkali-halides [9]. As far as the production efficiency is concerned, it is below 10^{-6} for the direct slow μ^+ while it is larger than 10^{-2} for thermal Mu. Therefore, if we find an ionization method with an efficiency larger than 10^{-2}, the resultant thermal μ^+ becomes far stronger than the direct μ^+ production.

After a series of long discussions, we have employed laser two-photon ionization as the most suitable method for the thermal Mu ionization, where the Mu is excited from 1S to 2P by 122 nm photons followed by the absorption of photons with wavelength shorter than 366 nm. The advantage can be summarized as follows: high ionization efficiency can be expected since with an intense pulsed laser source a

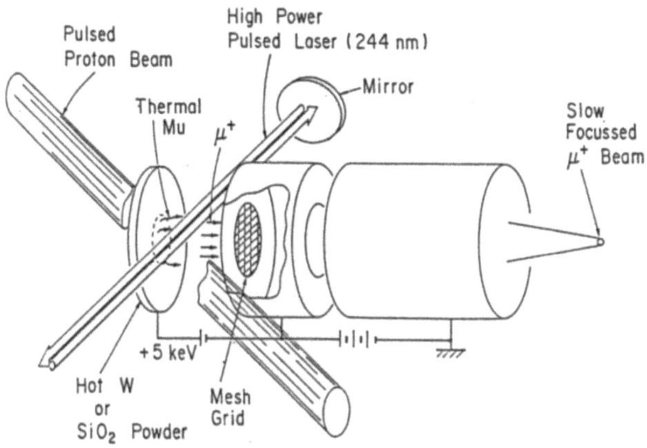

Fig. 2. Schematic picture of intense slow μ^+ production by laser ionization of the thermal Mu produced at the hot W of primary proton beam target.

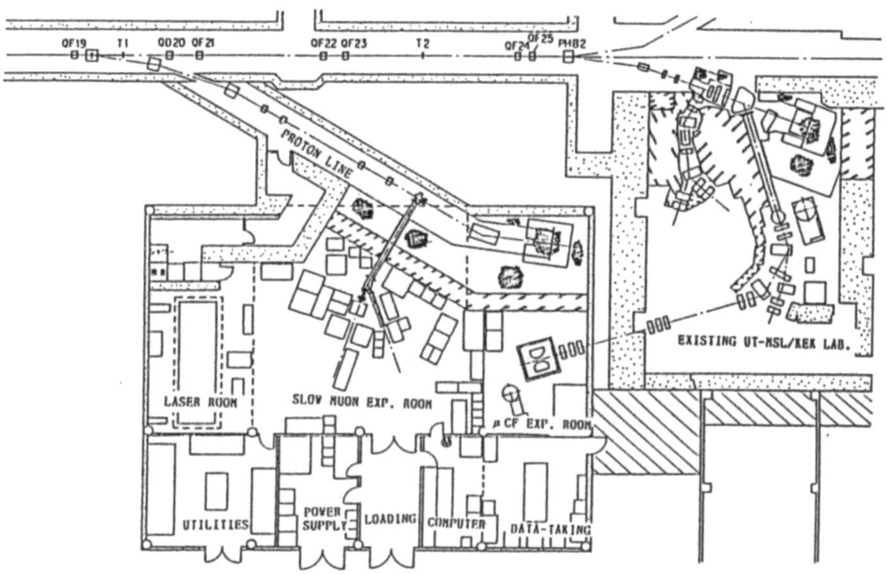

Fig. 3. Layout of the new laboratory of ultraslow μ^+ facility at UT-MSL/KEK.

full ionization is possible for the Mu in a volume of $(5 \text{ mm})^2 \times (\text{a few cm})$. No further momentum spread is expected due to recoil motion and the μ^+ extraction optics following the ionization process is not disturbed. New laser technology is just available for the intense pulsed light source for both 122 nm and around 366 nm. Here, we must emphasize the advantage of sharply pulsed μ^+ beam available at UT-MSL/KEK [10] as well as at RAL (Rutherford Appleton Laboratory), which is just suitable for coupling with the pulsed laser source.

Now we come to the final step of the intense slow μ^+ production; the intense thermal Mu production and the following laser ionization at the production target of the primary proton beam. The basic idea of the new project is depicted schematically in Fig. 2. The project was under construction as the extension program of the UT-MSL/KEK facility. The new laboratory is now under construction including a dedicated primary proton beam line. The schematic picture of the new laboratory is shown in Fig. 3. The installation of the dedicated proton line enables us to have a flexibility not only on the thermal Mu producing material but also on the ionization method of thermal Mu without interfering with the other activities at KEK-BSF (KEK Booster Synchrotron Utilization Facility) facility which, otherwise, are located downstream of our target station.

The slow μ^+ yield was estimated for hot W placed at the proton beam line. For this purpose, a dependence of the yield on the thickness of the W foil was calculated by a Monte-Carlo method in the two types of geometry, forward production and backward production, with the help of K. Ishida (RIKEN, Institute of Physical and Chemical Research). The absolute yield of the slow μ^+ (N_μ) can be expressed in terms of thermal Mu yield (N_{Mu}), ionization efficiency of Mu (ε_{ion}) and collection efficiency (ε_{coll}) as follows;

$$N_\mu = N_{Mu} \cdot \varepsilon_{ion} \cdot \varepsilon_{coll} \qquad (1)$$

where N_{Mu} can be represented with proton current (I_p), cross-section of the stopping π ($\sigma_{stop\pi}$), π to μ conversion rate inside the target ($\eta_{\pi\mu}$) and thermal Mu production efficiency (ε_{Mu}),

$$N_{Mu} = I_p \cdot \sigma_{stop\pi} \cdot \eta_{\pi\mu} \cdot \varepsilon_{Mu} \qquad (2)$$

By using the numbers obtained for $\sigma_{stop\pi}$ $\eta_{\pi\mu}$ by the Monte-Carlo method which is the stopping muon yield within 100 mg/cm2 from the surface (2×10^{-6} for 1 g/cm^2 W) and the other reasonable numbers ($N_p = 3 \times 10^{13}$ p/s, $\varepsilon_{Mu} = 0.04$, $\varepsilon_{ion} = 0.05$, $\varepsilon_{coll} = 0.2$) we obtain the yield of slow muon as 30,000 μ^+/s at the present UT-MSL/KEK with $I_p = 6\mu$A, such a value can be increased up to 5×10^5 μ^+/s when the system is installed at JHP with $I_p = 100$ μA.

In order to obtain the high value of ε_{ion}, the new laser system with all solid-state laser elements are now under development at UT-MSL with the helps of Y. Miyake (UT-MSL), T. Kuga (ISSP, Institute for Solid State Physics of the University of Tokyo) and S. Chu (Stanford). The 122 nm Lyman α light will be produced by the method recently developed by Marangos et al [11]; two-photon resonant excitations of $4p^5 5p$ level of atomic Kr with 212 nm light followed by the subtraction of 823 nm light. The required wave-lengths of 212 nm and 823 nm are all produced in the Nd:YAG (Nd doped Y-Al garnet) pumped Ti:Sapphire laser. At the same time 355 nm light is produced by the 3rd harmonics of the Nd:YAG laser.

The construction of the new building was started in June 1990 and was to be completed in February 1991. All the components for the new proton beam line, hot W target under ultra high vacuum, extraction electrostatic optics, clean room for the laser set-up were to be ready in April 1991. The first slow μ^+ beam was to be produced, hopefully, in May 1991.

3. SLOW μ^- BEAM

The production of slow μ^- beam by the re-emission method is not easy compared to the slow μ^+. All the injected μ^- in condensed matter are captured to the atomic nuclei to form the muonic atom. Except for the emission of neutral (μd) discovered recently [12], it is almost impossible for us to obtain the slow μ^- reemitted from the surface of the μ^- stopping material.

In order to overcome this essential difficulty, a new idea was recently proposed by the present author [3] based upon the phenomena of muon catalyzed fusion (μCF) usually happening in high density and high T concentration D-T mixture. The principle is as follows (see Fig. 4) : (a) the disappearance of the core nuclei of ^{5}He at the instant of the fusion reaction inside the (dtμ) molecule can produce a liberation of 12 keV slow μ^- ; (b) this liberation process is known to be repeated up to 150 times during the μ^- life time ; (c) after successive liberation processes of slow 10 keV μ^- we can expect that a significant fraction of the stopping μ^- inside the solid D-T layer target can be re-emitted from the target surface.

Fig. 4._ Concept of the slow μ^- production via muon catalyzed fusion.

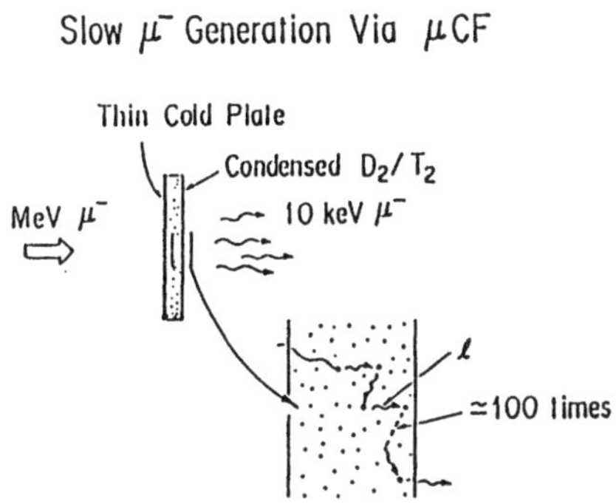

The conversion efficiency (ε_μ) can be estimated to be the ratio of range of the 12 keV μ^+ (0.3μm) versus that of the incoming μ^+ with the energy, say, of 1 MeV (0.9 mm) and the multiplication factor due to number of the μCF cycles ($\sim \sqrt{150}$), giving $\varepsilon_\mu \cong 0.04$.

For a more realistic estimate of ε_μ, the kinetics in μCF must be taken into account. The diffusion length of the neutral ($d\mu$) or ($t\mu$) gives us a significant correction to the value mentioned above. On the other hand, the kinetics of the μCF is one of the central problems to be solved related to the subjects like excited state transfer of (μd) to (μt) or effect of non-thermalized ($d\mu$) and/or ($t\mu$) states, etc. Thus, the proposed slow μ^- production experiment can be considered as the tool to understand the kinetic in the μCF phenomena. Another important correction is the energy spectrum of the liberated μ^- after the fusion reaction inside the ($dt\mu$). The spectrum was calculated by Mueller et.al [13]. The efficiency of the conversion rate was calculated by using Monte-Carlo calculation with the help of K. Ishida (RIKEN) and P. Strasser (UT-Nucl. Eng.). There, the effective diffusion length of ($d\mu$) and ($t\mu$) was taken as a free parameter and the thickness of the D-T layer was considered to be a variable parameter. In summary, by optimizing the thickness against the unknown diffusion length of ($d\mu$) or ($t\mu$), the production efficiency of μ^- below 12 keV could become in the range of a few %.

The realization of the slow μ^- generation will be confirmed by placing a momentum analyzing magnetic spectrometer after the solid D-T target. Actually, the axial focussing magnetic spectrometer which contains a sealed D-T gas and cooling device is under preparation to be used for the first test experiment.

There are several proposals for how to increase the production efficiency of the slow μ^- from MeV μ^-. As described for the slow μ^+, it is possible to place the D-T target assembly close to the production target or even right on the proton beam, where a small (≅1 %) probability of the $\pi^- \to \mu^-$ decay is expected to occur in the D-T target. Another remarkable idea was proposed with the help of G.M. Marshall (TRIUMF, Tri-University Meson Facility). Let us consider that the MeV μ^- be injected in a thick (1 mm) H_2 layer with a small T_2 impurity (around 1000 ppm) to which a thin D-T or even D_2 layer is attached. There, during the slowing down of the injected μ^-, because of the famous Ramsauer effect appearing in the scattering cross-section for (μt) + p, the 3 eV (μt) may have an extremely long range. Thus, the injected μ^- will be fully converted into (μt) which will then be efficiently converted to the slow μ^- at the attached D_2 layer.

4. FUTURE SCIENCE WITH SLOW μ^+ AND SLOW μ^-

Once either slow μ^+ or slow μ^- is produced, a large amount of new science will be realized which may range from fundamental physics to its application fields. Some of the novel subjects are as follows.

(1) Surface sciences with slow μ^+

The intense and high quality polarized slow μ^+ will be used to explore new aspects of material surfaces such as the electronic structure around hydrogen-like impurities on the surface, catalytic chemical reactions of Mu on surface, surface magnetism etc.

(2) Intense production of thermal Mu and QED (Quantum Electrodynamics)

Once we have the keV μ^+ source, thermal Mu can be produced in a controlled way ; complete conversion can be expected from keV μ^+ to thermal Mu on either hot W or SiO_2 surface. Intense and localized thermal Mu can be used for the improved measurements of QED, e.g. 1S → 2S resonance.

(3) Nuclear particle physics with slow μ^-

The μ^- capture will be studied on ultra-low pressure hydrogen gas. The PCAC (Partially Conserved Axial-Vector Current) problem will be solved by the separate measurements on the nuclear capture rate in ortho and para (μp) state.

(4) Improved compact μCF reactor

Intense 10 keV μ^- obtained from the method described above can be collected and injected into a small volume of D-T mixture, providing a compact energy source of μCF. At the same time, the high stopping density of μ^- may produce a new condition for the creation of molecular states with two negative muons. The enhanced fusion rates in (μ^- μ^- d ^{3}He) might open a new concept of the μCF energy source.

(5) Formation of $(\mu^+\mu^-)$ atom

The reaction of $(\mu^+e^-) + \mu^-$ at keV range can efficiently produce a new atomic state of $(\mu^+\mu^-)$ atom. The further advanced QED testing will become possible with this [14].

The author acknowledges contributions from the following colleagues to the creation of the ideas mentioned here, Drs A.P. Mills Jr., K. Ishida, M. Leon, G.M. Marshall and Mr. P. Strasser. Collaborations of Drs S. Chu, Y. Miyake, T. Matsuzaki, T. Kuga, Y. Murata and A. Matsushita are also appreciated. Encouragements from Drs. T. Yamazaki, T. Nishikawa and M. Oda are also acknowledged.

References

[1] K. Nagamine and A.P. Mills, Jr., Los Alamos Report LA-10714-C P. 216 (1986).
[2] K. Nagamine, Atomic Physics 10, 221 (1987).
[3] K. Nagamine, Proc. Japan Academy 65, 225 (1989).
[4] D. Taqqu, private communications.
[5] A.P. Mills, Jr., J. Imazato, K. Nagamine, et al., Phys. Rev. Lett. 56, 1463 (1986).
[6] G.A. Beer et al., Phys. Rev. Lett. 57, 671 (1988).
[7] A. Matsushita et al., contribution to μSR90 (Oxford) (1990).
[8] S. Chu, A.P. Mills, Jr., A. Yodoh, K. Nagamine et.al., Phys. Rev. Lett., 60, 101 (1988).
[9] D.A. Harshman et al., Phys. Rev. Lett. 56, 2850 (1986).
[10] K. Nagamine, Hyperfine Interactions, 8, 787 (1981).
[11] J.P. Marangos, et al., to be published (1990).
[12] G.M. Marshall et al., to be published (1990).
[13] B. Mueller et al., Phys. Rev. 40A, 2839 (1989).
[14] K. Nagamine, Proceeding of the Third LAMPF II Workshop (eds. J. C. Allred et al.). LA9933-C, Los Alamos, p.258 (1983).

Part II

Condensed Matter Studies with Negative Muons

Light Particles in Solids – μ^- in Semiconductors

K. Maier

Max-Planck-Institut für Metallforschung, Institut für Physik,
Postfach 80 06 65, W-7000 Stuttgart 80, Fed. Rep. of Germany

Crystallographic sites of muons and pions are determined using channelling effects of the charged decay particles. By means of time differential techniques at pulsed beams the fate of the particles in the crystal is followed by measuring the channelling profile as a function of time between implantation and particle decay. With negative muons μ^-SR experiments were performed in Si between room temperature and 6 K. The muon spin precession in an applied magnetic field of 0.04 T disappeared below 20 K. A zero-field measurement at 6 K revealed a μ^- spin precession frequency of 650 MHz. The muonic atom represents an aluminium acceptor in the silicon matrix, its electronic state is responsible for the μSR signal. A possible influence of the γ recoil produced by the X-ray cascade is discussed.

1. Introduction

In this contribution, light particles are positrons, muons and pions. Light means that the particle mass is small compared to the proton mass ($m_\pi \approx 1/7 m_p, m_\mu \approx 1/9 m_p, m_{e^+} \simeq 1/2000 m_p$). Over the past two decades these particles have become well-established both in solid state physics and chemistry [1,2]. After implantation in the sample with kinetic energies from several keV to a few MeV, they are thermalized in times which are short compared to their lifetimes τ in the sample ($\tau_{e^+} \approx 10^{-10}$s, $\tau_\mu = 2,2 \cdot 10^{-6}$s, $\tau_\pi = 2,6 \cdot 10^{-8}$s). Now positrons in the sample are in a bound state like the electrons, but we have to note that positive pions and positive muons with thermal energies behave in the lattice like hydrogen isotopes. They occupy well-defined interstitial sites, and their diffusion is determined by quantum effects and in many experiments, pions and muons are used to study the behaviour of hydrogen in materials independent of all solubility problems.

Positrons and positive muons are also very sensitive defect probes. One ppm vacancies in a metal suffices to trap a measurable amount of positrons and muons during their lifetime.

In semiconductors and insulators the light positive particles may capture an electron and form a hydrogen-like atom (Ps = Positronium, Mu= Muonium, Pi = Pionium) which may undergo chemical bonding with the lattice atoms. Two types of Mu have been discovered in many materials:
a free extremely mobile normal Mu atom and an immobile atom Mu* localized in a bond centred position [3,4] Almost all the information we gained about hydrogen in crystalline semiconductors was obtained with the "light isotope", whereas the solubility in Si and Ge is too small for a direct experimental observation of free hydrogen. Nevertheless, the influence on acceptors was underestimated in the past.

As far as the techniques are concerned we can say that positron annihilation and μ^+SR have become well-established. In general they are not sensitive to the position

© Springer-Verlag Berlin, Heidelberg 1992

of the probe particle in the lattice. Channelling with decay particles is another new technique to obtain this information [5,6,7].

The application of negative muons in semiconductors allows us to get new insight into the behaviour of acceptors within the lifetime of the muonic atom.

The behaviour of *negative muons* implanted into condensed matter is radically different from that of μ^+ [8,9]. The electrostatic attraction between the nuclei and the μ^- leads to the incorporation of the μ^- in 1s orbits. Owing to the large mass of the μ^- the spatial extent of the muonic 1s orbital is very small compared with that of the electronic 1s orbitals. This means that in its ground state a "muonic atom"of charge number Z appears to the electonic system like an "ordinary atom" of charge number Z-1.

In the case of Si (Z=14) capture of μ^- in a 1s orbital leads to the formation of quasi-aluminium (Z=13). The particular interest of this is that the introduction of the quasi-Al impurity takes place almost suddenly and that its effects on the electronic system may be studied on the μs time scale.

2. Site spectroscopy with π/μ and μ/ν decay

The use of channelling of the decay products of pions and muons gives direct information on occupied lattice sites. According to

$$\pi^+ \rightarrow \mu^+ + \nu \tag{1}$$

$$\mu^+ \rightarrow e^+, \nu\bar{\nu} \tag{2}$$

$$\mu^- \rightarrow e^-, \nu\bar{\nu} \tag{3}$$

the pions and muons decay into muons, positrons and neutrini. When emitted along low-index crystal directions or planes the decay particles experience the steering forces of the periodic arrangements of the crystal atoms. Therefore, when emerging from the crystal surface, the decay particles show channelling or blocking in their angular distribution, depending on whether, at the moment of their decay, the pions or muons were located at mid-channel positions or at positions inside lattice rows. The intensity distribution of the decay particles measured at a large distance from the crystal is a "finger-print"of the decay sites occupied by the parent particles (Fig. 1). In addition, the spatial extension of the wave function (probability distribution) of the pions and muons around the occupied lattice sites may be derived from the angular width of the muon yield profiles [10].

With pulsed pion beams at the Paul Scherrer Institute we can use a time-differential technique. The channelling pattern is a snapshot of the lattice site occupied at the moment of the decay. The fate of the pions in the crystal is followed by measuring the muon channelling profile as a function of the time interval between pion implantation and pion decay (pion age). With this technique, we measured the dynamic behaviour of the pion in some metals, e.g., in pure copper we observed the pion at octahedral interstitial sites at short times after implantation ($0 - 2 \cdot 10^{-8}$s) between 10 and 400 K (Minimum muon yield along the $\langle 100 \rangle$ directions). However, only below room temperature pions which remain longer than $2 \cdot 10^{-8}$s in the crystal remain on octahedral sites. Above room temperature, the muon profile changes from dip to a maximum which is compatible with metastable tetrahedral site occupation [11].

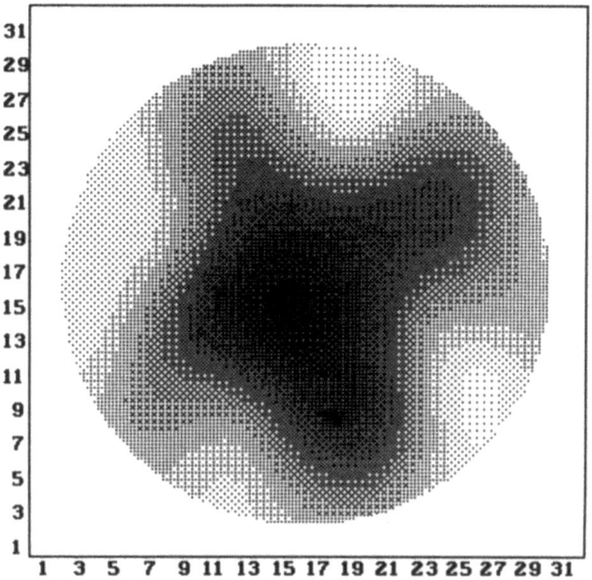

Figure 1
Two dimensional decay muon plot intensity from an iron single crystal aligned in the ⟨100⟩ direction. (Black spot = axial channelling, dark cross = planar channelling of the planes {110} angular scale: one channel = 0,01 degrees.

3. μ^- in Semiconductors

When an incoming μ^- has been slowed down to kinetic energies in the eV range, the formation of a "muonic atom" starts with the capture of a μ^- in a high-lying excited muonic orbital. This process is accompanied by the ejection of an Auger electron. Subsequently, the μ^- cascades down to its 1s orbital, strongly disturbing the electronic shells. Depolarization of the μ^- takes place due to spin-orbit coupling and due to spin-flip transitions during the cascade [12]. Typically only about 18 % of the initial polarization remains when the muon has reached its ground state [13,14]. The entire rearrangement process takes place in a time that is short compared with the typical μSR timescale. Above $Z \approx 7$, the lifetime of μ^- in muonic atoms decreases rapidly with increasing Z, thus giving a signature of the element involved (Fig. 2). In Si (Z=14) the lifetime is 8×10^{-7} s [15].

Another signature is provided by characteristic X-rays emitted during the cascade (Fig. 3).

By means of μ^-SR-experiments in coincidence with X-Ray measurements the local magnetic fields near the nuclei of each element in alloys and compounds ("x-ray triggered μ^-SR") may be measured.

In Si the replacement of a lattice atom of charge number Z= 14 by a muonic atom of effective charge number Z=13 leads to the introduction of an acceptor (Al). The acceptor level close to the edge of the valence band (E_A = 57 meV for Al in Si (Fig. 4) disappears again when the captured muon disappears.

During the μ^- lifetime the acceptor level may be occupied by an electron. This process can be followed quantitatively by observing the interaction between the magnetic moment of the unpaired electron and that of the μ^- by means of μ^-SR.

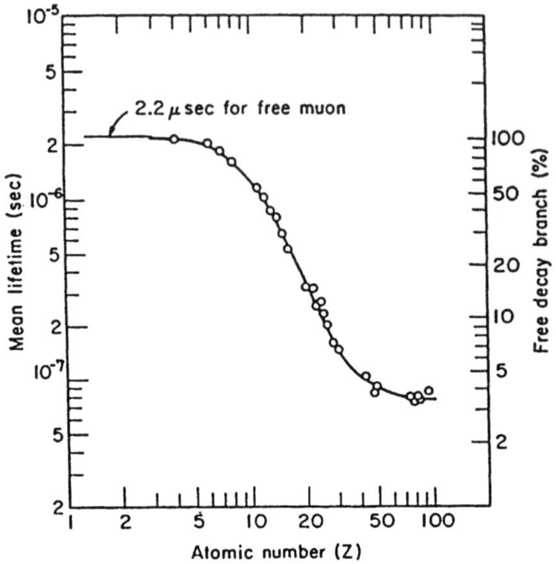

Figure 2
Muon lifetime and free decay branch as a function of the atomic number.

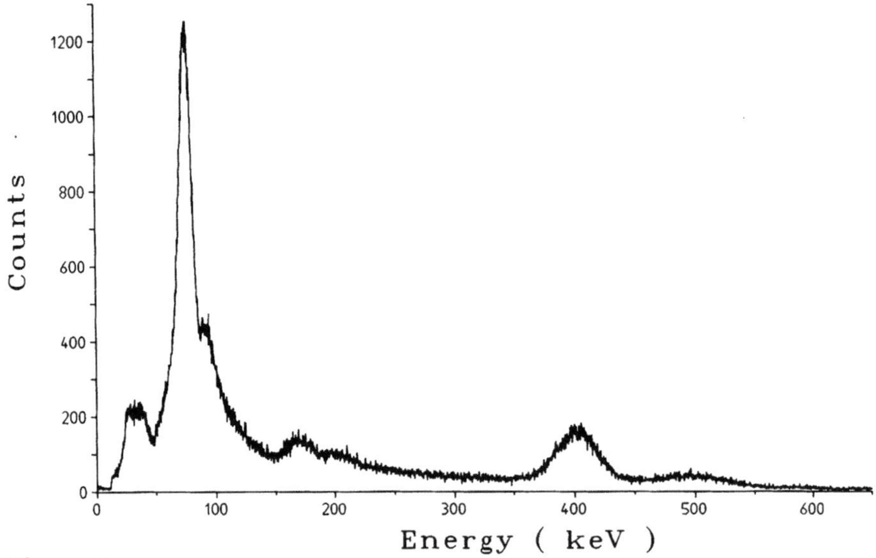

Figure 3
Characteristic muonic X-rays as measured with a 4×1 inch NaI scintillator. For charge numbers Z < 20 the energy resolution is good enough to distinguish between different elements. The energy sprectrum shows a large X-ray line at 76 keV, emitted by the muonic transition from n=3 to n=2 in Si and a smaller peak at 411 keV due to the transition n=2 to n=1. The peaks at 34 keV and 171 keV are caused by muons stopped in the beryllium moderator and by the capture of muons at fluorine in the BaF_2 stop detector, respectively.

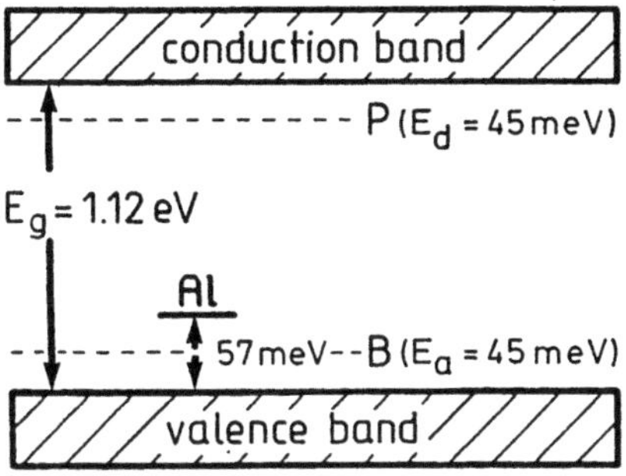

Figure 4
Energy levels of phosphorous P, aluminium Al and boron B in Si. Ea and Ed are the distances from valence *viz.* conduction band.

We must keep in mind, however, that there is no guarantee that the muonic atom remains on its lattice site. It is true that the γ-recoil energy does not suffice to displace the atoms permanently in a perfect environment. (In Si, the emission of a 411 keV photon results in a recoil energy of 3.1 eV). This may be different during the transit of the μ^- to its 1s orbital since the electronic structures and hence the bonding of the atom is strongly disturbed.

4. Experimental Techniques

The experiments were performed on three different silicon crystals between room temperature and 6 K, viz. n-doped Si ($2.5 \cdot 10^{14}$cm^{-3} phosphorus) [16], p-doped Si ($1.4 \cdot 10^{14}$cm^{-3} boron), and intrinsic high purity Si (room temperature resistivity $2 \cdot 10^3\Omega$cm). Natural Si possesses a 95 % abundance of isotopes with zero nuclear magnetic moment, so that further depolarization due to dipolar interaction is negligible.

Two types of experiments were performed using the μE1 beam at the Paul Scherrer Institute (Villigen, Switzerland), viz. μ^-SR experiments in an applied field B_{appl} = 40 mT and zero-field measurements. For the latter an apparatus with high time-resolution was constructed. We used pilot-U plastic scintillators coupled directly to XP-2020 phototubes. By means of constant-fraction discriminators acting on the fast-rising signal of a dynode a time resolution of 130 ps (FWHM) was achieved for prompt events from beam electrons.

5. Results and discussion

In each run of the measurements with $B_{\text{appl}} = 0.04$ T more than $6 \cdot 10^7$ events were accumulated in the four electron-counter telescopes surrounding the sample. The count rates measured in telescope No. i were fitted to

$$\begin{aligned} N(t) = {} & N_1 \exp(-t/\tau_1)[1 + A_1 e^{-\lambda_1 t}(\cos\omega_1 t + \phi_1)] \\ & + N_2 \exp(-t/\tau_2)[1 + A_2 e^{-\lambda_2 t}(\cos\omega_2 t + \phi_2)] \\ & + B_s(\cos\omega_s t + \phi_s) + U, \end{aligned}$$

where τ_1 denotes the decay time of the μ^- in the sample and A_1, λ_1, ω_1, and ϕ_1 are

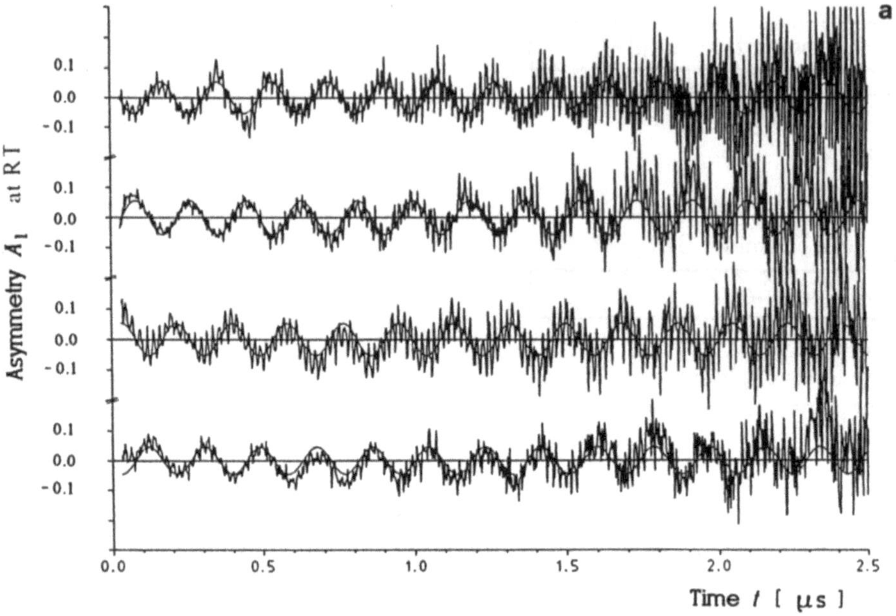

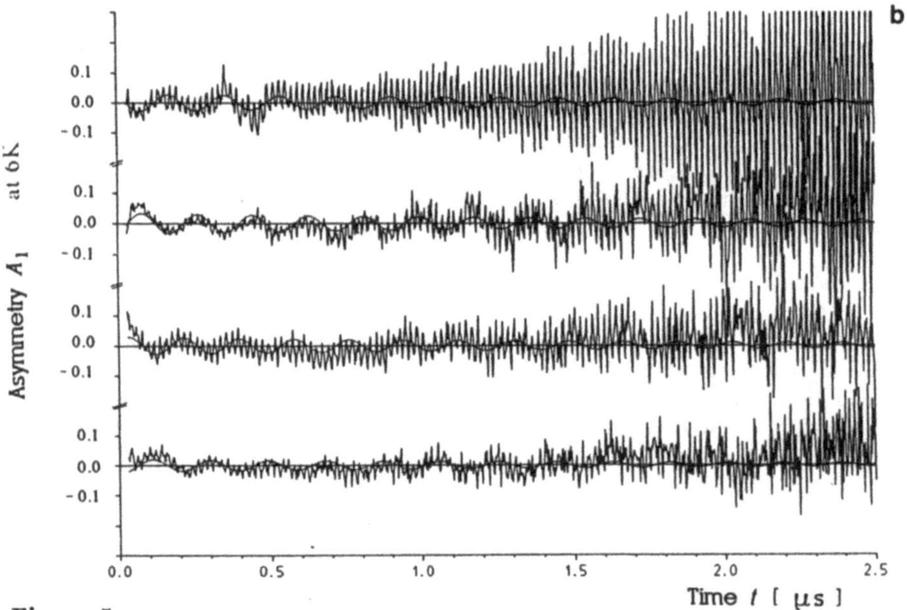

Figure 5

μ^-Sr signal of n-doped silicon (sample Si n1, $2.5 \cdot 10^{-3}$cm-3 phosphorous) in an applied magnetic field of 0.04 T at room temperature (a) and at 6 K (b) after background stubtraction and correction for the finite muon lifetime as measured in four electron-counter telescopes surrounding the sample.

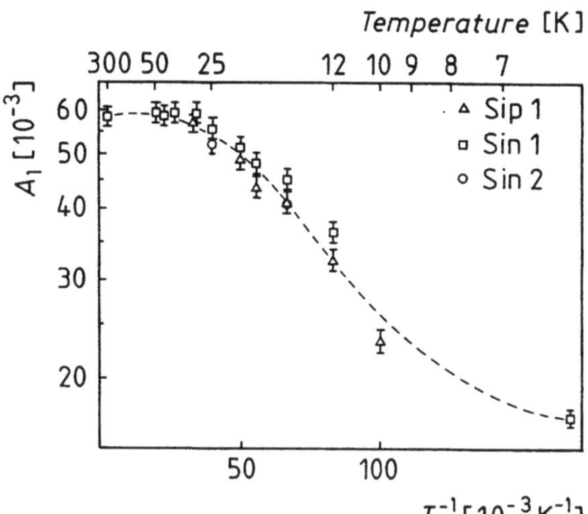

Figure 6
Arrhenius plot of the average symmetries A_1 measured in an external magnetic field of 0.04 T on samples Si p1 (p-doped, $1.4 \cdot 10^{14}\text{cm}^{-3}$ boron), Si n1 (n-doped $2.5 \cdot 10^{14}\text{cm}^{-3}$ phosphorous) Si in (intrinsic, $2 \cdot 10^3 \Omega\text{cm}$). The straight line fitted to the data below 30 K corresponds to an activation enthalpy of $\approx$ 1 meV.

the asymmetry, the damping rate, the circular frequency, and the phase of that signal. The subscript 2 refers to the muons stopped in the moderator. B_s, ω_s, ϕ_s account for the background caused by the beam electrons. U represents the time-independent background.

Fig. 5 shows the spin precession signal of μ^- in Si as measured in the four counter telescopes at room temperature (a) and 6 K (b) with $B_{\text{appl}} = 0.04$ T.

The damping rate λ_1 is less than 100 kHz in the entire temperature range. In Fig. 6 the temperature dependence of the weighted average A_1 of the asymmetries A_1^i is shown for all experiments with $B_{\text{appl}} = 0.04$ T. The type of doping seems to have no significant influence on A_1. An Arrhenius fit to the data below 30 K (the straight line in Fig. 6) yields a very low activation enthalpy of about 1meV. From the fact that in spite of the decrease of the asymmetry A_1 at temperatures below 30 K no significant change in the damping rate λ_1 is observed we conclude that the process responsible for the variation of A_1 is terminated after less than 10^{-9}s.

At 6 K a first measurement in zero field revealed a peak in the Fourier spectrum at 650 MHz (Fig 7). At room temperature no frequency could be resolved.

From hyperfine splitting in EPR measurements on Al-doped silicon we obtain a hyperfine constant of 1.3 GHz [14]. This value, scaled with the ratio between the magnetic moment of the muon and the nuclear moment of aluminium, leads to a hyperfine constant of 3.2 GHz. However, the "nucleus"of the muonic quasi-Al is significantly larger than the aluminium nucleus, so the lower frequency observed in our experiment is also reasonable.

The simplest explanation of the variation of A_1 with temperature would be the temperature dependence of the acceptor ionisation, but with the small activation enthalpy deduced from Fig. 3 we do not get a reasonable value for the energy of the Al level in the bandgap. This is obvious in the case of n-doped Si when the Fermi level is above the aluminium acceptor level at all temperatures. However, there is a substantial difference between μ^-SR at muonic Al-acceptors and classical experiments

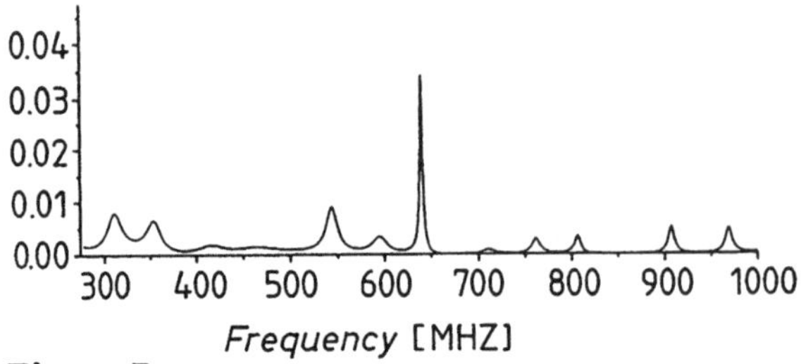

Figure 7
Fourier-spectrum of the μ^+SR signal measured in zero magnetic field at 6 K in intrinsic silicon (sample Si int).

in Al-doped silicon crystals: in contrast to the classical methods, the μ^-SR technique starts to be sensitive immediately after production of the Al-impurity.

Al on interstitial sites may be responsible for the temperature-dependence of the μSR amplitude. At interstitial sites the Al will have Al^{2+} configuration with an unpaired electron. When the muon cascades down to the ground state, a 411 keV γ-quantum is emitted. Under normal conditions the recoil energy of 3.1 eV would not be sufficient to displace Si atoms from their lattice sites. As most of the outer electrons are removed by the muon capture the threshold energy for displacement is presumably lowered to such an extent that quasi-Al interstitials are produced. From the experiments we deduce that only below 20 K are the Frenkel-pair configurations sufficiently long-lived to give rise to a detectable μ^- signal.

In order to exploit the full potential of the present technique further experiments are necessary, e.g. measurements on more heavily doped Si crystals. Experiments to investigate the anisotropy of the displacement process using pairs of aligned γ-detectors and decay site spectroscopy appears to be particularly promising.

Acknowledgements

The author is indebted to his colleagues R. Abela, V. Claus, M. Koch, K. Maier, J. Major, A. Seeger, W. Sigle, W. Staiger, W. Templ, and E. Widmann of the Max-Planck-Institut für Metallforschung and PSI for numerous discussions, and suggestions and last but not least to the 24-hour-per-day support during the experiments. The pion and muon experiments are supported by the Bundesministerium für Forschung und Technologie, Bonn, Fed. Rep. of Germany, under contract no. 03-SE2STU-0.

References

[1] A. Seeger, in: Muons and Pions in Materials Research, Eds. J. Chappert, R. Grynszpan, Elsevier Publishers B.V. 251 (1984)

[2] P.J. Schultz, K.G. Lynn, Rev. Mod. Phys. **3**, 701 (1988)

[3] B.D. Patterson, Rev. Mod. Phys. **60**, 69 (1988)

[4] T. L. Estle, Hyperfine Interact. **64** 525 (1990)

[5] K. Maier, in: Hyperfine Interact. **17** 3, (1984)

[6] H. Simmler, P. Eschle, H. Keller, W. Kündig, W. Odenmatt, B.D. Patterson, B. Pümpin, I.M. Savic, J.W. Schneider, U. Straumann, P. Truöl, Hyperfine Interact. **64** 535 (1990)

[7] G. Flik, J.N. Bradbury, D.W. Cooke, R.W. Heffner, M. Leon, M. A. Paciotti, M.E. Schneider, K. Maier, H. Rempp, C. Boekema, J.J. Reidy and H. Daniel, Phys. Rev. Lett. **57**, 563 (1986)

[8] H. Keller, R.F. Kiefl, H.P. Baumeler, W. K ündig, B.D. Patterson, J. Imazato, K. Nishiyama, K. Nagamine, T. Yamazaki, and R.I. Grynszpan, in: Hyperfine Interact. **31** 461-466 (1986)

[9] J. Imazato, K. Nagamine, T. Yamazaki, B.D. Patterson, E. Holzschuh, and R.F. Kiefl, in: Phys. Rev. Lett. **53** 1849 (1984)

[10] W. Staiger, E. Widmann, S. Connell, G. Fabritius, K. Maier, and A. Seeger, in: Hyperfine Interact. **64** 701 (1990)

[11] G. Fabritius, R. Feldmann, G. Flik, T. Gaul, V. Heinemann, D. Herlach, M. Krenke, K. Maier, A. Seeger, W. Staiger, and E. Widmann, in: Nuc. Instr. and Meth. **B33**, 49 (1988)

[12] K. Nagamine, in: Hyperfine Interact. **6** 347 (1979)

[13] T. Yamazaki, K. Nagamine, S. Nagamiya, O. Hashimoto, K. Sugimoto, K. Nakai, and S. Kobayashi, in: Physics Scripta **11** 133 (1975)

[14] G.D. Watkins, in: Radiation Damage in Semiconductors, P. Baruch (Ed), Dunod, Paris, 333 (1965)

[15] T. Yamazaki, in: Muons and Pions in Materials Research, J. Chappert and R.I. Grynszpan (ed), Elsevier Publishers B.V., 1984, p. 241

[16] M. Koch, K. Maier, J. Major, A. Seeger, W. Sigle, W. Staiger, W. Templ, E. Widmann, R. Abela, V. Claus, M. Hampele, and D. Herlach, in: Hyperfine Interact. **65** 1039 (1990)

Condensed Matter Studies with Polarized Bound Negative Muons

K. Nagamine

Meson Science Laboratory, University of Tokyo (UT-MSL),
Hongo, Bunkyo-ku, Tokyo, Japan and
Institute of Physical and Chemical Research (RIKEN), Wako, Saitama, Japan

Abstract: Experimental advances in solid state physics research using polarized negative muons (in the ground state of muonic atoms) are reviewed. The main subjects are studies of μ^- hyperfine interactions in magnetic materials, electronic structure around oxygen, and some exotic bonding effects, etc. Basic principles and distinctive features of the μ^-SR method are also presented. Futhermore, possible future developments are briefly discussed.

1. INTRODUCTION

The essence of the μ^-SR technique is to measure the precession and relaxation of the spin of the polarized μ^- bound in the ground state of muonic atoms [1,2,3]. As a probe of condensed matter, the bound μ^- behaves as a dilute (Z-1) quasi nucleus which has a spatial distribution extended substantially beyond the nucleus. Since the μ^- wave function is known or calculable, one may expect to explore the electronic structure just at the quasi nucleus, which may appear as a hyperfine anomaly. At the same time, the μ^-SR can be applied to probe the electronic structure around the even-nuclei to which no access exists for NMR. Also, the local exotic bonding formed by the implanted (Z-1) core "nucleus" might be a new subject to be explored.

The earliest μ^-SR experiments were carried out by the Russian group starting in 1958 [4]. They measured the residual μ^- polarization in various materials, with emphasis on the depolarization mechanism during atomic capture or cascade. Then, aiming to measure the relativistic correction on magnetic moment of the bound μ^-, Hutchinson et al. performed μ^-SR experiments on some insulators and metals and discussed, for the first time, a change of the microscopic local magnetic field at the muon due to the surrounding electronic structure, so-called Knight shifts in metals [5]. In 1973, Yamazaki et al. started a program at Berkeley with the measurement of the hyperfine anomaly effect as an ultimate goal. Then, during these last 15 years, various important experimental results have been obtained for the subjects mentioned above.

However, at present, the μ^-SR is still in its infancy, particularly when we compare its accomplishment to the vast accumulation of data in μ^+SR experiments. The main reason is the inherent difficulties of measuring μSR pattern with smaller polarizations and in shorter time windows.

2. FEATURES OF BOUND μ^- PROBE

The muonic atom has a fine structure splitting which is in most cases larger than the natural width, and the capture process for an energetic μ^- is spin-dependent so that there is a strong depolarization. When the μ^- is captured into the atomic state of J = L + 1/2, the population parameters of substates m are given by, for example, Mann and Rose [6] based upon the assumption that the direction of the muon momentum is completely randomized at the time of capture. The statistical tensor $B_1(J)$ exhibiting the polarization of J state is given as:

 © Springer-Verlag Berlin, Heidelberg 1992

$$B_1(J) = (J+1)/3J \times B_1(1/2), \quad \text{for } L + 1/2$$
$$= - J/(3J+3) \times B_1(1/2), \quad \text{for } L - 1/2 \qquad (1)$$

where $B_1(1/2)$ corresponds to the muon beam polarization. The muon passes through the fine structure multiplets where the depolarization is expressed by geometrical coefficients called U-coefficients. We require the calculation of these changes in polarization through the entire cascades. For almost all the even nuclei, the residual polarization P_I when μ^- reaches at the $1s_{1/2}$ state is roughly 1/6.

When muon is captured to Z nuclei, a large number of electrons are ejected from the μ^-Z atoms during earlier stages of the muonic cascade where the Auger effect is dominant. Either during or subsequent to the end of the cascade, the μ^-Z system will be repopulated by surrounding electrons. In most solids, particularly, in metals, the time required for repopulation is short enough compared to that for the muonic cascade (10^{-14} to 10^{-12} sec) so that the rearrangements of electrons might be completed without causing any additional depolarization. Thus, an electronic state corresponding to (Z-1) quasi nucleus is formed before the start of the μ^-SR measurement. In some insulators, however, the time for refilling might be relatively long so that the unpaired spin of atomic electrons may cause a significant μ^- depolarization. By taking analogy with radiation damage experiment, it is suggested that the μ^-Z system might be displaced towards an interstitial location due to Coulomb repulsion of highly ionized states after Auger transition. Further investigations are required concerning these arguments.

For the studies of compound materials containing various elements, the capture probability is roughly due to the so-called Fermi-Teller law with corrections due to electronic structures. It is remarkable that the μ^-SR signal from the different elements can be distinguishable. Because of a difference in nuclear capture rate, the μ^- lifetime (τ_μ) decreases monotonically against the increase of Z value of the nuclei to which μ^- is bound [1,2]; thus, τ_μ for (μ^-O) is 1.6 μs, while τ_μ for Cu is 200 ns, etc. On the other hand, because of this nuclear capture process, the muon free decay branch is substantially reduced. Therefore, along with the reduced polarization, the μ^-SR is more difficult to carry out compared to the μ^+SR.

At the same time, the magnetic moment of the bound μ^- at the ground state of muonic atom is subject to the relativistic correction; 9×10^{-4} for the bound μ^- at the ground state of (μ^-O), while 1 % for Zn, etc. Therefore we are required to know the value of this correction before we obtain a precise number of absolute frequency shift.

3. SPATIAL DISTRIBUTION OF HYPERFINE INTERACTION OBSERVED IN μ^- HYPERFINE ANOMALY

The spatial distribution of the bound μ^- at 1s state for a point nucleus is characterized by the muon Bohr radius (260/z fm). The nuclear finite size effect is most important for heavier Z elements. In Fig. 1, we present realistic charge distributions of core nucleus, bound μ^- and 1s electrons for μ^-Ni and μ^-Pb. As is clearly seen, compared to the hyperfine field felt by the (Z-1) nucleus, a substantially different hyperfine field is expected for the bound muon for the following reasons: (a) the magnetization distribution of the bound μ^- ($\rho_\mu(r)$) is entirely different from that of the (Z-1) nucleus ($\rho_N(r)$) in the region where the 1s electron density has a significant radial dependence (Bohr-Weisskopf effect); (b) the electron density ($\rho_e(r)$) might be significantly different because of the different charge distribution of the core (Breit-Rosenthal-Crawford-Schawlow effect). In the case of no orbital contribution for (Z-1) nuclei, the degree of the hyperfine anomaly ε in H_{hf} for the inner 1s electron is defined as

$$1 + \varepsilon = H_{hf}(\mu^-Z)/H_{hf}(Z-1) \qquad (2)$$

The hyperfine anomaly Δ may be calculated using the form:

$$\Delta = \varepsilon_{\mu} - \varepsilon_{N}, \tag{3}$$

$$\varepsilon_{\mu,N} = \int 4\pi r^2 \rho_{\mu,N}(r) \quad \rho_e(r)/\rho_e(0) \quad dr - 1. \tag{4}$$

Here ε_{μ} (ε_N) is the muonic (nuclear) hyperfine anomaly relative to a hypothetical point-like nucleus. Note that for a point-like probe (muon or nucleus) ε is zero by definition. As an extreme case adequate for the light elements, one can estimate ε by using the following assumptions [3]: (i) $\rho_e(r)$ as well as $\rho_{\mu}(r)$ are represented by hydrogen-like 1s-wave functions; (ii) $\rho_N(r)^{\mu}$ is constant inside the nuclear sphere and zero outside; and (iii) $\rho_e(r)$ has the quadratic form $\rho_e(r) = \rho(0)[1 - Zr^2/(a_0R_0)]$ for $r >> R_0$. Taking these assumptions into account the above integrals yield:

$$\varepsilon_{\mu} = -3m_e/m_{\mu} = -1.45\ \%, \quad \varepsilon_N = -3/5(R_0/a_0)Z$$
$$= -0.13\ \%\ (Z = 25). \tag{5}$$

With these estimates one obtains for the total hyperfine anomaly

$$\Delta = \varepsilon_{\mu} - \varepsilon_N = -1.3\ \%. \tag{6}$$

Once we can detect the different value from these simplest estimates, the extended magnetization distribution of the muonic atom can be used to probe the electron spin density outside the nucleus. In particular, for a magnetic ion, the electron spin-density distribution may be different from the charge-density distribution due to the core polarization mechanism; the polarized outer 3d electron causes a change in the spatial distribution of the inner s electrons through the s-d exchange interaction.

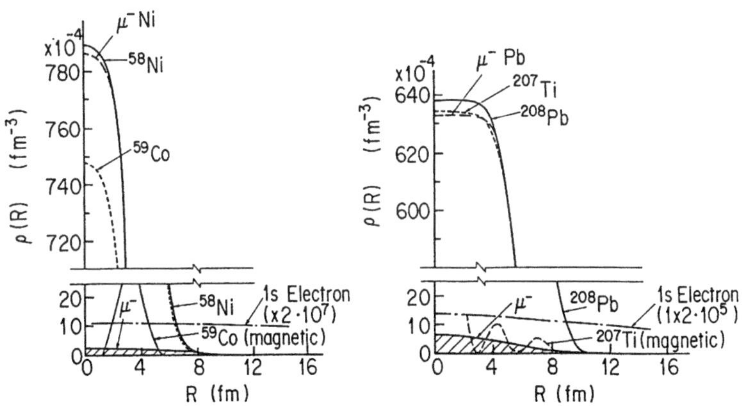

Fig. 1 Density distributions of nuclear charge (Z), bound μ^- at ground state of muonic atoms and μ^-Z core charge for ^{58}Ni (left) and ^{208}Pb (right). Calculated density distributions of 1s electrons for μ^-Z core are also shown. In order to see the hyperfine anomaly effect, density distributions of core nuclear charge and magnetic moment are also shown for the corresponding (Z-1) nucleus (^{59}Co and ^{207}Tl).

3.1 Paramagnetic Transition Metals

a) Paramagnetic Shift at Si site in MnSi

In the μ^-SR work by the Tokyo-Los Alamos group at the Los Alamos Meson Physics Facility a successful observation was made for the μ^- paramagnetic shift at Si sites in MnSi [7] which is weakly ferromagnetic below 29 K [8]. The μ^- paramagnetic shift was measured from 60 K to 300 K under 3.6 kG. The $K_\mu - \chi$ plot yields a hyperfine field of -58.7(11) kG/μ_B after 0.3 % relativistic correction expected for (μ^-Si). The nuclear hyperfine field has been measured by NMR method on the ^{29}Si site in the ordered phase and on the ^{55}Mn site in both the ordered and paramagnetic phases. In order to see the hyperfine anomaly effect, $H_{hf}(\mu^-Si)$ should be converted to $H_{hf}(Al)$. The conversion coefficient was taken as the ratio of the 3s electron density at the origin (0.96). Thus we obtained an estimate of the hf anomaly [2]:

$$\varepsilon = H_{hf}(\mu^-Si)/H_{hf}(Al) - 1 = -0.06(2). \tag{7}$$

b) Paramagnetic Shift in Pd

Similar experiments have been performed by the same group on paramagnetic transition metal of Pd [9]. The results can be compared to the hyperfine field on Rh in Pd. Thus, $H_{hf}(\mu^-Pd)/H_{hf}(Rh\underline{Pd})$ was obtained to be 0.64(5). The results may indicate the effect of spatial distribution of the electron spin density in Pd. The detaild theoretical studies have been done by using the unrestricted Hartree-Fock method by Freeman et al., which can explain only a part of the difference [10].

3.2 Hyperfine Anomaly in Ferromagnetic Ni and Fe

For μ^-Ni in ferromagnetic Ni metal, the hypefine field for Co in Ni metal is known to be - 118 kOe. Imazato et al. [11], succeeded in observing a precession frequency at Swiss Institute for Nuclear Research, using a high time resolution apparatus. The Fourier transforms of the observed time spectrum are shown in Fig. 2. The precession amplitude in the μ^-SR spectrum was consistent with the value expected after various corrections. Therefore, this peak was a dominant one and was assigned as coming from the substitutional site. Comparing with the experimental data of Co in Ni, they obtained

$$\varepsilon(\mu^-Ni) = -2.3(5)\ \%. \tag{8}$$

This value is significantly larger in magnitude than the simple estimate. This infers that the electron spin density has a larger gradient than the charge-density distribution itself. This seems to support the inner-shell polarization mechanism. Actually, Freeman et al. [10] calculated the hyperfine anomaly in this case and obtained $\varepsilon(\mu^-Ni)$ = -2.5 %, which is in agreement with the experiment.

Then, the SIN-Tokyo group extended their study to ferromagnetic Fe [12]. Like the case of μ^-Ni, the corresponding nuclear hyperfine field, namely that of ^{55}Mn in ferromagnetic Fe, was known from NMR experiments, which show a peculiar temperature dependence which was explained by Jaccarino et al. From the observed clear μ^- precession signals at various temperatures, the μ^- local fields were deduced and plotted against T/T_c and compared with the nuclear hyperfine field of ^{55}Mn. A remarkable proportionality has been revealed. The hyperfine anomaly thus deduced is $\varepsilon(\mu^-Fe)$ = -1.2 %, which is significantly smaller in magnitude than the μ^-Ni value. This is surprising, because there should be no difference between μ^-Ni and μ^-Fe, as far as the core polarization mechanism is concerned. This discrepancy may suggest a need of somewhat different mechanisms in the origin of the hyperfine fields.

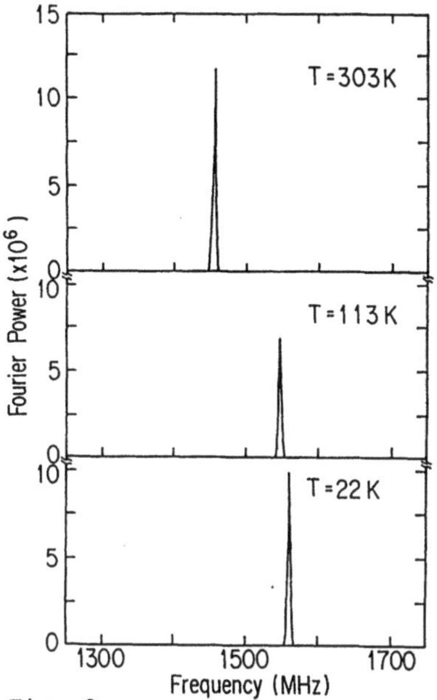

Fig. 2
Frequency spectra for negative-muon spin precession in single-crystal Ni in zero applied field as a function of temperature.

4. PROBING OXYGEN ELECTRONIC STRUCTURE WITH (μ^-O)

4.1 Magnetic Oxides

As already mentioned earlier, μ^-O has a lifetime longer than that of μ^- trapped by most metal ions of magnetic oxides, the μ^-SR signal can be selectively separated. Thus, the paramagnetic shift corresponding to the "quasi nitrogen" placed at the O-site can be studied.

The paramagnetic shift of K_μ (μ^-O) observed in MnO at room temperature is 1.16(21) % ($K_\mu = \delta H/H$) [13]. This value is about 1/3 of K_N (^{17}O) (3.2 %). Half of the difference was ascribed to the decreased electron density at μ^- in the probable ground state of $(Mn^{2+})_5(\mu^-O)^{3-}(Mn^{3+})$ compared to that at oxygen. The other half was explained by taking into account the admixture of $(\mu^-O)^{2-}(Mn^{2+})_6$ excited state into the ground state [14].

4.2 High T_c Superconductors

Since the discovery of high T_c superconducting oxides, the origin of the superconducting mechanism has been the central problem in both theoretical and experimental studies. For almost all the high T_c superconductors, pure materials such as La_2CuO_4 are insulators with magnetic orderings. A doping of impurity causes the hole conductivity and the transition from insulator phase to metallic, as well as superconductor one. The correlation of magnetic interaction and superconducting mechanism have been suggested by many theoretical and experimental studies. However, a consistent picture has not been achieved yet in theoretical interpretations. In order to obtain a break-through, the studies of electronic structure around the O-site are the most important among many possible experiments.

Generally speaking, the μ^-O is one of the most important and useful tools for high T_c superconductors. Even though many other microscopic methods were applied such as μ^+SR, NMR, Nuclear Quandru-

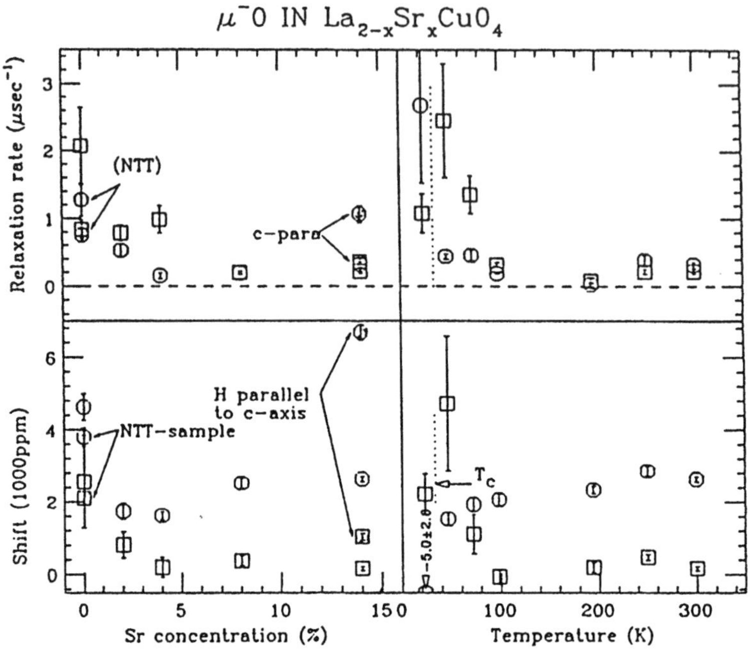

Fig. 3
The observed frequency shift for μ^- in LaSrCuO superconductors as a factor of Sr concentration (x) at room temperature and as a function of T for x = 0.14.

pole Resonance, the Mössbauer Effect and neutron scattering, only μ^-SR for μ^-O and enriched ^{17}O NMR can probe directly the O-site. Although some NMR experiments on high T_c materials with enriched ^{17}O have been carried out, the μ^-O experiment is another useful method to probe the oxygen site in high T_c materials. Especially, we can apply the μ^-O method for various samples without causing any damage in the samples. The strong advantage over the NMR experiment is an easy access to the various single crystals in the case of μ^-SR. Since in high T_c materials the basic interaction appearing in the mechanism of superconductivity is expected to be very anisotropic, the use of single crystal is inevitable for clear interpretation of the experimental results. Also the lack of the quadrupole moment of the μ^-O means a pure magnetic probe and the interpretation of the experimental results is much simpler.

The LaSrCuO system is one of the most theoretically studied systems and also the only system among high T_c superconducting oxide for which a high quality single crystal sample can be obtained. In addition, it has the simplest basic structure of Cu-O plane of high T_c superconductors. Thus, μ^-SR experiments were carried out for the (μ^-O) paramagnetic shift in the $La_{2-x}Sr_xCuO_4$ and its temperature as well as crystalline axis dependence [15].

The experiment was done at the upgraded M9B superconducting muon channel of TRIUMF [16]. The samples used are all single crystalline $La_{2-x}Sr_xCuO_4$; lower Sr concentration samples were made at RIKEN using a flux method (x = 0,4,6,8%), and higher Sr concentration samples were prepared at the University of Yamanashi using the travelling solvent floating zone method (x = 11-15%. named the KOSHU sample). For the absolute value of paramagnetic shift at room temperature we used the μ^-C signal from graphite and the μ^-O signal from pure water as references.

The observed shifts in spin precession frequency under 6.7 kG external transverse field, are summarized in Fig. 3 in the form of the x-dependence at room temperature and T-dependence for x = 0.14. This analysis was done by two-component fitting based upon a clear beating pattern seen in the μ^-SR time spectrum. There, equal population of μ^- for planar oxygen (O in the CuO plane forming Cu-O-Cu network) and apical oxygen (O on the top or bottom of the CuO plane) is assumed.

As shown in Fig. 3, the observed μ^- shifts at room temperature have two characters: a small shift and a large one. The large shift has a weak temperature dependence, while a small one has a diverging temperature dependence around T_c. The crystal anisotropy dependence of the shift along the c-axis as well as a-/b- axis may tell us which one is associated with either O in the plane or O in the chain. The measurement is under preparation for "KOSHU". Thus the μ^-SR is now ready to tell us clearly the electronic structures of the two types of oxygens in high T_c superconductors.

5. CHEMICAL EFFECTS IN (Z-1) PROBES

The (μ^-Z) atomic states produce a microscopic local defect with the charge state of (Z-1), where the response of the host atomic lattice can be sensitively probed or studied by the negative muon itself. So far, the following two experiments have demonstrated the usefulness of the μ^- probe; a) the state of μ^-C = B in HOPG (Highly Oriented Pyrolytic Graphite) and its intercalation compounds to probe the nature of bonding against the change of bonding number as well as the additional introduction of the intercalated atoms; b) the effect of B type substitutional impurity to probe the stability of Peierls distorted lattice of polyacetylene $(CH)_x$.

5.1 Graphite and Its Intercalation Compound

When a negative muon (μ^-) is introduced into highly oriented pyrolytic graphite (HOPG) and its second-stage intercalation compound of rubidium (HOPG-Rb), the μ^- bound with the ground state of a spinless ^{12}C nucleus forms a muon-nucleus system μ^-C which behaves as a boron-like pseudo-nucleus. These chemically "anomalous" compounds provide a unique opportunity for studying the nature of the chemical bonds, when we carry out the high precision measurements of the μ^- spin precession frequencies. Since the μ^-C in the two-dimensional carbon network is probably bonded by sp^2 hybrid orbitals to the three nearest-neighbor carbon atoms, the paramagnetic nuclear shielding at the μ^-C site is expected to be anisotropic. Thus, measurements of the frequency anisotropy enable us to study the nature of the valence bonds. Furthermore, the anisotropy could be changed by intercalation, since the σ electrons as well as the π electrons could be influenced by intercalation.

The experiment was carried out at SIN by the Tokyo group [17]. Fig. 4 shows the observed precession frequency as a function of the angle between the c-axis and the direction of the applied magnetic field. The results are summarized as follows: (1) the shift is anisotropic for HOPG with $\delta(90) - \delta(0) = -325$ ppm, and (2) intercalation enhances the negative shift by -310 ppm for the $\delta(0)$ component, while it produces a slight positive change of + 15 ppm for the $\delta(90)$ component. Hereafter, in most cases we will refrain from using the absolute value of δ because of the systematic errors in ω_0 arising, for example, from the g-factor correction or from uncertainty in the absolute value of the external magnetic field. Here, the shift of the precession frequency can be known to originate mainly from the nuclear shielding due to the interaction of the nucleus with electron orbital motions (the chemical shift) [17]. The anisotropy is given solely by the paramagnetic contribution, $\delta^{(p)}$. In addition, it is highly likely that $\delta^{(d)}$ is almost unaffected by the intercalation because it causes the electron density of the μ^-C to increase by only 1 %. Consequently, if we let C and Rb represent HOPG and HOPG-Rb, respectively, the expression

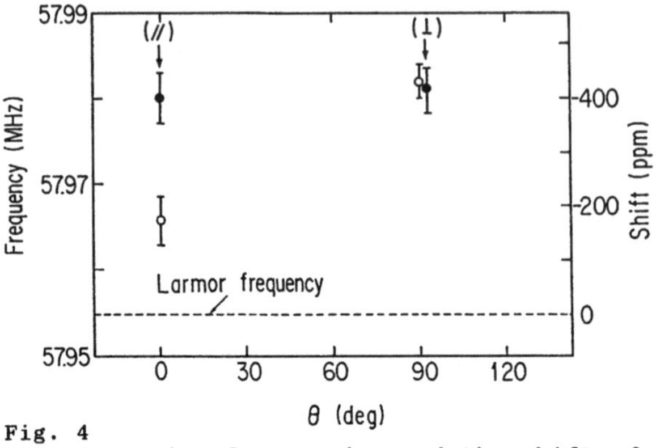

Fig. 4
The precession frequencies and the shifts δ for HOPG (ɸ) and HOPG-Rb (⧫) are shown as a function of the angle between the c-axis of the samples and the external field H_0.

$$\delta(Rb) - \delta(C) = \delta^{(p)}(Rb) - \delta^{(p)}(C) \qquad (9)$$

provides an adequate approximation for the shifts.

Then, $\delta^{(p)}(0,90)$ in a p-bonded system can be related to $P_u(0,90)$ representing the "unbalance" of the valence electrons in the 2p orbitals centered on the μ^-C [17]. In the case of HOPG, the μ^-C-C system in the carbon network can be treated as a pseudo-molecule containing a tri-coordinated boron atom bonded to three identical unsaturated hydrocarbons, since the μ^-C can be regarded as a localized boron impurity. In such a molecule, the B-C bond is formed from sp^2 hybrid orbitals involving the partially filled p_z orbital whose electron density p_{zz} is in the range of 0.09-0.11. When the μ^-C-C bond in HOPG is formed from the ideal sp^2 hybrid orbitals, $P_u(0)$ and $P_u(90)$ are given by 0 and 2 - p_{zz}, respectively. From the observed shift $\delta(90)$ - $\delta(0)$ = -325 ppm one obtains p_{zz} = 0.1 which is comparable with the measured and calculated values for trico-ordinated organoboron compounds mentioned above.

$P_u(90)$ depends mainly on the electron density of the p_z orbitals while $P_u(0)$ is determined exclusively by the nature of the σ orbitals. Therefore, the effects of intercalation on the π and the σ bonds can be studied by the paramagnetic shifts of $\delta(90)$ and $\delta(0)$ due to intercalation. Observed results suggest that the electron density of the p_z orbital increases by 0.1 per one μ^-C upon intercalation; this trend agrees with the filling of the vacant π bond of graphite upon intercalation. On the other hand, the intercalation of Rb causes the $\delta(0)$ component to shift by -310 ppm, corresponding to an increase of $P_u(0)$ by 1.8. This finding shows that the relative contribution of the P_x orbital to the P_y orbital deviates from the ideal sp hybridization, otherwise $P_u(0)$ = 0. That is, the σ bonds are distorted from the three-fold symmetry when HOPG is intercalated by Rb.

5.2 Probing the Stability of the Peierls Distorted Lattice with μ^-C Impurity

The organic semiconductor $(CH)_x$ is well known as a typical example of the so-called Peierls distorted lattice with two isomers; trans-$(CH)_x$ and cis-$(CH)_x$. It is quite interesting to learn the stability of these lattice structures against the presence of different substitutional impurities. As the most simple and sensitive method, we have introduced μ^-C = B impurity and intended to detect the response

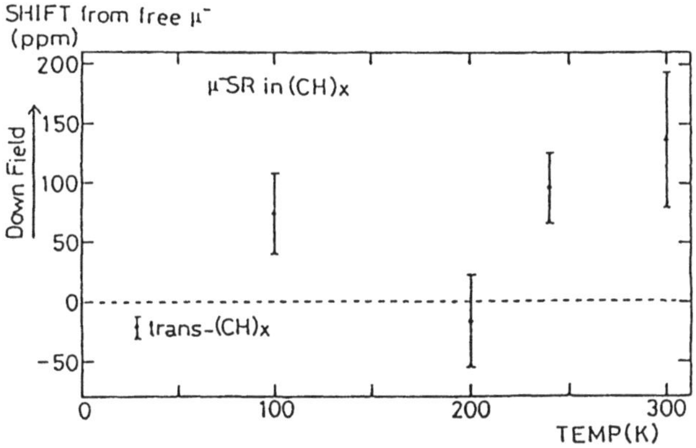

Fig.5. Temperature variation of the shift for μ^- in trans-polyacetylene.

of the host lattice by the change of the local magnetic field at (μ^-C) seen as a Knight shift [18].

For the precise detection of the expected small Knight shift we employed the 500 MHz muon spin resonance spectrometer developed at UT-MSL/KEK [19]. The results are summarized in Fig. 5: a frequency shift was detected with reference to the μ^+SR resonance and with a correction to the magnetic moment of (μ^-C). At room temperature, no significant shift exists for cis-$(CH)_x$, while it is -360 ppm for trans-$(CH)_x$. The large shift in trans-$(CH)_x$ at room temperature disappears below 200 K. All these results suggest the Peierls distorted lattice which is stable in cis-$(CH)_x$ and in trans-$(CH)_x$ at low temperature becomes unstable in trans-$(CH)_x$ above 200 K producing, possibly,a metallic state at room temperture.

6. FUTURE DIRECTIONS

Consistent progress in the μ^-SR studies on condensed matter can be expected for the near future. With the help of the advanced technology for μ^- beam production, the following experiments which have not yet been done will be realized soon;

a) longitudinal relaxation measurements to study a dynamical relaxation phenomena of the (μ^-Z) probe;

b) level crossing resonance studies for the precise determination of the hyperfine fields on the (μ^-Z) probe.

At the same time, a recently proposed method of ultra-slow μ^- beam production [20] will open an entirely new era for the μ^-SR studies, particularly towards the characterization of materials surface.

The author acknowledges Professor T. Yamazaki for continuing collaborations and fruitful discussions during these 15 years since the initiation of the μ^-SR work. The experiments described here have been carried out in collaboration with many scientists to whom the author would like to express sincere thanks; in particular to Drs. K. Nishiyama, J. Imazato, T. Kondow, Y. Kuno, S. Nagamiya, O. Hashimoto, R.S. Hayano, B.D. Patterson, E. Torikai, H. Kojima, I. Tanaka, H. Kitazawa and S. Ikehata. Helpful discussions with Professors H. Kamimura, A.J. Freeman and J. Kanamori are also acknowledged. This work was supported in part by the Grant-in-Aid for Scientific Research from the Ministry of Education, Science and Culture.

References

[1] T. Yamazaki, K. Nagamine, S. Nagamiya, O. Hashimoto, K. Sugimoto, K. Nakai and S. Kobayashi, Physica Scripta 11 (1975) 133.
[2] K. Nagamine, Hyperfine Interactions 6 (1979) 747.
[3] T. Yamazaki, in: Muons and pions in materials research, eds. J. Chappert and R.I. Grynszpan (North-Holland, Amsterdam, 1984) ch. 14.
[4] V.S. Evseev, T.N. Majebov, V.S. Roganov and M.V. Fronmasieva, Pisma, JETP 27 (1978) 249.
[5] D.P. Hutchinson, J. Menes, G. Shapiro and A.M. Patlach, Phys. Rev. 131 (1963) 1362.
[6] R.A. Mann and M.E. Rose, Phys. Rev. 121 (1961) 293.
[7] R.S. Hayano, Y.J. Uemura, J. Imazato, N. Nishida, K. Nagamine, T. Yamazaki, Y. Ishikawa and H. Yasuoka, J. Phys. Soc. Japan, 49 (1980) 1773.
[8] R.S. Hayano, Y.J. Uemura, J. Imazato, N. Nishida, T. Yamazaki, H. Yasuoka and Y. Ishikawa, Phys. Rev. Lett. 41 (1978) 1743.
[9] T. Yamazaki, R.S. Hayano, Y. Kuno, J. Imazato, K. Nagamine, S.E. Kohn and C.Y. Hung, Phys. Rev. Lett. 42 (1979) 1241.
[10]A.J. Freeman, J.V. Mallow, J.P. Desclaux and M. Weinert, Hyperfine Interact. 1719 (1984) 865.
[11]J. Imazato, K. Nagamine, T. Yamazaki, B.D. Patterson, E. Holzschuh and R.F. Kiefl, Phys. Rev. Lett. 53 (1984) 1849.
[12]H. Keller, R.F. Kiefl, Hp. Bauneller, W. Kundig, B.D. Patterson, J. Imazato, K. Nishiyama, K. Nagamine, T. Yamazaki and R.I. Grynszpan, Hyperfine Interactions 31 (1986) 461.
[13]S. Nagamiya, K. Nagamine, O. Hashimoto and T. Yamazaki, Phys. Rev. Lett. 35 (1975) 308.
[14]H. Kamimura and Y. Natsume, Hyperfine Interactions 6 (1979) 357.
[15]K. Nishiyama, K. Nagamine, H. Kitazawa, K. Katsumata, E. Torikai, H. Kojima and I. Tanaka, Contribution to μSR 90 (1990).
[16]K. Nagamine, K. Nishiyama, Y. Miyake, K. Fukuchi, T. Matsuzaki, J.L. Beveridge, G.E. Clark, W. Anderson, G.M. Marshall and J. Doornbos, to be published.
[17]T. Kondow, Y. Kuno, R. Kadono, J. Imazato, K. Nagamine and T. Yamazaki, Phys. Lett. 110A (1985) 319.
[18]S. Ikehata, M. Oda, K. Nagamine, K. Nishiyama and H. Shirakawa, Synthetic Metals 28(1989) D275.
[19]K. Nishiyama, T. Azuma, K. Ishida, T. Matsuzaki, J. Imazato, Y. Yamazaki and K. Nagamine, Hyperfine Interactions 32 (1986) 88.
[20]K. Nagamine, Proc. Japan Academy, 65 (1989) 225.

Part III

Low Energy Fusion

Dense Deuterium and Deuterium–Muon Systems

N.W. Ashcroft[1]

Laboratory of Atomic and Solid State Physics and Materials Science Center, Cornell University, Ithaca, NY 14853, USA

Abstract. The states of dense deuterium are treated starting with the fundamental highly symmetric Hamiltonian where all interactions are entirely Coulombic. Such a system is known to be unstable to deuteron pairing and crystalline ordering at low densities; experimentally it has been probed to densities an order of magnitude higher than normal solid density. At densities somewhat higher than this deuterium is predicted to undergo a dynamic instability to a monatomic metallic state passing on its way through a paired band-overlap phase. Some of the pairing characteristics are immediately transferable to the states formed when positive or negative muons are introduced into dense deuterium.

1. Introduction

Because of its possible importance in the understanding of muon catalysed fusion, the system to be treated is deuterium, in condensed phase, and under relatively high compression. Much of the physics we shall be discussing is altered little by the substitution of hydrogen (protium) for deuterium. In either system the Hamiltonian is known exactly, at least for length and energy scales appropriate to condensed matter physics. This statement continues to be true if the system is extended to include positive and negative muons, as we shall do below.

Let N electrons (fermions, spin 1/2) have coordinates $\vec{r}_i, \sigma_i$ and a compensating assembly of deuterons (bosons, spin 1) have coordinates $\vec{R}_j, s_j$: for either system let $\hat{\rho}_\alpha^{(1)}(\vec{r})$ be the one-particle density operator (i.e., $\sum_i \delta(\vec{r}_i - \vec{r})$, for the electrons). Then the non-relativistic Hamiltonian is spin independent, and if the system is macroscopic and confined to volume V it has the form

$$H = \hat{T}_e + \hat{T}_d + \tfrac{1}{2}\int_V d\vec{r} \int_V d\vec{r}\,' v_c(\vec{r} - \vec{r}\,')\{\hat{\rho}_{ee}^{(2)}(\vec{r},\vec{r}\,') - 2\hat{\rho}_e^{(1)}(\vec{r})\hat{\rho}_d^{(1)}(\vec{r}\,') + \hat{\rho}_{dd}^{(2)}(\vec{r},\vec{r}\,')\} \tag{1}$$

where $v_c(\vec{r} - \vec{r}\,') = e^2/|\vec{r} - \vec{r}\,'|$ is the basic Coulomb interaction, and

$$\hat{\rho}_{\alpha\alpha}^{(2)}(\vec{r},\vec{r}\,') = \hat{\rho}_\alpha^{(1)}(\vec{r})\hat{\rho}_\alpha^{(1)}(\vec{r}\,') - \delta(\vec{r} - \vec{r}\,')\hat{\rho}_\alpha^{(1)}(\vec{r})$$

is the two particle density operator reflecting the original presence of pairwise interactions in the problem. Note that from its fundamental statistical interpretation

$$< \hat{\rho}_d^{(1)}(\vec{r})\hat{\rho}_d^{(1)}(\vec{r}\,') >$$

is a two-particle correlation measuring the probability density of finding a deuteron in

[1] Work supported by the National Science Foundation.

 © Springer-Verlag Berlin, Heidelberg 1992

$d\vec{r}\,'$ at $\vec{r}\,'$ and another in $d\vec{r}$ at $\vec{r}$. The properties of this function very much reflect pairing tendencies. The Hamiltonian (1) has considerable symmetry: the subtlety of condensed matter physics is illustrated immediately by the fact that in spite of the simplicity of (1) the enormity of N ($\sim 10^{23}$) leads to cooperativity of phenomena involving states in which the symmetry is broken in most remarkable ways. (This we assured from the known states of (1) at low densities.)

The statement just made holds true for an even simpler system. Let $\hat{\rho}_d^{(1)}(\vec{r}\,)$, the one-particle density operator for positively charged deuterons, be replaced by a constant and rigid (non-dynamic) background density say $\rho(= N/V)$. Then (1) becomes

$$H = H_{eg} = \hat{T}_e + \tfrac{1}{2}\int_V d\vec{r}\int_V d\vec{r}\,' v_c(\vec{r}-\vec{r}\,')\left\{\hat{\rho}_{ee}^{(2)}(\vec{r},\vec{r}\,') - 2\hat{\rho}_e^{(1)}(\vec{r}\,)\rho + \rho^2\right\} \tag{2}$$

which is the much studied interacting electron-gas problem [1]. It constitutes part of the analysis of very high density deuterium (see below): its states are also known to break the translational and rotational symmetry of (2) (for example, in the transition at low densities to a Wigner crystal). In a very real sense (1) is the simplest Hamiltonian that can be written down for condensed matter: as it happens, it conforms for mass 1 (m_h) to the most abundant form of matter in the universe.

2. States of Dense Deuterium

Given what we know about relatively simple forms of matter (the alkali metals might be an example) it is far from clear that without hints from nature we would easily predict the states of (1). To be convinced of this, let us begin with the problem described by (2) and place N electrons in a neutralizing background at an average density *far in excess* of any alkali metal; in fact we will choose a density much closer to that of metallic Be whose density, characterized by the standard linear measure r_s $\left(\frac{4\pi}{3}r_s^3a_0^3 = V/N\right)$ has a value $r_s \simeq 1\cdot 87$. At such densities the states of (2) are known to be deeply within range of quite normal metallic behavior; instabilities of magnetic and structural character certainly appear, however, at higher values of r_s.

Now keeping the average electron density fixed we begin the formation of dense deuterium by replacing the static uniform background (ρ) by fully dynamic deuterons ($\hat{\rho}_d^{(1)}(\vec{r}\,)$), passing therefore from (2) to (1). We are now invited to predict, for example, the ground state of the new system, given that the former dense electronic system, was highly metallic. It is a reasonable guess that our predictions will be quite far from the mark. The particular density ($r_s = 1\cdot 87$) under consideration can actually be related to a known physical state: it corresponds to a phase of deuterium which has been compressed by almost a factor of 5 over normal one atmosphere solid density. Since compressions of an entire order of magnitude have now been reported, it is clear that this proposed state has been achieved by experiment. What is known [2] about the state is that

(a) it is insulating, not metallic
(b) it is crystalline (and not translationally invariant, as is (1))
(c) it is paired (in the ordinary sense of molecules)
(d) it is rotational.

In the last case (d), deuterium is close to becoming orientationaly ordered at this density: however, its companion, hydrogen, is very far from this point. We certainly know that a single isolated molecule has rotational states; what the observation argues is that when the molecules are in exceedingly close proximity, the self-consistently determined crystal-fields are apparently quite small compared with the rotational constants, and as a consequence angular momentum remains secure as a reasonably good quantum number. This property is strongly linked to observation (c), which also tells us how remarkably robust the pairing tendency is, in spite of intermolecular separations that have been reduced by a factor of 2 compared with ordinary solid deuterium. It is generally accepted that this pairing must give way to monatomic order, at sufficiently high density: but exactly when is still not known. However, the ensuing monatomic state certainly begins to conform much more to the simplicity of (1).

As noted, the observation of pairing is already an indication of broken symmetry: observation (b) of crystalline order indicates a further breaking. The known crystal structures of hydrogen and deuterium do change with density [3]. At the highest densities for which they have both been measured, deuterium and hydrogen are both hexagonal close packed (two pairs of deuterons per primitive cell). However, from the continuity of Raman active phonons [4] there is reasonable evidence to suppose that the hexagonal phase may actually persist to exceedingly high pressures.

Observation (a), that the replacement of the uniform positive background by discrete and dynamic nuclear charge changes the system from highly conducting to insulating, demonstrates the importance of cooperative electronic structure. In the simplest picture, that of independent electrons, this is band-structure, a term normally applied to perfect crystalline arrangements. But given (d), we are presented with a non-trivial complication. Crystalline band-structure remains definable in the presence of nuclear vibrations, where it appears in the zero-phonon approximation. However the equivalent for a highly rotational problem remains to be defined [5].

2.1 The High-Density Environment: Background

Hydrogen is normally placed in Group I of the periodic table, as befits an element with a single electron entering a new shell. Its companion elements in this group are the alkalis, which in the condensed phase and under normal conditions are monatomic metals, and with an electronic structure which is quite simple. However, under the same conditions condensed hydrogen has properties that are quite remote from those of the alkali metals. Its behavior far more resembles the Group VII elements, into which category it might also be placed by virtue of its proximity (lacking one electron) to a closed shell. Like the halogens in this group, deuterium and hydrogen are strongly paired covalent molecular elements at low densities. They also form rotational liquids and solids, and they retain both their rotational character and their pairing to surprisingly high densities [3,5].

From this we might surmise that for a first clue to the likely electronic behavior of hydrogen in a high density environment we should examine the dense halogens. The case of solid iodine is interesting because it is known already that application of pressure does lead to a metallic state, but in a structure where diatomic ordering is preserved [6]. Unlike hydrogen, this solid is not rotational. From Bloch theory of independent electrons, molecular pairing guarantees that a necessary condition for insulating behavior – an even number of electrons per primitive cell – is automatically satisfied. However, this is obviously not a sufficient condition as witnessed by the numerous cases of divalent metals; for these metallic behavior results from the overlap of bands which would otherwise by filled, and inert. From the tight-binding rendition of Bloch theory [7] we most readily see the growth of the width of bands with increasing density and the corresponding shrinkage of gaps of forbidden states between such bands.

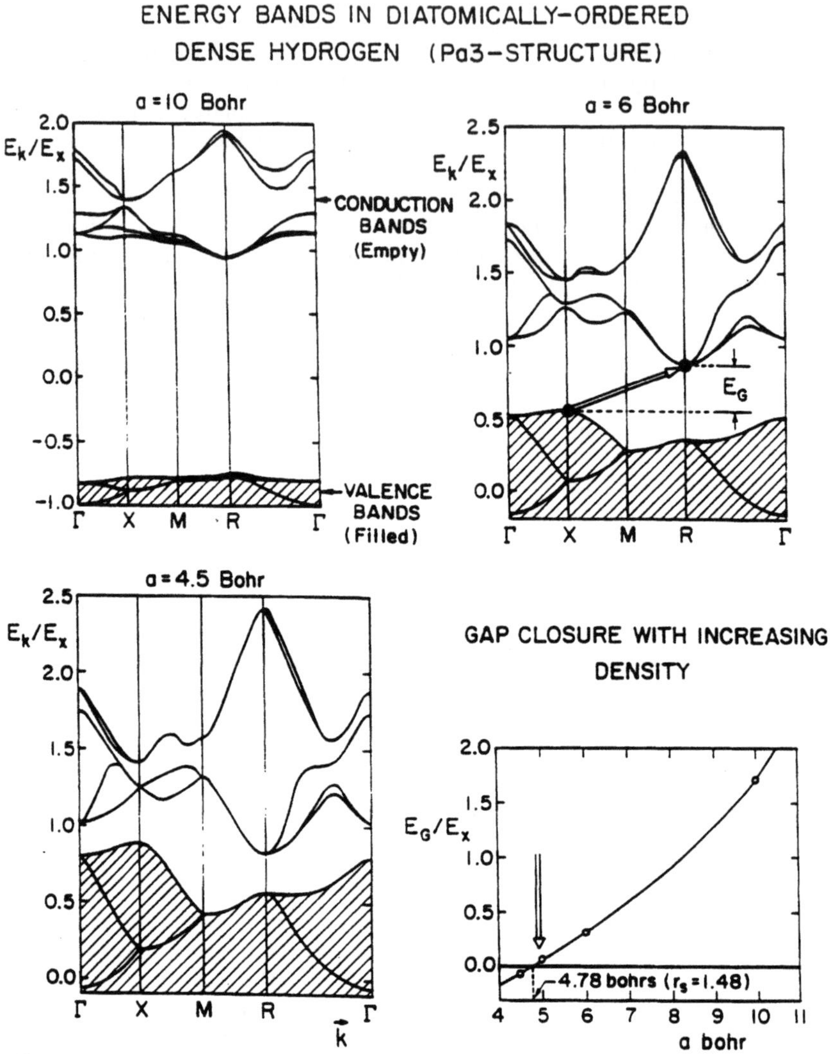

Fig. 1

Band gap closure with pressure as demonstrated for the rotationally ordered Pa3 structure in hydrogen. The lower right panel shows the indirect gap E_G as a function of lattice constant (From Ref. [8]).

With sufficient increase in density the gap between any adjacent pair of bands can be made to shrink to zero. (Fig. 1 shows an example of this for the $\alpha - N_2$ or $Pa3$ structure of dense hydrogen or deuterium [8].) When the bands do cross, the prospect of a paired metal arises providing always that from considerations of the total energy of the system the paired state is actually favored over any other competing structure. An obvious competitor is a monatomic phase.

This particular route to metallization is apparently circumvented by the alkalis which, upon formation of a condensed state of matter, proceed directly to a monatomic, paramagnetic, and metallic phase. With a single electron per primitive cell, Bloch theory guarantees such a metallic state. This is by no means the case when electron-electron interactions, treated only in an average way in the Bloch picture, are restored. Indeed the monatomic hydrogen lattice was the very example used by Mott [9] to argue that a transition from a high density alkali-metal-like paradigm to a lower density isostructural insulator was inevitable, and it could not be continuous. Here the electron spin (not directly evident in (1)), is crucial as later elucidated by the notion of spin-split bands and their eventual description by Hubbard and related strong coupling models [1]. Mott's problem confronts exceedingly important many-particle physics; however, the key structural assumption, namely that of a strictly monatomic lattice, is not pertinent to the problem here. The deuterons are subject to no such constraint; they are free to pair, and at low densities they do. The resulting pairs may still reflect the quantum symmetry of the constituents (the spin of nuclei and electrons) through the low density classifications of ortho-para differences [3].

2.2 The High-Density Environment

Returning to (1), we transform to reciprocal space and with $q = 0$ terms removed by charge neutrality, we obtain in the thermodynamic limit

$$H = \hat{T}_e + \frac{N}{2V} \sum_{q\neq 0} v_c(q)\left(\frac{1}{N}\hat{\rho}_e^{(1)}(\vec{q})\hat{\rho}_e^{(1)}(-\vec{q}) - 1\right) \tag{3a}$$

$$+ \quad \hat{T}_d + \frac{N}{2V} \sum_{q\neq 0} v_c(q)\left(\frac{1}{N}\hat{\rho}_d^{(1)}(\vec{q})\hat{\rho}_d^{(1)}(-\vec{q}) - 1\right) \tag{3b}$$

$$-\frac{N}{V} \sum_{q\neq 0} v_c(q)\left(\frac{1}{N}\hat{\rho}_e^{(1)}(\vec{q})\hat{\rho}_d^{(1)}(-\vec{q})\right) \tag{3c}$$

where $\hat{\rho}_d^{(1)} = \sum_j e^{i\vec{q}\cdot\vec{R}_j}$, $\hat{\rho}_e^{(1)} = \sum_i e^{i\vec{q}\cdot\vec{r}_i}$, and $v_c(q) = 4\pi e^2/q^2$. In exactly the same way (2) becomes

$$H_{eg} = \hat{T}_e + \frac{N}{2V} \sum_{q\neq 0} v_c(\vec{q}\,)\left(\frac{1}{N}\hat{\rho}_e^{(1)}(\vec{q}\,)\hat{\rho}_e^{(1)}(-\vec{q}\,) - 1\right) \tag{4}$$

whose ground state energy per particle is

$$E_0/N = <\hat{T}_e> + \frac{1}{16\pi^3} \int_{0+} d\vec{q} v_c(q)(S_{ee}(\vec{q}\,) - 1) \tag{5}$$

where $S_{ee}(\vec{q}\,)$ is the static structure factor of an interacting electron system placed in a static uniform compensating background:

$$S_{ee}(\vec{q}\,) = \frac{1}{N} < \hat{\rho}_e^{(1)}(\vec{q}\,)\hat{\rho}_e^{(1)}(-\vec{q}\,) > -N\delta_{q,0}. \tag{6}$$

By comparing (4) and (3a) we see immediately that the original problem (1), can be viewed as the synthesis of the interacting electron-gas problem (3a) the corresponding "interacting deuteron gas" problem (3b), and the coupling between two such problems (3c). As noted above, in the standard metallic range $2 \leq r_s \leq 6$, the states of (2) or (3a) have continuous symmetry and are strongly metallic. However, at much lower densities ($r_s \sim 100$) the states of (2) are crystalline [10].

It was stated earlier that at sufficiently high density, deuterium will become metallic assuming thereby a simple structure as do the other Group I elements. For such a phase, a simplified treatment of (3) already sets the important scales of energies and densities. To see this we first ignore the inhomogeneities in the electron system: then the ground state energy is approximately that of (3a) augmented by the electrostatic energy of an array of point positively charged deuterons, to which finally is added the dynamic energies $< T_d >$ of the deuterons. In atomic units $(e^2/2a_0; a_0 = \hbar^2/m_e e^2)$ the energy of the continuous paramagnetic phases of (3a) are, per particle,

$$E_0/N = \tfrac{3}{5}\,\alpha^2/r_s^2 - 3\alpha/2\pi r_s + E_{corr}(r_s) \tag{7}$$

where $\alpha = (9\pi/4)^{1/3}$. The first two terms constitute the Hartree-Fock result; the remainder is the correlation correction beyond statistical correlation, which is a slowly varying function of r_s. The electrostatic (Madelung) term is readily estimated from a structure independent argument which holds that N deuterons with point positive charges and situated with reasonable homogeneity in a uniform compensating background will have an energy which is approximately N times the energy of a single sphere with a deuteron at its center and uniform charge within a radius $r_s a_0$. If the spheres were truly non-overlapping this result would follow immediately from Gauss's theorem; it is therefore only valid to within structure dependent multipole corrections. The energy of a single sphere is $-e^2 \int_0^{r_s a_0} 4\pi^2\, dr \left\{1 - (r/r_s a_0)^3\right\}\rho/r = (-9/5)(e^2/2a_0 r_s)$ and actually differs remarkably little from the known structural values for the Madelung energy (for face centered cubic 1·792 replaces 9/5). It follows that at the level of electron gas and Madelung energies the energy is approximately

$$E/N = \tfrac{3}{5}\,\alpha^2/r_s^2 \; - (\tfrac{3\alpha}{2\pi} + \tfrac{9}{5})/r_s + E_{corr}(r_s) \tag{8}$$

and this has a minimum at $r_{sm} \simeq 1\cdot 63$. At the corresponding density, the energy per particle has the value

$$E/N \sim -0\cdot 831 + E_{corr}(r_{sm}). \tag{9}$$

Typical values of E_{corr} are also $\sim -0\cdot 1$ Ry and as noted are quite slowly varying with r_s in this range. Non-uniformity of the electron gas leads to additional binding, at about the same level of $\sim -0\cdot 1$ Ry. This is referred to as band structure energy, and it can be determined by response techniques providing the response functions of the electron-gas problem defined by (3a) are known [11,12].

Opposing binding is $< \hat{T}_d >$: this too is easily estimated from the neutral sphere construction already used to determine the Madelung energy. We now imagine an assembly of N Einstein oscillators, in which each deuteron in displaced *at random* by a small amount δ from the center of each neutral sphere. The potential of any deuteron is then $(e^2/2a_o rs)((\delta/r_s a_o)^2 - 3)$, under an assumption that random multipole corrections from neighboring spheres cancel on average. The obvious harmonic nature of this result immediately fixes the Einstein frequency: it is

$$\omega_E = \omega_p/\sqrt{3} = (4\pi\; Ne^2/3m_d V)^{1/2}$$

and the corresponding zero point energy is thus (in Rydbergs)

$$< \hat{T}_d > /N \approx (m_e/m_d)^{1/2}\; 3\sqrt{3}/r_s{}^{3/2}, \tag{10}$$

or $\sim$ 0.02 Ry. This corresponds to an Einstein temperature of about 4000°K, and it turns out to be a non-trivial energy in the issue of binding of the metal as a metastable state. This follows from (9) and the subsequent proximity of the total energy to the free atomic state.

2.3 Metallic Muonium

The last point can be made a little more strongly by considering a hypothetical metallic state of positive muons and again a compensating assembly of electrons. Since the muon can be viewed as a light isotope of hydrogen, the Hamiltonian of any assumed macroscopic [13] state is (at least briefly)

$$H = \hat{T}_e + \frac{N}{2V}\sum_{q\neq 0} v_c(q)\Big(\frac{1}{N}\hat{\rho}_e^{(1)}(\vec{q})\hat{\rho}_e^{(1)}(-\vec{q}) - 1\Big) \tag{11a}$$

$$+ \ \hat{T}_\mu + \frac{N}{2V}\sum_{q\neq 0} v_c(q)\Big(\frac{1}{N}\hat{\rho}_\mu^{(1)}(\vec{q})\hat{\rho}_\mu^{(1)}(-\vec{q}) - 1\Big) \tag{11b}$$

$$-\frac{N}{V}\sum_{q\neq 0} v_c(q)\Big(\frac{1}{N}\hat{\rho}_\mu^{(1)}(\vec{q})\hat{\rho}_\mu^{(1)}(-\vec{q})\Big). \tag{11c}$$

So the exercise just performed for deuterium carries through again, the major changes being just those resulting from the change in mass of the positive nucleus (the mass ratio being 0.0564). The estimate of $< T_\mu > /N$ now rises to 0.084 Ry [14] (nearly 1.1 ev), which is competitive with the band structure energy, but is opposing binding. It is also large enough to affect estimates of where, in density, such a transitory system would find its metastable point. Notice that this argument assumes that the states appropriate to (11b) are, in the absence of electrons, essentially a "muon Wigner crystal" at $r_s \approx 1{\cdot}63$. Since the Wigner crystal for electrons occurs at $r_s \sim 100$ [10] and since there is a length scale reduction of order 200 between the problems represented by (11a) and (11b), this assumption is borderline. At $r_s \sim 1$ the "muon gas" in its ground state would be close to quantum melting. Note that for this system the Born-Oppenheimer parameter $(m_e/m_\mu)^{1/4} = 0.264$ so that non-adiabatic corrections will also be significant.

2.4 Dense Molecular Deuterium

It may be of interest to consider an experiment in which dense deuterium is exposed to an intense beam of negative muons. To understand the behavior of any ensuing clusters of muonic atoms, we need first to examine the standard case of (electronically) paired deuterium. We are therefore once again viewing the states of (1) but now at densities somewhat lower than $r_s = 1.63$, and also at densities where the system is known to be diatomically ordered. The dynamics of the deuterons under these conditions are still clearly delineated into internal pair-motions of a localized vibron (and even roton) character, and a collective center-of-mass motion of optic and acoustic phonon character. Within the standard Born-Oppenheimer separation of time scales, the deuteron dynamics are controlled at reasonably low temperatures by a ground state electronic trace of (1). The result is an effective Hamiltonian

$$H_d = \hat{T}_d + \mathcal{V}(\{\vec{R}_j\}; V) \tag{12}$$

where $\mathcal{V}(\vec{R}_j; V)$ is an instantaneous many-body potential energy. Its volume dependence is fundamental since the volume appears directly in the boundary condition on the electron problem (over whose states an electron trace is being taken). This volume is under the control of the experimentalist; in fact diamond anvil cells have achieved

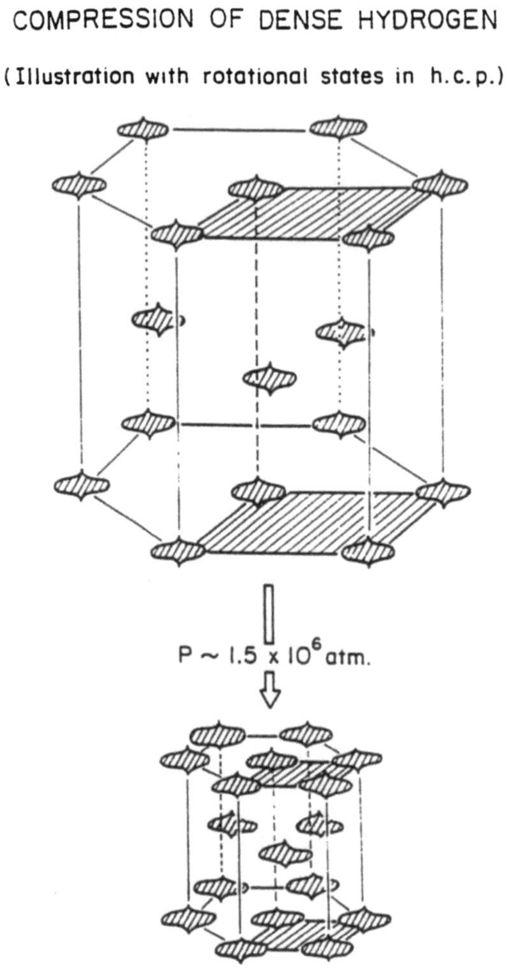

Fig.2
Approximate scale of compression of hydrogen accomplished by P~1·5 Mbar (150 GPa) illustrated for a hexagonal close packed structure. The objects located on each site represent basis pairs of hydrogen or deuterium in rotational states that are not specified.

volume compressions amounting to a reduction in linear dimension of over a factor of 2. Figure 2 shows the approximate scale of this reduction for a hexagonal structure and for unspecified rotational states. The simplest physical manifestation of such a change is an expected transfer of electron density from the "bond" (or intra-molecular potential) of the deuteron pair, to the interstitial region between pairs. Pairs themselves also participate in bonding (initially of the fluctuating multipole or Van der Waals type) which is weak at low densities, but becomes progressively stronger at higher compression as the pairs are driven together. This simple intuitive division between intra- and inter-molecular bonding is normally expressed through a decomposition of the many-particle function into pair and multi-center interactions. The simplest approximation is to write it as

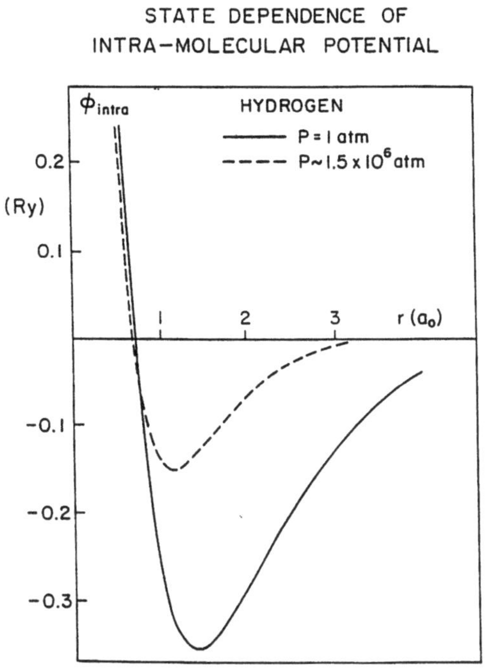

Fig. 3
State dependence of the intra-molecular potential in dense hydrogen (From Ref. [17]).

$$\mathcal{V}(\{\vec{R}_j\};V) \doteq \sum_{intra} \phi\ (V) + \sum_{inter} \Phi\ (V) \tag{13}$$

where the state dependence arising from electron transfer is now directly revealed in the intra-molecular (ϕ) and inter-molecular (Φ) pair potentials. At very low densities ϕ obviously approaches the standard intra-molecular potential of free deuterium or hydrogen molecules.

To illustrate the general effects of increasing density on, say vibron dynamics [15], it is only necessary in a first approximation to modify standard model descriptions of ϕ to incorporate the density dependence mandated by (12). The simplest representation incorporating the known attractive and repulsive regions of typical intra-molecular potentials is the Morse function: including the state-dependence it is written

$$\phi(r;r_s) = D(r_s)\left\{e^{-a(r_s)(r-r_m(r_s)} - 2e^{-2a(r_s)(r-r_m(r_s))}\right\} \tag{14}$$

which is well known to lead to solutions for vibron and roton states in closed form [16]. We focus, in particular on the vibron states, since it is these that are probed by Raman-(inelastic light-) scattering in experiments that have been carried out at high pressures on both deuterium and hydrogen. From the arguments advanced above, $\phi(r;r_s)$ as given by (14) is common to both; however the vibron and roton levels will differ on account of the obvious mass differences. This is clear from the form of the primary excitation probed by the Raman effect where the excitation energy (in Ry) is

$$\hbar\omega(r_s) = 2(Da^2/m^*)^{1/2} - (a^2/m^*) \tag{15}$$

with m^* the reduced mass. From this it is apparent that at fixed density (or r_s) the measurement of $\hbar\omega(r_s)$ for <u>two</u> isotopes can fix $a(r_s)$ and $D(r_s)$. Figure 3 shows the

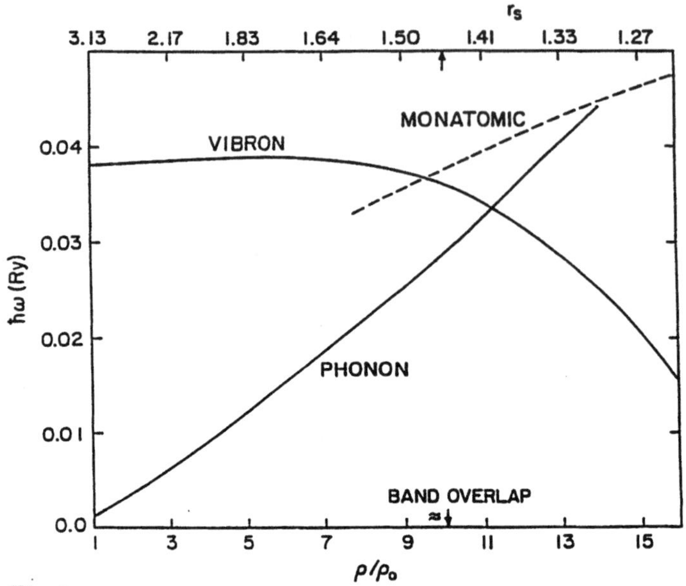

Fig. 4

Projected trends in vibron and zone boundary acoustic phonons showing their confluence at ρ/ρ_0. This probably sets an upper bound on pairing stability [17]. Band overlap indicates the density where in a paired state semi-metallic behavior begins.

results of such a procedure [5]. The picture of electron transfer leading to an expected weakening of the bond is confirmed. Corresponding to this weakening, the vibron frequency is dropping quite precipitously [5,17] with density at the highest pressures measured. In contrast to this the transfer of charge to the interstitial region is strengthening the system elastically and in confirmation the equation of state, which has been measured for the hexagonal structure to approximately 5-fold compression [18], shows a rapid power-law rise pressure with density. This information is also useful because with an assumption of reasonable elastic isotropy it is possible to obtain an estimate of the zone-boundary acoustic phonon frequency. This is the "anti-phase" companion of the vibron for an orientationally ordered structure, and it is apparent that should the two frequencies coincide, the force constants (derived from the potentials), would also equalize. This could only happen in a symmetric (monatomic) structure, so that a "one-sided" criterion for the instability of pairing can be obtained from a search for the confluence of vibron and zone-boundary frequencies. Figure 4 shows the results of such an extrapolation, indicating an approximate bound on density below which molecular pairs are dynamically stable of about 11 times normal. This is somewhat above the current estimates of the density for band overlap, which is in the vicinity of 9 times normal. So far as rotational states are concerned, orientational order appears to be established in dense deuterium at about 5-6 fold compression; in hydrogen the rotational states persist to much higher densities (the rotational constant is twice what it is in deuterium) but where, in density, orientational order might eventually be established is not yet clear.

2.5 Molecular Muonium

In Section 2.3 it was noted that a hypothetical state of metallic muonium, the analog of a group I metal, would have a minimum in its equation of state at about $r_s \sim 1.63$. However, because of the zero-point energy of the positive muons at such a density, it appeared unlikely that such a state would bind. Accordingly, the more likely state (if a dense assembly of muons and electrons could ever be formed at all) is a paired state. Apart from rising non-adiabatic corrections, the arguments leading to (12) go through as before, except that the kinetic energy operator refers now to muons ($\hat{T}_\mu$). Pairing is again expected, the basic unit being two muons and two electrons. Because of the radical change in mass, however, the energy scales for vibrons and rotons, are again very different. The vibron energy is immediately estimable from (14) since at low densities (with a and D adjusted to the deuterium vibron) this potential will again apply to the light muon isotope. A vibron scale of some 0.1 Ry (12,000 K) is expected: this is also over an electron volt, and as a consequence the binding energy of the muonium molecule (compared with deuterium) will be reduced by the same amount. The rotational constant ($\hbar^2/m_\mu^* \, r_m^2$) will be larger by a factor of 18 compared to the same quantity in deuterium, and this indicates immediately that if a muonium molecule is formed in a condensed state, it is quite likely to be highly rotational.

3. Muons in Dense Deuterium or Hydrogen

We turn now to three component systems, namely positive and negative muons in an environment provided by dense hydrogen on deuterium (the case of deuterons in a background formed entirely of negative muons will be left to last). We shall assume that the systems are again canonical, and neutral overall. This essentially requires that some external agency provides a source or sink of electrons. Though the density of muons is likely to be low compared with the volume average of $< \hat{\rho}_\alpha^{(1)}(\vec{r}) >$ ($N_\mu << N_d$) we will continue with the notation introduced earlier: thus

$$\hat{\rho}_\mu^{(1)}(\vec{r}\,) = \sum_i^{N_\mu} \delta(\vec{r} - \vec{r}_{\mu i})$$

is the one-particle density operator for muons.

3.1 Positive Muons in Deuterium

For this system the Hamiltonian is the straightforward extension of (1):

$$H = \hat{T}_e + \hat{T}_d + \hat{T}_\mu + \tfrac{1}{2}\int_V d\vec{r} \int_V d\vec{r}\,' \; v_c(\vec{r} - \vec{r}\,')\Big\{\hat{\rho}_{ee}^{(2)}(\vec{r}, \vec{r}\,') + \hat{\rho}_{dd}^{(2)}(\vec{r}, \vec{r}\,') + \hat{\rho}_{\mu\mu}^{(2)}(\vec{r}, \vec{r}\,') \\ - 2\hat{\rho}_e^{(1)}(\vec{r})\hat{\rho}_d^{(1)}(\vec{r}\,') + 2\hat{\rho}_d^{(1)}(\vec{q})\hat{\rho}_\mu^{(1)}(\vec{r}\,') - 2\hat{\rho}_\mu^{(1)}(\vec{r})\hat{\rho}_e^{(1)}(\vec{r}\,')\Big\}. \qquad (16)$$

The single length scale implied in (16) is important. First, since all charges are point charges, the energetics of neutral ensembles of electrons, deuterons, and muons are constrained by the virial theorem. Second, the electronic densities at either the deuterons or the mesons are rigorously constrained by the cusp theorem [19]. Since $N_\mu << N_d$ we expect from the arguments given above that for $r_s > 1.6$ the deuterons will be electronically paired; for $r_s < 1.6$ the system will be metallic in the electronic sense, either with pairing (the band overlap case) or without (the monatomic case). Any remaining electrons might be expected to bind to the positive muons. At low densities this is possible, the ensuing muonium appearing as substitutional 'impurities,' but probably quite diffusive on account of the low overall mass. Because of this diffusion, eventual pairing will occur with (a) lone deuterium atoms, the result being an even more extreme hybrid of the HD molecule (once again the vibron and roton energies of this couple will be strikingly different from either H_2 or D_2), or (b) another atom of muonium, the result being 'molecular' muonium which was discussed in Sect. 2.5. However, as the density of the system increases, the electrons bound in this system will more and more significantly interfere with the band electrons. Note that if such a molecule can be found, it will have extremely energetic vibron and roton states. As density increases it is also susceptible to a dynamic instability, and by applying the arguments of Sect. 2.4 this instability is expected before that of deuterium. At metallic densities the positive muon becomes an extreme case of the problem of hydrogen in metals: it will be a highly mobile quantum diffuser.

3.2 Negative Muons in Deuterium

The Hamiltonian for this system is also straightforward, namely

$$H = \hat{T}_e + \hat{T}_d + \hat{T}_\mu + \tfrac{1}{2}\int_V d\vec{r}\int_V d\vec{r}\,'\; v_c(\vec{r}-\vec{r}\,')\Big\{\hat{\rho}^{(2)}_{ee}(\vec{r},\vec{r}\,') + \hat{\rho}^{(2)}_{dd}(\vec{r},\vec{r}\,') + \hat{\rho}^{(2)}_{\mu\mu}(\vec{r},\vec{r}\,')$$
$$- 2\hat{\rho}^{(1)}_e(r)\hat{\rho}^{(1)}_d(\vec{r}\,') - 2\hat{\rho}^{(1)}_d(q)\hat{\rho}^{(1)}_\mu(r'\ ) + 2\hat{\rho}^{(1)}_\mu(\vec{r})\hat{\rho}^{(1)}_e(\vec{r}\,')\Big\}. \qquad (17)$$

The situation here appears more interesting: the correlation function $< \hat{\rho}^{(1)}_d(\vec{r})\hat{\rho}^{(1)}_\mu(\vec{r}\,') >$, in particular, will reflect characteristics similar to $< \hat{\rho}^{(1)}_d(\vec{r}\,)\hat{\rho}^{(1)}_e(\vec{r}\,') >$ whose behavior we have already inferred. To begin with, most deuterons will again be electronically paired; however, in addition to forming mesic atoms $(m_\mu D)$ deuterons can also be paired by the negative muons as has already been noted by Zel'dovich and Gerstein [20]. For example, the direct analog of D_2 will be an object bound at a separation of about $(1.4/207)a_o$. A somewhat more intriguing case is the following: consider two deuterons bound by a single negative muon – the analog of D_2^+ in the electronic case. Such an object appears to an electron as a point charge augmented by a relatively small permanent dipole (probably highly rotational, by the arguments given earlier). A single additional electron would yield a 'molecule' whose electronic excitation spectrum should not be too different from ordinary atomic D. Now bind *two* D_2^+ molecules with a *pair* of electrons: the resulting molecule has two electrons in a spin singlet state, and its "nuclei" are composite objects of two deuterons and negative muon each. In the simplest approximation the intra-molecular potential will again be given by (14) from which it is straightforward to conclude that spectroscopy on such a molecule will reveal vibrational and rotational structure at much lower values than found in D_2. Similar arguments obviously apply to the molecule in which hydrogen replaces deuterium. We note in passing that the overall spin statistics of those molecules (the equivalent of the standard ortho-para distinctions) are also interestingly different.

Finally note that for the $(\mu - D)_2$ molecules, the issue of whether their environment is insulating or metallic is probably of little practical significance. It is of far more interest to consider, again at least briefly, the transitory states of the neutral system described by

$$H = \hat{T}_d + \hat{T}_\mu + \tfrac{1}{2}\int_V d\vec{r}\int_V d\vec{r}\,'v_c(\vec{r}-\vec{r}\,')\Big\{\hat{\rho}^{(2)}_{\mu\mu}(\vec{r},\vec{r}\,') - 2\hat{\rho}^{(1)}_\mu(\vec{r})\hat{\rho}^{(1)}_d(\vec{r}\,') + \hat{\rho}^{(2)}_{dd}(\vec{r},\vec{r}\,')\Big\} \qquad (18)$$

that is, Coulomb matter comprised of deuterons and negative muons, alone. Here all length scales are gauged by $a_0/(m_\mu/m_e)$. At low densities and on such scales, the deuterons are paired. We note immediately that the scale of *inter*molecular attraction is set by internal fluctuation: the long-range fluctuating dipole interaction is given by $-\Delta(\alpha/r^3)^2$ where Δ is a typical internal excitation energy, and α is the (muonic) polarizability. It is clear that Δ is increased by two orders of magnitude over the normal energies found in D_2; on the other hand α is decreased by six orders of magnitude. So compared with normal Van der Waals attraction, $D_{\mu 2}$ molecules are attracting with energies 10 order of magnitude lower than usual. In a very real sense a gas of such objects would be "ideal." It is likely that a liquid phase will not exist; quantum zero point fluctuations at low temperatures will overwhelm any residual attraction. On the other hand, only modest pressures will be required at low temperatures to pack these molecules. The ensuing density will be 300,000 times that of normal deuterium. Note finally that the fluctuation argument as applied to a single mesic atom in a condensed matter environment leads to a reduction in fluctuation based attraction which is 5 orders smaller than normal Van der Waals attraction. The analogy to a neutron, both at short range and at long range, is therefore very apt.

Perhaps the ultimate state of Coulomb matter will be the one formed in the metallic modification of the system just described. From the arguments given in Section 2.2 this should occur at $r_{\mu s} = (1 \cdot 63)a_0(\frac{m_e}{m_\mu})$, i.e., approximately $2 \cdot 4 \times 10^6$ times the density of normal deuterium (at which density there should again be a minimum in the equation of state). It is energetically unfavorable with respect to the molecular state; but is it bound relative to free muonium atoms $(\mu - D)$? It appears so [21], and it also appears that at such densities the deuterons will be in a state of continuous symmetry. This is immediately interesting from the fusion point of view because the *overlap* of deuterons in such a *fluid* state (a quantity which is scalable from the known electron equivalent [22]) is now quite substantial.

References and Footnotes

1. G.D. Mahan, "Many Particle Physics" (Plenum, N.Y., 1983), Ch. 5.
2. See, for example, H.E. Lorenzana, I.F. Silvera, and K.A. Goettel, Phys. Rev. Lett. **64**, 1939 (1990).
3. I. Silvera, Rev. Mod. Phys. **52**, 393 (1980).
4. R.J. Hemley, H.K. Mao, and J.F. Shu, to be published.
5. N.W. Ashcroft, Phys. Rev. B **41**, 10963 (1990).
6. K. Takemura, S. Miromura, O. Shimomura, and K. Fujii, Phys. Rev. Lett. **45**, 1881 (1980).
7. N.W. Ashcroft and N.D. Mermin, "Solid State Physics" (Holt Saunders, N.Y., 1976).
8. C. Friedli and N.W. Ashcroft, Phys. Rev. B **16**, 662 (1977).
9. N.F. Mott, Proc. Phys. Soc. A **62**, 416 (1949).
10. D.M. Ceperley and B.J. Alder, Phys. Rev. Lett. **45**, 566 (1981).
11. E.G. Brovman and Yu. Kagan, Sov. Phys. - JETP **30**, 721 (1970); E.G. Brovman, Yu. Kagan, and A. Kholas, Sov. Phys. - JETP **34**, 1300 (1972).
12. J. Hammerberg and N.W. Ashcroft, Phys. Rev. B **9**, 409 (1974).
13. Or perhaps more realistically a "metallic" cluster formed by beam cooling techniques. Low energy muon beams are discussed by Nagamine in the Proceedings of this Workshop. For general references on muons in condensed matter, see "Muon Physics," Vernon W. Hughes and C. S. Chu, eds. (Academic Press, N.Y., 1977), Volumes I, II, and III.

14. The Einstein frequency rises to well above 10,000 K, and this sets the scale for any phonon mediated pairing mechanism for electrons. Superconductive pairing in such a remote system would have a scale about 3 times what is expected for metallic hydrogen.
15. Strictly speaking the argument applies to a vibron band (see Ref. 5).
16. See, for example, S. Flügge, P. Walger, and A. Weiguny, J. Mol. Spectros. 23, 243 (1967).
17. R. Hemley and H.K. Mao, Phys. Rev. Lett. **61**, 857 (1988).
18. V.P. Glazkov, S. P. Besedin, I.N. Goncharenko, A.V. Irodava, I.N. Makarenko, V.A. Somenkov, S.M. Stishov, and S. Shil'stein, JETP Lett. **47**, 763 (1988).
19. T. Kato, Comm. Pure Appl. Math. **10**, 151 (1957).
20. Ya. B. Zel'dovich and S.S. Gerstein, Sov. Phys. Usp. 3, 593 (1961). See also K. Fukustima and F. Iseki, Muon Catalyzed Fusion 1, 225 (1987).
21. N.W. Ashcroft, to be published.
22. K. Moulopoulos and N.W. Ashcroft, Phys. Rev. B **41**, 6500 (1990).

Basic Concepts of Muon Catalyzed Fusion

A. Scrinzi

Institute for Medium Energy Physics, Austrian Academy of Sciences, Boltzmanngasse 3, A-1090 Vienna, Austria

Present address: Kurtchatov Institute for Atomic Energy, Moscow

Abstract. We give an overview over the basic concepts of muon catalyzed fusion in theory and experiment. With the examples of μCF in pure deuterium and deuterium–tritium mixtures the present level of theoretical understanding of experiments is illustrated. The most important open questions in the field and possible applications are summarized.

1. Introduction

The idea of muon catalyzed fusion (μCF) is simple: after insertion of a negative muon μ^- into a mixture of deuterium and tritium the muonic molecular ion $(dt\mu)^+$ is formed. Due to the big mass of the muon $m_\mu = 207\,m_e$ the nuclei of this molecule are by a factor of $\sim$ 200 closer to each other than in the ordinary hydrogen ion H_2^+. This drastically enhances the probability for t and d to tunnel through the Coulomb barrier, thus leading to virtually instantaneous t–d fusion

$$(dt\mu)^+ \longrightarrow {}^4He^{++} + n + \mu^- + 17.6\,MeV\,. \qquad (1)$$

In most cases the μ is released after the nuclear reaction to form another molecule $(dt\mu)^+$. Analogous reactions occur with all hydrogen isotopes, as long as at least one of the nuclei is a heavy isotope.

Soon after the possibility of this process was realized [1] quantitative considerations [2, 3] exposed the two fundamental bottlenecks of the process: low muonic molecule formation rates seemed to limit the number of fusions to at best a few within the muon lifetime of $\tau_\mu = 2.2\,\mu s$, and the probability of losing the muon through binding to the fusion product He^{++} would further reduce the efficiency of the catalytic process, in particular for the $dp\mu$ and $dd\mu$ systems primarily discussed at that time.

With initial hopes for energy production crushed, during the 1960s muon μCF was studied as an important background in $d\mu$ weak capture experiments [4], until in 1966 unexpectedly high rates of $(dd\mu)^+$ formation with a strong temperature dependence were observed [5]. It was realized that there exists a so-called *resonant molecular formation* mechanism [6], which enhances the rates by 2 orders of magnitude. This mechanism requires the existence of an extremely weakly bound state of the muonic molecule. In the course of a large theoretical effort [7, 8, 9] the existence of such states of $dd\mu$ and $dt\mu$ was shown and their wave functions were calculated. The prediction of extremely high formation rates [10] was soon afterwards confirmed by experiment [11]. The last ten years have been devoted to achieving quantitative understanding of μCF .

The catalytic cycle comprises phenomena from atomic, molecular, and nuclear physics (Fig. 1): first, a muonic $d\mu$ or $t\mu$ atom is formed by *Coulomb capture* of the μ in a highly excited state ($n \approx 14$) [12, 13], whose binding energy roughly equals the energy of the

 © Springer-Verlag Berlin, Heidelberg 1992

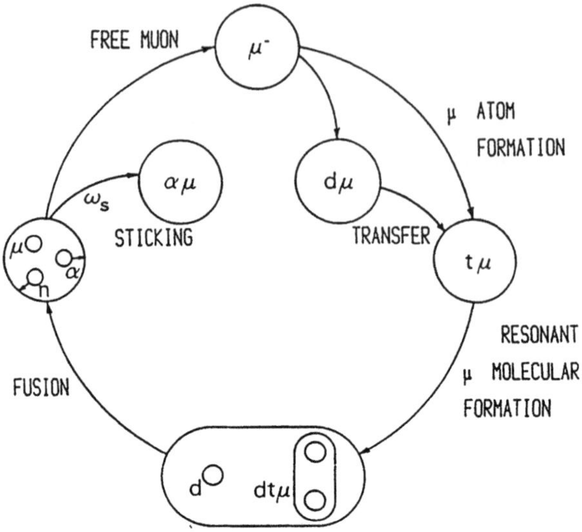

Figure 1: Basic μCF cycle in td mixtures

$^1\Sigma$ electron. The details of the subsequent atomic cascade [14], which is made up from radiative and Auger transitions, as well as elastic and inelastic scattering of the muonic atom determine the distribution of the muonic atom kinetic energies [15], although with a duration of $\sim 10^{-12}\,s$ (in liquid hydrogen) these processes are not directly observable. Because of the large isotopic difference between the ground state energies of $t\mu$ and $d\mu$ (48 eV) molecular formation is preceded by *transfer* of the μ from $d\mu$ to $t\mu$. With a rate of $\sim 300\,\mu s^{-1}$ in liquid D_2–T_2 mixtures this process crucially enters the duration of the cycle. Next, through the resonant mechanism the hybrid molecule $[(dt\mu)dee]$ (or $[(dt\mu)tee]$) is formed, in which the weakly bound first excited (v=1) state of angular momentum J=1 (denoted by $(dt\mu)_{11}$) serves as one nucleus of a hydrogen like molecule. Experimentally observed formation rates lie between 300 and 600 μs^{-1} [17, 16]. After a fast Auger cascade ($\sim 10^{-12}s$) [18, 19] to a J=0 state of $dt\mu$ (either to the first excited Jv=01 or ground Jv=00 state) fusion occurs. Due to the high $^4He^{++}$ kinetic energy (3.5 MeV) after the reaction there is only a small probability for *sticking* of the μ to He^{++} of about $\omega_s \sim 0.4\%$ [20, 16]. This value is the *initial sticking* after the nuclear reaction reduced by *reactivation* [21, 22] of the muon through $(\mu He)^+$ scattering on the surrounding hydrogen molecules. The free μ enters a new cycle.

The numbers given above allow a quick estimate of the number of fusions per μ : The inverse of the fusion yield Y_n is given by the probability of μ loss in one cycle. With μ decay and sticking as the main losses this becomes

$$Y_n \approx (\frac{\tau_c}{\tau_\mu} + \omega_s)^{-1} \approx (\frac{0.005}{2.2} + 0.004)^{-1} \approx 160 \qquad (2)$$

where we estimated the time for one cycle by $\tau_c \approx \tau_{transfer} + \tau_{formation}$. Indeed, up to 150 fusions per μ were observed in experiments [20, 23].

While μCF in D_2–T_2 mixtures is most efficient in terms of *cycle rate* ($\lambda_c := \tau_c^{-1}$) and energy gain per fusion, all other mixtures of hydrogen isotopes have been studied as well (see Ref. [24]). The corresponding nuclear reactions are

$$
\begin{array}{llll}
t+t & \longrightarrow & {}^4He+n+n & Q=11.3\,MeV \\
t+p & \longrightarrow & {}^4He+\gamma & Q=19.8\,MeV \\
d+d & \longrightarrow & \left\{ \begin{array}{l} {}^3He+n \\ t+p \end{array} \right. & \begin{array}{l} Q=3.3\,MeV \\ Q=4.0\,MeV \end{array} \\
p+d & \longrightarrow & {}^3He+\gamma & Q=5.5\,MeV
\end{array}
\tag{3}
$$

At present the $dd\mu$ system, in which the resonant mechanism was discovered, is the best understood of all. There resonant molecular formation has been verified in great detail [25] allowing the experimental determination of the $(dd\mu)_{11}$ binding energy with an accuracy of $0.3\,meV$ or $10^{-7}\,m.\,a.\,u.$ (muonic atomic units $\hbar=e^2=m_\mu=1$, muonic Rydberg energy $2.7\,keV$).

The remaining systems $tt\mu$, $tp\mu$, and $dp\mu$ do not possess weakly bound states and, consequently, they only have small, *nonresonant* rates of molecular formation. They are studied both as important background processes in the $dt\mu$ cycle and for their own sake, e.g. they show exotic nuclear reactions like radiationless pt fusion with μ conversion [29, 30].

2. Molecular Formation

Initially, unaware of the weakly bound states of $dt\mu$ and $dd\mu$, only *nonresonant formation*

$$
t\mu + D_2 \longrightarrow [(dt\mu)de]^+ + e^- \tag{4}
$$

was considered, where the conversion electron carries away the $dt\mu$ binding energy of about 30 to 300 eV.

The tiny binding energy of the excited $(dt\mu)_{11}$ state $\epsilon_{11} \approx 0.6\,eV$ can be absorbed by rovibrational excitations of the hybrid $[(dt\mu)dee]$ molecule leading to the more efficient process

$$
(t\mu)_F + (D_2)_{K_i,v_i=0} \longrightarrow [(dt\mu)_{11,S}dee]_{K_f v_f} (=: MD_{K_f v_f})\,. \tag{5}
$$

Here the quantum number F=0 and 1 denotes the $t\mu$ hyperfine states ($\uparrow\downarrow$) and ($\uparrow\uparrow$), respectively, K_i is the rotational quantum number of D_2, which will always be in the vibrational ground state $v_i = 0$ at target temperatures below $\sim 3000\,K$. S labels the $dt\mu$ hyperfine states and $K_f v_f$ are the rovibrational quantum numbers of the hybrid complex.

Since the energy spectrum of $[(dt\mu)dee]$ is approximately discrete, reaction (5) can occur only at a discrete set of relative kinetic $t\mu$–D_2 energies. These *resonance energies* are given by

$$
\begin{aligned}
\epsilon_{if} &= \epsilon_{11} + \delta\epsilon_{dt\mu}(S) + E_{MD}(K_f, v_f) \\
&\quad -\epsilon_{t\mu} - \delta\epsilon_{t\mu}(F) - E_{D_2}(K_i, v_i = 0)
\end{aligned}
\tag{6}
$$

with the MD and D_2 rovibrational energies E_{MD} and E_{D_2}, the $(t\mu)_{1s}$ and $(dt\mu)_{11}$ binding energies $\epsilon_{1s}, \epsilon_{11}$ and the corresponding hyperfine corrections $\delta\epsilon_{t\mu}(F)$ and $\delta\epsilon_{dt\mu}(S)$. In muonic systems the hyperfine energies are quite sizable, bigger than the fine structure splitting. For the $dt\mu$ system with a value of $\sim 200\,meV$ the hyperfine splitting is comparable to the binding energy $\epsilon_{11} \approx 600\,meV$.

The rate for process (5) can be written as [31]

$$
\lambda_{FK_i,SK_f}(T) = 2\pi\, N\, f(\epsilon_{if}, T)\omega_i(K_i)\, W_{FS}\, |V_{if}|^2 \tag{7}
$$

where $\omega_i(K_i)$ is the occupation probability of the D_2 rotational state K_i. The distribution of the $t\mu$ kinetic energies $f(\epsilon_{if}, T)$ depends on the target temperature T, although $t\mu$ will not always be completely thermalized [32, 33]. N is the density of the target molecules, $|V_{if}|$ the transition matrix element for the given resonance and W_{FS} a spin overlap factor.

The typical separation of the individual resonances is given by the rotational quanta of MD and D_2, which are of the order $4 \cdot K_{i,f}(K_{i,f}+1)\, meV$. Therefore for accurate predictions at very low temperatures ($\sim 50\, K$) the resonance energies have to be known to the accuracy of $1\, meV$.

3. μCF in Pure Deuterium

As a first illustration of the concepts discussed above and of some experimental techniques we present theoretical and experimental results for μCF in pure deuterium. Apart from being interesting as a background process in the deuterium–tritium mixture and in weak $d\mu$ capture experiments, this system is sufficiently simple to allow detailed understanding of all processes involved. Like $(dt\mu)_{11}$ there exists a weakly bound $(dd\mu)_{11}$ state with binding energy $\epsilon_{11} \approx 2\, eV$, which causes resonant $dd\mu$ formation.

The dd system differs from td in some specifics:

The hyperfine splittings of $48.5\, meV$ for $d\mu$ and $24\, meV$ for $dd\mu$ are significantly smaller than for $t\mu$ and $dt\mu$. Incidentally, there are some resonances for $dd\mu$ formation from the $d\mu(F = 3/2)$ hyperfine state at very low ($\sim 4\, meV$) energies, while, due to hyperfine splitting, the first resonances for $d\mu(F = 1/2)$ appear only at $\sim 28\, meV$. Consequently, formation from $d\mu(F = 3/2)$ dominates at low target temperatures ($\lesssim 200\, K$)

The inverse reaction of *backdecay* of the hybrid molecule

$$[(dd\mu)_{11}dee]^* \longrightarrow d\mu + D_2 \tag{8}$$

plays an important role in the $dd\mu$ system. While $(dt\mu)_{11}$ quickly stabilizes through Auger transitions to lower states, the $(dd\mu)_{11}$ state is long–lived, because dd exchange symmetry forbids $\Delta J = 1$ transitions. Fusion from $J = 1$ states is slow due to the centrifugal barrier between the nuclei. Thus backdecay rates exceed the sum of fusion and deexcitations rates, the *effective fusion rate* $\tilde{\lambda}_f = 460\, \mu s^{-1}$ [34] by a factor of ~ 3.

The dd μCF reaction kinetics is given in Fig. 2. The dashed arrows in the kinetics indicate the backdecay process. Initial formation is reduced by the fraction of backdecay and a backdecay contribution $\lambda_{back}^{\frac{3}{2}\frac{1}{2}}$ corresponding to the path $d\mu(\frac{3}{2}) + D_2 \rightarrow [(dd\mu)dee] \rightarrow d\mu(\frac{1}{2}) + D_2$ is added to the hyperfine transition rate. Since backdecay is almost instantaneous compared to molecule formation and hyperfine transitions, the kinetics can be described by *effective rates* (denoted by $\tilde{\lambda}$'s, solid arrows). Clearly, the effective formation rates increase with the fusion rate, while $\lambda_{back}^{\frac{3}{2}\frac{1}{2}}$ decreases with increasing fusion rate.

Initially the $d\mu$ hyperfine states are populated statistically, i.e. 1/3 in $F = 1/2$ and 2/3 in $F = 3/2$. Molecular formation starts predominantly from the $F = 3/2$ state at rates of $\tilde{\lambda}_{dd\mu}^{\frac{3}{2}} \sim 4\, \mu s^{-1}$ in liquid D_2. (Formation from $F = 1/2$ is 2 orders of magnitude smaller at target temperatures $< 90\, K$ and reaches $\tilde{\lambda}_{dd\mu}^{\frac{1}{2}} \sim 2\, \mu s^{-1}$ at $T \sim 150\, K$.) At the same time the $F = 3/2$ state is depopulated by *hyperfine transitions* to the $F = 1/2$ state at a rate of $\tilde{\lambda}_{d\mu}^{\frac{3}{2}\frac{1}{2}} \sim 35\, \mu s^{-1}$ in liquid D_2, i. e. much faster than through molecular formation. At low temperatures the process of inverse hyperfine transitons (rate $\tilde{\lambda}_{d\mu}^{\frac{1}{2}\frac{3}{2}}$) is of little importance.

After depopulation of the $F = 3/2$ state molecular formation continues at low intensity from $d\mu(F = 1/2)$, which decreases at the rate of μ decay. The probability of μ sticking to

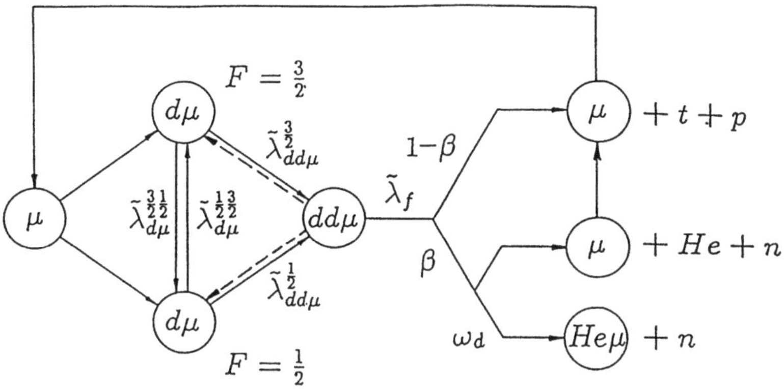

Figure 2: Reaction kinetics of μCF in pure deuterium.

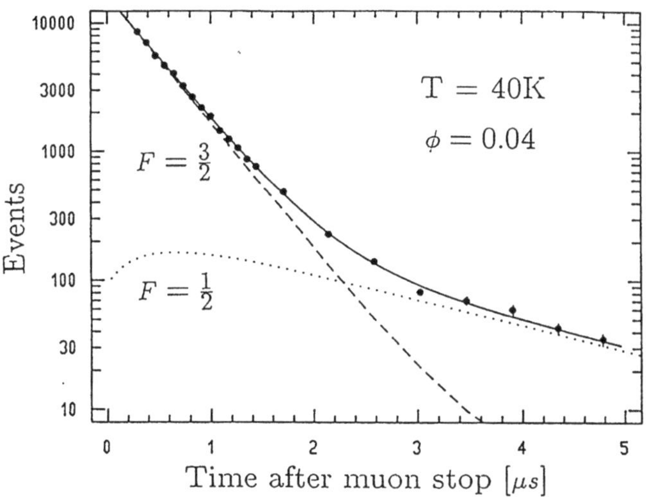

Figure 3: Intensity of dd fusion neutrons after μ stop. PSI experiment at temperature 40 K and D_2 density of 5% of liquid hydrogen. From Ref. [25]

^3He after dd fusion $\omega_d = 12.2\%$ is comparatively high because of the smaller recoil energy of ^3He. Nevertheless it plays little role in the cycle, since the main loss is the competition of μ decay with $dd\mu$ formation.

A characteristic time spectrum of fusion neutrons after μ stop in the D_2 target obtained in PSI experiments is shown in Fig. 3. The data were taken at temperature $40\,K$. Target density was 5% of liquid hydrogen, which proportionally slows down the molecular formation and hyperfine transiton rates. The structure discussed above is clearly visible: initially high fusion neutron intensity from $F = 3/2$ molecular formation, then fast depopulation of $d\mu(F = 3/2)$, and finally molecular formation from $d\mu(F = 1/2)$, which decays approximately with the μ lifetime $\tau_\mu = 2.2\,\mu s$.

From such spectra the rates of the kinetics of Fig.2 can be extracted unambigously. Fig.4 shows the effective rates obtained in the PSI experiment [25]. $\tilde{\lambda}^{\frac{1}{2}}_{dd\mu}$ is nonresonant at low target temperatures, picking up only at higher temperatures. The solid line is the

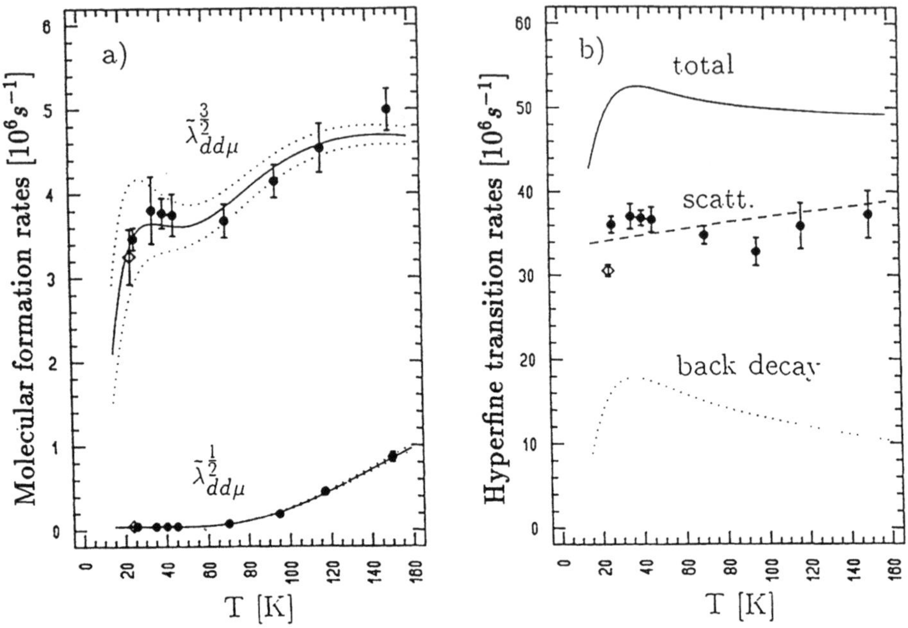

Figure 4: **$dd\mu$ formation rates (a) and hyperfine transition rates (b) from Ref.[25]. In (a) solid lines are best fits, dotted lines are obtained with ϵ_{11} shifted by $\pm 1\,meV$. In (b) the total theoretical hyperfine transition rate (solid line) is the sum of the scattering (dashed) and backdecay (dotted) contributions.**

best fit using the theory discussed above. The main fit parameters are the $(dd\mu)_{11}$ binding energy ϵ_{11} and the effective fusion rate $\tilde{\lambda}_f$. As discussed above, varying the effective fusion rate $\tilde{\lambda}_f$ allows us to adjust the overall size of the effective formation rates. The shape of the curve $\tilde{\lambda}^{\frac{3}{2}}_{dd\mu}$, in turn, is extremely sensitive to variations of ϵ_{11} due to the near threshold resonances. The dotted lines result, if one shifts ϵ_{11} by $\pm 1\,meV$. For this reason it was possible to determine ϵ_{11} very accurately:

$$\epsilon_{11} = -1966.1(3)\,meV\,. \tag{9}$$

From the same set of data hyperfine transition rates were extracted (Fig. 4). There the comparison between theory and experiment is less satisfactory: theoretical expectations exceed the observed rates by about 40%. The observed rate is the sum of the backdecay contribution $\lambda^{\frac{3}{2}\frac{1}{2}}_{back}$ mentioned above and a contribution $\lambda^{\frac{3}{2}\frac{1}{2}}_{scatter}$ due to the inelastic scattering process

$$d\mu(F=3/2) + D_2 \xrightarrow{\lambda^{\frac{3}{2}\frac{1}{2}}_{scatter}} d\mu(F=1/2) + D_2\,, \tag{10}$$

which has been calculated in a simplified manner as $d\mu + d$ scattering. Since on the scale of muonic energies this is very low energy scattering, nearly linear temperature dependence was obtained. The temperature dependence of the backdecay contribution $\lambda^{\frac{3}{2}\frac{1}{2}}_{back}$ reflects the resonant character of the molecular formation process. The data indeed seem to hint at a superposition of both effects.

The possible reasons for the disagreement can be listed as follows: the theoretical $d\mu+d$ cross-section is too high, the influence of the D_2 molecular structure in $d\mu + D_2$ scattering is significant, or the backdecay contribution is much smaller, which would require a much higher effective fusion rate $\tilde{\lambda}_f$ than calculated. Which (or which combination) of the above-mentioned effects is responsible for the discrepancy cannot be decided at present: $d\mu + d$ scattering calculations seem to be well converged [26], taking into account the D_2 molecular structure rather seems to enhance the hyperfine rate [28] and the fusion rate required to suppress backdecay would be ~ 2 times larger than present theoretical values. This is one of the most appealing theoretical questions in μCF , because its solution would give the final and comprehensive confirmation of the theory of resonant formation, fusion rate calculations and calculations of $d\mu + d$ inelastic scattering.

4. The Quantum Mechanical 3 Body Problem

The accurate experimental $dd\mu$ binding energy ϵ_{11} represents a challenge to theory. At the beginning of the theoretical efforts was the task of determining the purely Coulombic $dt\mu$ and $dd\mu$ binding energies. This is a difficult problem, not only because of the accuracy required ($\lesssim 1\,meV$ compared to the muonic Rydberg of $2.7\,keV$), but in particular because at a mass ratio $m_\mu/m_{deuteron} = 1/20$ any Born–Oppenheimer or adiabatic approximation is inadequate and a genuine 3 body problem has to be solved.

Here the first breakthrough was achieved with an expansion in 2 center Coulomb wave functions [10], which at that time allowed the determination of the binding energies of the weakly bound J=1,v=1 states with an accuracy of $\approx 20\,meV$.

During the last few years the desired accuracy of better than $1\,meV$ was reached in a number of large scale variational calculations using up to 3000 basis functions of exponential [35, 36, 37], Gaussian [38] or Hylleraas [39, 40] type. Fig. 5 gives the spectra of the

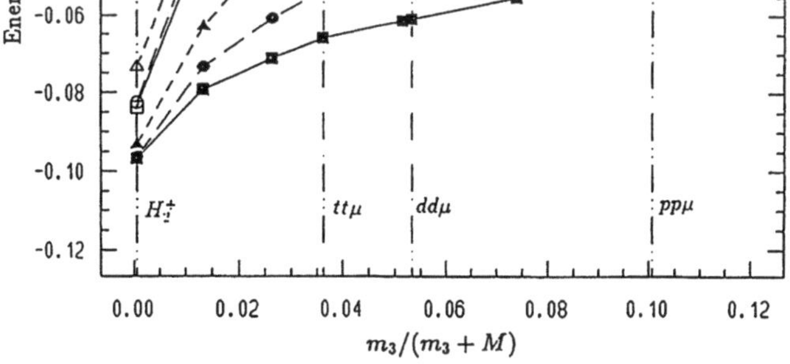

Figure 5: Spectra of the 3 body Coulomb problem for 2 identical particles with mass M and the third particle with m_3 for various mass ratios. Squares... J=0, circles... J=1, triangles... J=2. Empty symbols indicate first excited states. Units $\hbar = e^2 = M^{-1} + m_3^{-1} = 1$.

systems with identical nuclei. There one can see what a remarkable coincidence the existence of the $(dd\mu)_{11}$ (and analogously the $(dt\mu)_{11}$) state is. At present, the bigger problems in calculating ϵ_{11} are the various corrections: relativistic corrections, vacuum polarization, interaction with surrounding electrons etc. [41, 42, 43]. Recently theoretical results for the $(dd\mu)_{11}$ binding energy have converged to $-1966.0\,meV$ with estimated $0.5\,meV$ uncertainty [44], which impressively agrees with the experimental value of $-1966.1 \pm 0.3\,meV$. Thus μCF in pure deuterium could be used to confirm a variety of techniques used in the 3 body Coulomb problem and corrections.

5. μCF in Deuterium Tritium Mixtures

This system attracted most attention both from theory and experiments, because it possesses the highest cycle rates and a high yield of fusion energy. At the same time the system is quite complex and difficult to handle. In experiments radiation safety requires careful tritium handling. From the theoretical point of view, the kinetics analogous to Fig. 2 involve more states $(t\mu, d\mu, [(dt\mu)dee], [(dd\mu)dee], \ldots)$ multiplying the number of kinetic rates. Moreover, due to low $t\mu$ scattering cross–sections $t\mu$ thermalization is no longer instantaneous compared to formation and transfer rates. In particular, *excited state transfer* $(d\mu)^* + t \rightarrow d + (t\mu)^*$ significantly enters the cycle rates.

These many physical parameters have to be extracted from relatively few experimental observables. Instead of the nice two–component curve of Fig. 3, the typical neutron time spectrum in dt–μCF is in most cases (see below) a straight exponential

$$n(t) = \lambda_c\, e^{-\lambda_n t}\,. \tag{11}$$

The decay rate $\lambda_n := \lambda_0 + W\lambda_c$ is composed of the μ decay rate λ_0 and the cycle rate times *raw sticking* W, i.e. sticking ω_s plus losses in the side cycles with $dd\mu$, $tt\mu$ and scavenging of μ by He. The cycle rate λ_c depends on the physical parameters in the way [45]

$$\lambda_c^{-1} = \tau_c = \frac{q_{1s} c_d}{\lambda_{dt} c_t} + \frac{3/4}{\lambda_{t\mu}^{10} c_t + \Lambda_{dt\mu}^{1}} + \frac{1/4 + (3/4)\chi}{\Lambda_{dt\mu}^{0}}\,, \tag{12}$$

i.e. the cycle time τ_c is the sum of times spent in the individual stages of the cycle. The first term is for transfer: q_{1s} is the fraction of $d\mu$ atoms, which reach the $1s$ state without excited state transfer, λ_{dt} is the rate of $(d\mu)_{1s} \rightarrow (t\mu)_{1s}$, c_d and c_t are the deuterium and tritium concentrations, respectively. Next is the time spent in the $t\mu(F=1)$ hyperfine state with the $t\mu(F=1) \rightarrow t\mu(F=0)$ transition rate $\lambda_{t\mu}^{10}$ and the molecular formation rate from this state $\Lambda_{dt\mu}^{1}$. And finally the time spent in $F=0$ with formation rate $\Lambda_{dt\mu}^{0}$ and $\chi = \frac{\lambda_{t\mu}^{10} c_t}{\lambda_{t\mu}^{10} c_t + \Lambda_{dt\mu}^{1}}$. The rates $\Lambda_{dt\mu}^{F} = c_{DT}\lambda_{(dt\mu)t}^{F} + c_{D_2}\lambda_{(dt\mu)d}^{F}$, $F = 0, 1$ are the sum of the formation rates on the molecules DT and D_2 weighted with the corresponding molecular concentrations c_{DT} and c_{D_2}.

Although a wide range of experiments has been performed up to temperatures of $800\,K$ [17, 16, 20, 23], at densities between 0.01 and 1.2 of liquid hydrogen, at all relative isotopic concentrations $c_t : c_d$, and even varying the relative molecular concentrations $D_2 : DT : T_2$, up to now not all basic parameters could be extracted unambiguously.

Nevertheless resonant $dt\mu$ formation has been qualitatively confirmed. The mere fact of very high $(200\,\mu s^{-1})$ temperature dependent cycle rates like in $dd\mu$ confirms resonant formation. Moreover, it was found that, at fixed ratio $c_t : c_d$ the cycle rate depends on the ratio of the molecular concentrations $D_2 : DT : T_2$ [23]. This is due to the fact that the resonances $t\mu + D_2 \rightarrow [(dt\mu)dee]$ are quite different from $t\mu + DT \rightarrow [(dt\mu)tee]$.

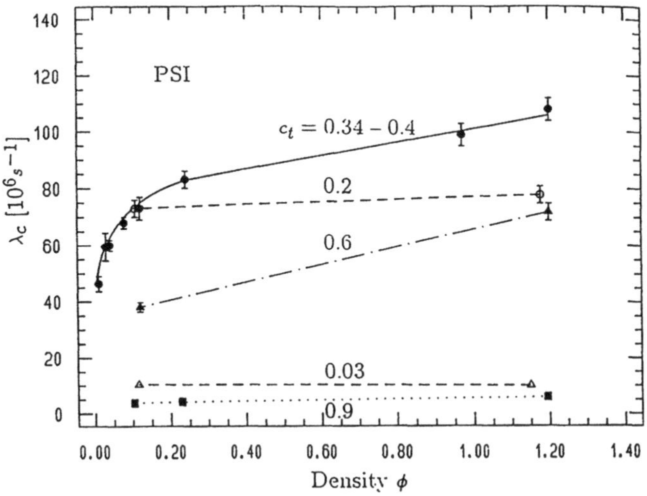

Figure 6: Nontrivial density dependence of the cycle rates of μCF in td mixtures for various tritium concentrations c_t [23].

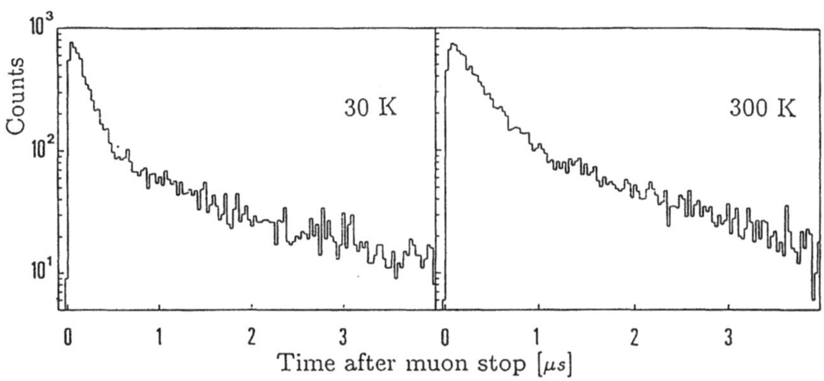

Figure 7: "Hot transients" in the intensity of td fusion neutrons in low density gas [24].

However, there are several poorly understood observations on td:
First, contrary to initial theoretical predictions, the cycle rates do not decrease at very low temperatures, although there seem to be no resonances for molecular formation at low energy. The common hypothesis to explain this behaviour is that a very strong resonance lying some meV below threshold, i.e. at negative $t\mu$–D_2 kinetic energies, has finite width with the tail reaching up to positive energies. However, quantitative statements are hard to make, because shape, width, and exact location of the resonance have not been determined.

Secondly, a nontrivial density dependence of the cycle rates was observed [20, 23]. Fig. 6 gives the cycle rates divided by the target density over the full range of densities at various tritium concentrations. Horizontals would mean trivial, i.e. linear density dependence. The general explanation for the deviations from linearity are three and many body effects in molecular formation, or, putting it differently, again collisional broadening of the resonances.

Finally, in experiments with low density td gas (1% of liquid hydrogen) [46] peaks were observed at short times after μ stop (Fig. 7). These peaks are temperature dependent, broadening with increasing temperature. The peaks correspond to processes faster than transfer and molecular formation. The phenomenon is qualitatively explained as a manifestation of the $t\mu$ deceleration process, where the $dt\mu$ formation rates are very high for $t\mu$ kinetic energies above thermal energies.

6. Sticking

As the limiting factor in the $dt\mu$ cycle sticking has been most carefully studied experimentally as well as theoretically. There seem to be few means of influencing sticking on the scale of ordinary chemical energies, since the $dt\mu$ initial state, the fusion reaction and $^4He\mu$ binding involve quite different scales of energy (muonic, nuclear) and are largely independent of the surroundings.

The calculation of initial sticking after the nuclear reaction is quite straightforward once accurate $dt\mu$ wave functions are available: initial sticking ω_0 is the overlap of the $dt\mu$ wave function for td distance =0 (assuming a sudden nuclear reaction at this point) with the wave function of a fast (3.6 MeV) recoiling $^4He\mu$ in a bound state. The majority of calculations of initial sticking use this approach with minor modifications to account for effects of the nuclear forces and the finite size of the nuclei [42, 47, 48, 49]. Results agree within 1% of ω_0.

This initial sticking is reduced by the reactivation factor R, when μ may be shaken off in collisions of $^4He\mu$ with target molecules:

$$\omega_s = (1 - R)\omega_0 \tag{13}$$

Reactivation introduces but a slight density dependence into sticking, since the number of collisions till stopping of $^4He\mu$ is independent of density. Present theoretical values are $\omega_s \approx 0.6\%$.

The first experimental determinations of sticking exploit the relation Eq. 11: with λ_c determined from the intensity of fusion neutrons raw sticking W can be extracted from the decay rate λ_n. After correcting for the side cycles and other losses, one obtains ω_s. There is no agreement between the experiments [16] and [20]: in the former an approximately constant value of $\omega_s = 0.45\%$ was observed, while the latter shows a strong density dependence with values varying from 0.3% to 1.2%.

An ambitious experiment [51] tried to directly measure the ratio of fusion products $He : \mu He$ from a low density td target, which was separated from the detector by MYLAR foils. First results of $\omega_s \approx 1\%$ are planned to be improved using a pulsed μ beam.

The last generation of sticking experiments unites target and detector. With pure deuterium gas in an ionization chamber sticking in this system has been determined sucessfully [52]. The experimental value of 12.2(3)% sticking in the He–n branch of the dd fusion reaction agrees very well with theory [53]. In D_2–T_2 the small value of ω_s and the β activity of tritium make the analogous experiment much more complicated. Currently data from such an experiment at PSI are being analyzed.

Reactivation may open the way for higher μCF efficiency. It was remarked [54] that in plasma of 100 eV deceleration of μHe would be reduced, thus enhancing the chance for reactivation, which in turn would reduce sticking to almost 0.

7. Applications

The physical break–even for energy production by means of μCF requires us to extract more energy than spent on μ production. Estimates of the cost of the μ under most favourable conditions vary between 3 and 8 GeV beam energy/μ . Muon catalyzed fusion under conditions comparable to present day experiments will give a maximum of 200 fusions/μ with a release of a total of 3.5 GeV/μ , where sticking is the fundamental limiting factor. Clearly, counting conversion losses and costs of tritium production, μCF is far from producing energy.

In any case, μCF can serve as an intense source of 14 MeV neutrons, even if not energy positive. A hybrid reactor using μCF neutrons to breed Pu was proposed in [55]. The complexity of the system and the undesirable aspects of fission technology make alternatives attractive, exotic as they may be.

One line of thought tries to exploit the smaller sticking due to higher reactivation in plasma. Unfortunately, in plasma the resonant molecular formation mechanism discussed above cannot work. This problem might be overcome [54] by using an inhomogenous mixture of "cold" ($T < 2\,eV$) drops of T_2–D_2 in a plasma of $\sim 100\,eV$, where $dt\mu$ molecules would be formed quickly in the drops, while reactivation in plasma would reduce sticking. The required plasma parameters ($T \sim 100\,eV$, density 20% of liquid hydrogen, 5 times the density of liquid hydrogen in the drops) are claimed to be attainable in Z–pinches. Muons have to be supplied in an intense pulsed beam with bunches of the order $10^{14}\,\mu$ in $10^{-7}s$.

Acknowledgement

I would like to thank G. Benedek and H. Schneuwly for the opportunity to participate in this International Workshop on Exotic Atoms in Condensed Matter. Support by the Austrian Science Foundation is gratefully acknowledged.

References

[1] Frank F.C., Nature 160: 525 (1947)

[2] Sakharov, A.D., Report,Phys.Inst.Acad.Sci. U.S.S.R. (1948)

[3] Zeldovich, Ya. B., Gerstein, S.S., Dokl.Akad. Nauk. USSR 95:493 (1954)

[4] Zavattini, E., in *Muon Physics*, ed. V. W. Hughes, C. S. Wu. New York: Academic (1975), 2:219

[5] Dzhelepov, V. P., et al., Sov. Phys. JETP 50:1235 (1966)

[6] Vesman, E. A., Sov.Phys.JETP Lett. 5:91 (1967)

[7] Matveenko, A.V., Ponomarev, L.I., Zh. Eksp. Teor. Fiz. 58:1640 (1970)

[8] Vinitsky, S.I. et al., preprint JINR P4–10336, Dubna (1976)

[9] Gerstein, S.S. and Ponomarev, L.I., in *Muon Physics*, ed. V. W. Hughes, C. S. Wu. New York: Academic (1975), 3:141

[10] Vinitsky, S.I. et al., Sov. Phys. JETP 47:444 (1978)

[11] Bystritski, V.M. et al., Sov.Phys.JETP 49:232 (1979)

[12] Korenman, G.Ya., Sov.J. Nucl. Phys. 32:472 (1980)

[13] Cohen, J.S., Phys. Rev. A27:167 (1983)

[14] Markushin, V.E., Sov.Phys.JETP 53:16 (1981)

[15] Kraiman, J.B. et al., P.R.L. 63:1942 (1989)

[16] Breunlich, W.H. et al., Phys.Rev.Lett. 58:329 (1987)

[17] Anderson, A.N., in *Muon–Catalzyed Fusion, Sanibel Island*, ed. S.E. Jones, J.Rafelski, H.J. Monkhorst (AIP Conf. Proc. 181). New York: Am. Inst. Phys. (1989), p. 145

[18] Vinitsky, S.I. and Ponomarev, L.I., Sov. J. Nucl. Phys. 20:310 (1975)

[19] Scrinzi, A. and Szalewicz, K., Phys. Rev. A39:2855 (1989)

[20] Jones, S.E. et al., Phys.Rev.Lett. 56:588 (1986)

[21] Cohen, J.S., P.R.L. 58:1407 (1987)

[22] Markushin, V.E., Muon Catalyzed Fusion 3:395 (1988)

[23] Petitjean, C. et al.,Muon Catalyzed Fusion 2:37 (1988)

[24] Breunlich, W.H., Kammel, P., Cohen, J.S., and Leon, M., Ann. Rev. Nucl. Part. Sci., 39:311 (1989)

[25] Zmeskal, J. et al., Phys. Rev. A, to be published

[26] Melezhik, V.S. and Wozniak, J., Phys. Lett. 116A:370 (1986)

[27] Bracci, L. et al., in *Muon–Catalzyed Fusion, Sanibel Island*, ed. S.E. Jones, J.Rafelski, H.J. Monkhorst (AIP Conf. Proc. 181). New York: Am. Inst. Phys. (1989), p.214

[28] Adamczak, A., presented at the Intenational Conference on Muon Catalyzed Fusion, Vienna, 1990

[29] Hartmann, F.J. et al., Muon Catalyzed Fusion 2:53 (1988)

[30] Bogdanova, L.N. and Markushin, V.E., Muon Catalyzed Fusion 4:103 (1989)

[31] Faifman, M.P., Menshikov, L.I., and Strizh, T.A., Muon Catalyzed Fusion 4:1 (1989)

[32] Kammel, P., Nuovo Cimento Lett. 43:349 (1985)

[33] Cohen, J.S. and Leon, M., P.R.L. 55:52 (1985)

[34] Bogdanova, L.N. et al., Phys. Lett. 115B:171 (1982), errata: Phys. Lett. 167B:485 (1986)

[35] Frolov, A.M., Yad. Fiz. 44:1367 (1986)

[36] Korobov, V.I., Puzynin, I.V., and Vinitsky, S.I., Phys. Lett. 196B:272 (1987)

[37] Alexander, S.A. and Monkhorst, H.J., Phys.Rev. A 38:26 (1988)

[38] Kamimura, M., Phys. Rev. A38:621 (1988)

[39] Hu, C.Y., Phys. Rev. A 34:2536 (1986)

[40] Szalewicz, K., Monkhorst, H.J., Kolos, W., and Scrinzi, A., Phys. Rev. A36:5494 (1987)

[41] Bakalov, D., Muon Catalyzed Fusion 3:321 (1988)

[42] Kamimura, M., in *Muon–Catalzyed Fusion, Sanibel Island*, ed. S.E. Jones, J.Rafelski, H.J. Monkhorst (AIP Conf. Proc. 181). New York: Am. Inst. Phys. (1989), p. 330

[43] Scrinzi, A. and Szalewicz, K., Phys. Rev. A39:4983 (1989)

[44] Aissing, G., Bakalov D.D., and Monkhorst, H.J., Phys. Rev. A42:116 (1990)

[45] Leon, M., Phys.Rev.Lett. 52:605 (1984)

[46] Breunlich, W.H. et al., Muon Catalyzed Fusion 1:67 (1987)

[47] Bogdanova, L.N.,.et al., Nucl. Phys. A454:653 (1986)

[48] Hu, C.Y. and Kauffmann, S.K. Phys.Rev. A36:4135 (1987)

[49] Moszynsky, R., to be published

[50] Haywood, S.E., Monkhorst, H.J., and Szalewicz, K., Phys.Rev. A39:1634 (1988)

[51] Paciotti,M.A. et al., in *Muon–Catalzyed Fusion, Sanibel Island*, ed. S.E. Jones, J.Rafelski, H.J. Monkhorst (AIP Conf. Proc. 181). New York: Am. Inst. Phys. (1989), p. 330

[52] Balin, D.V. et al., Phys. Lett. 141B:173 (1984)

[53] Bogdanova, L.N., et al. Phys. Lett. 161B.1 (1985)

[54] Menshikov, L.I. and Shakirov, V.A., Sov. Phys. JETP 68:258 (1989)

[55] Petrov, Yu. V., Muon Catalyzed Fusion 1:351 (1987)

μ-Atomic and μ-Molecular Processes in Hydrogen Isotope Mixtures and Muon Catalyzed Fusion

S.S. Gerstein

Institute for High Energy Physics, Protvino, Russia

Abstract. This paper considers various μ-atomic and μ-molecular processes in hydrogen and its isotopes and muon catalyzed fusion.

Introduction. The behaviour of muonic atoms of hydrogen and of its isotopes is governed by their small sizes and electric neutrality. The size of the Bohr orbit of a muonic atom, $a_\mu = \hbar^2/m_\mu e^2 \cong 2.5 \cdot 10^{-11}$ cm, being much less than those of atom shells, the behaviour of muonic atoms of hydrogen in matter is similar to that of neutrons. Muonic hydrogen atoms, when penetrating atom shells, can approach close to other nuclei of matter to give rise to many interesting phenomena. These are as follows:

1) isotopic exchange, i.e. transitions

$$p\mu + d \rightarrow d\mu + p, \qquad (1a)$$

$$d\mu + t \rightarrow t\mu + d, \qquad (1b)$$

$$p\mu + t \rightarrow t\mu + p, \qquad (1c)$$

2) transitions between the levels of the hyperfine structure (h.f.s.) of muonic atoms coming about when they collide with other nuclei;
3) transitions of muons from muonic hydrogen atoms to admixture nuclei of heavier elements (Z), e.g.

$$p\mu + Z \rightarrow Z\mu + p; \qquad (2)$$

4) formation of molecules, or, putting it more precisely, of μ-molecular ions, ppμ, pdμ, ddμ, ptμ, dtμ, ttμ;
5) nuclear fusion reaction in various muonic molecules, except for ppμ, where the $pp \rightarrow de^+ \nu_e$ fusion is related to weak interactions;
6) nuclear fusion-induced re-injection of muons from muonic molecules and repetition of the whole previous chain of processes leading to muonic molecule formation and fusion.

The efficiency of these processes, in particular, of processes (4) and (5), is heavily dependent on the state of the matter they proceed in (density, temperature, isotope concentration, equilibrium or nonequilibrium composition of molecules, e.g. D_2, T_2, DT, etc.). It is just this circumstance which points to a possible relation of the μCF processes to solid state physics.

© Springer-Verlag Berlin, Heidelberg 1992

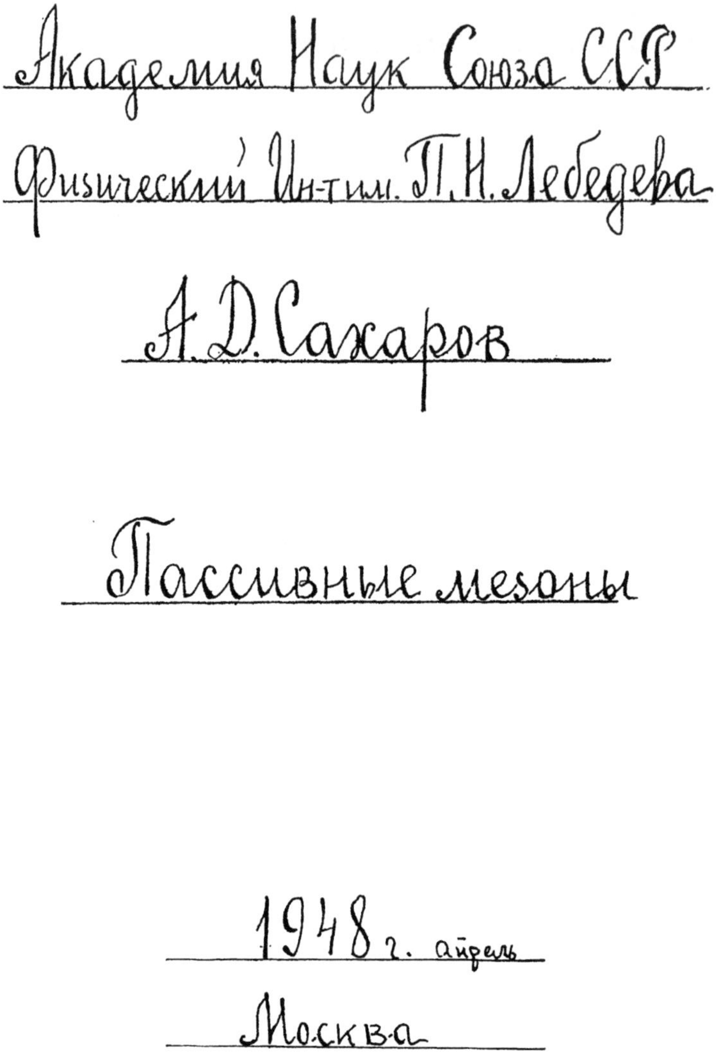
Академия Наук Союза ССР
Физический Ин-т им. П.Н. Лебедева
А.Д. Сахаров
Пассивные мезоны
1948 г. апрель
Москва

Front page of the hand-written report by A.D. Sakharov, Academy of Sciences of USSR, P.N. Lebedev Physical Institute, 1948 April, Moscow, entitled "Passive Mesons", where he expressed the idea of practical use of muon catalysis. This report has been published for the first time in English translation in *Muon Catalyzed Fusion* **4**, 235 (1991)

Though the possibility for processes (4)-(6) was first discussed more than 40 years ago [1-3], their deep study began only after L. Alvarez et al. discovered in 1957 μ-catalysis in pdμ and ddμ molecules [4]. The major attention in that study was given to the verification of the universal weak (V-A) interaction in the processes

$$\mu^- p \rightarrow n \nu_\mu, \tag{3a}$$

$$\mu^- d \rightarrow 2n \nu_\mu, \tag{3b}$$

whereas the μCF was studied spuriously because this phenomenon yielded some important information about which μ-atomic and μ-molecular states processes (3) take place in. Here I would like to note the brilliant research done by the groups from Chicago, the University of Columbia, CERN and Dubna with the would-be Nobel prize winners L. Lederman and C. Rubbia participating in them. A further interest in the μCF processes was related to the resonant ddμ-formation mechanism discovered by E. Vesman proceeding from the research work done by the Dubna team. This interest has particularly increased after L. Ponomarev and his noteworthy colleagues established the existence of a weakly bound level in the DDtμ molecule.

Its presence leads, with the resonant formation mechanism, to a high yield of fusion reactions d+t (about 10^2 events per muon in a condensed matter). The experimental corroboration of this prediction done at Los Alamos and SIN attracted much attention to μCF as a possible alternative nuclear energy source. In addition, the meson facilities put into operation at Los Alamos, SIN, TRIUMF, KEK combined up with the brilliant know-how of experimental physicists, made it possible to discover a number of new effects, e.g. the density effects, and also to make an experimental study of many fine details of processes (1)-(6). This could not have been thought of 30 years ago when the discussion of these processes just began.

Presently, several hundreds of papers are devoted to processes (1)-(6). Therefore I will refer mainly to reviews [5-11, 13] and to the μCF Journal where other references, if required, are available.

1. Muonic Atom Formation and Isotopic Exchange

In a condensed matter, e.g. hydrogen having a density of nuclei of $N_0=4.25\cdot 10^{22}$ cm^{-3}, relativistic muons are slowed down within $\sim 10^{-9}$ s due to ionization losses to experience further slowing down within $\sim 10^{-12}$ s and to be captured on the Coulomb orbits of excited muonic atoms $(n>(m_\mu/m_e)^{1/2}\sim 15)$ mainly due to the Wightman mechanism (adiabatic electron drift off the hydrogen nucleus screened by a muon).

Cascade de-excitation of the muonic atom formed takes place on higher orbits primarily due to the external Auger effect on the medium atoms and for $n\leq 3$ it proceeds via radiation transitions.

After transition into the ground state, the muonic hydrogen atoms diffusing in matter may, when colliding with the nuclei, experience elastic scattering and also processes (1)-(2). (In this Section I will talk about a pure mixture of hydrogen isotopes without admixtures of heavier nuclei, Z>1).

Since the masses presented in Fig. 1 differ, the binding energy of muonic atoms of different isotopes also differ. At all temperatures attainable, transition of a muon to a higher isotope is irreversible. Sakharov and Zel'dovich were the first to estimate in 1957 the order of magnitude of the effective cross-section for isotopic exchange (1) proceeding from dimension considerations. Since at small relative velocities v the cross-section for inelastic process (1) should, as in the case with neutrons, be inversely proportional to v, and the process itself proceeds when objects approach one another at a distance of about a_μ, the estimate of Sakharov and Zel'dovich yields

$$\sigma_{pd} = 4\pi a_\mu^2 \frac{v^*}{v} f, \tag{4}$$

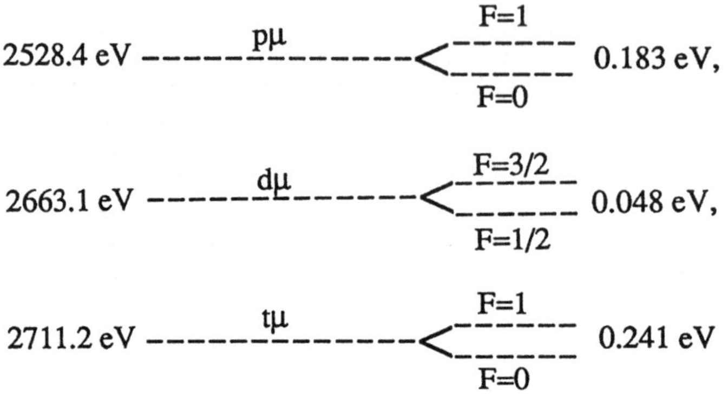

Fig.1: The μ-atom binding energies in the ground state and their hyperfine structure.

where v* is the specific velocity of scattering after transition and f is a factor of the order of unity. As it follows from the accurate calculations for processes (1a) and (1b), the order of this factor is really unity. However, for processes (1b) it is abnormally small, $f \sim 10^{-2}$. To have a guideline for the orders of the values of cross sections it might be helpful to use the value $\sigma = 4\pi a_\mu^2 \cong 0.8 \cdot 10^{-20}$ cm^2 as a natural scale. The probabilities for transitions (1a)-(1c),

$$\lambda_{ab} = N_o \sigma_{ab} v, \tag{5}$$

where "a" and "b" are taken to mean the isotope symbols, are presented in Table 1.

Table 1. The Main Characteristics of Muon Catalysis Processes.

Mesic molecule	ω_s	λ_m, s^{-1} *)	λ_f, s^{-1}	λ_{ab}, s^{-1} **)	Energy release Q (MeV)
pdμ	0.99	$5.8 \cdot 10^6$	$2.6 \cdot 10^5$	$1.7 \cdot 10^{10}$	5.4
ptμ	0.94	$6.8 \cdot 10^6$	$0.7 \cdot 10^5$	$0.7 \cdot 10^{10}$	20
ddμ	0.12	$\sim 4 \cdot 10^6$	$4.3 \cdot 10^8$	$3.7 \cdot 10^7$	3.6
dtμ	$0.43 \cdot 10^{-2}$	$\sim 3 \cdot 10^8$	$1.2 \cdot 10^{12}$	$2.8 \cdot 10^8$	17.6
ttμ	0.14	$3 \cdot 10^6$	$1.5 \cdot 10^7$	$1.2 \cdot 10^9$	11.3

*) Rates λ_m of muonic molecule formation are given at density φ=1 room temperature 300 K and are averaged over the internal variables of particles.

**) For symmetric systems, the spin-flip rates dμ(↑↑)→dμ(↑↓), tμ(↑↑)→dμ(↑↓) are given instead of rates λ_{ab} of the aμ+b → a+bμ process.

As is seen from these data, in HD and HT mixtures, actually the complete transition of a muon, within its lifetime $\tau_0 \cong 2\cdot10^{-6}$s, into a heavy isotope occurs when its concentration is about 0.1%. This just helped Alvarez and his colleagues to discover μCF in natural hydrogen containing a deuterium concentration of about 10^{-4}. As to a DT mixture, the tritium concentration required for this purpose is about two orders of magnitude higher. This is the reason why μCF cannot be applied practically though its efficiency is higher in this mixture. However, in a condensed matter, isotopic exchange from the excited states of muonic atoms proceeding simultaneously with their transitions into the ground state may possibly play an essential role in this case.

2. Transitions between Hyperfine Structure Levels and Elastic Muonic Atom Scattering

Zel'dovich pointed to a specific mechanism leading to a fast transition of muonic hydrogen atoms from the upper into lower state of the hyperfine structure (h.f.s.) when they collide with nuclei of the same isotope:

$$p\mu\,(F=1) \quad + p \rightarrow p + p\mu\,(F=0), \tag{6a}$$

$$d\mu\,(F=3/2) + d \rightarrow d + d\mu\,(F=1/2), \tag{6b}$$

$$t\mu\,(F=1) \quad + t \rightarrow t + t\mu\,(F=0). \tag{6c}$$

Here F is taken to mean the total spin of muon and nucleus. This mechanism consists in the transition of a muon from one nucleus to another having the opposite spin direction (see Fig. 2). It is not related to relatively weak spin-spin interactions but is governed by the exchange process probability being in this resonant case fairly large. The effective cross sections for transition (6) at low collision energies, $\varepsilon\sim10^{-3}$ eV, can be expressed through scattering lengths a_g and a_u corresponding to scattering in the symmetric (Σ_G) and antisymmetric (Σ_u) state of a muon, respectively, relative to two Coulomb centers. The probability of transition (6a) and (6b) is

$$\lambda_p = \lambda_{pp} = \frac{\pi}{4}\,(a_g^p - a_u^p)^2\, a_\mu^2 v_o N_p^o, \tag{7a}$$

$$\lambda_d = \lambda_{dd} = \frac{\pi}{3}\,(a_g^d - a_u^d)^2\, a_\mu^2 v_o N_d^o, \tag{7b}$$

respectively. Here $v_o = (2\Delta E_{h.f.s.}/M)^{1/2}$ is the velocity of muonic atom and nucleus scattering after transition and M is the relevant reduced mass. The probability of transition (6c) is found from a formula similar to (7a). Accurate

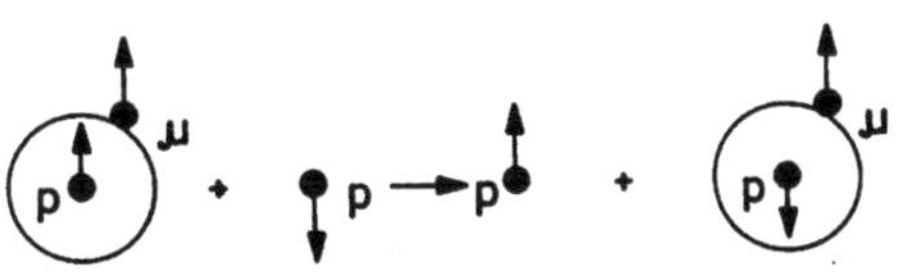

Fig. 2: The pμ(F=1)+p → pμ(F=0)+p transition mechanism.

calculation of the scattering lengths requires that the electron screening of nuclei be taken into account. The results of numeric calculations are presented in Table 1. The large difference between the values of λ_{dd} and λ_{tt} is explained by the scattering lengths a_g^d and a_u^d having the same sign and close values. The value of $\lambda_p = \lambda_{pp}$, not specified in Table1, is about one order of magnitude larger than that of λ_{tt}. This is explained by a_a^p having rather a large value and the a_g^p a sign opposite to that of a_u^p.

It is essential that processes (6) actually occur in nuclei in the molecules rather than on free nuclei. The spin flip of the nucleus remaining in the molecule should, in some cases, lead to para-ortho transitions in the molecule with the rotational level excited. The calculation of this effect shows that for isolated molecules the calculated transition velocity between the hyperfine structure levels varies but inessentially, by ~30%. Yet, one should not rule out the possibility that for a solid state, especially at high pressures, excitation of rotational transitions is slowed down, in which case the transition velocity between the h.f.s. levels may vary noticeably.

If the concentration of the same isotopes is sufficiently high, fast transition between the h.f.s. levels precedes either muonic atom formation or muon capture (3a,b). This leads to a number of observable effects:

1) In gaseous phase and under a pressure of some atm, when muonic molecule formation can be neglected, transition (6a) increases the probability of process (3a) about 4 times as compared with what would have been attained with a statistical distribution in the h.f.s. levels. This is due to the fact that for a weak (V-A)interaction the probability of process (3a) from the upper h.f.s. (F=1) state is negligible as against the probability for a muon to be captured from the F=0 state [9]. Similarly, the $d\mu(F=3/2) \rightarrow d\mu(F=1/2)$ transition increases the probability of process (3b) 3 times, which is also observed experimentally.
2) In condensed matter, transitions into the lower state of the h.f.s. of muonic atoms vary the distribution in the states of the h.f.s. of muonic molecules. This leads to the experimentally observed variation in the yield of nuclear reactions in the $pd\mu$, $pt\mu$ molecules and also to variation in the ratios between various reaction channels.
3) As is known from the experiments performed at SIN, the $d\mu(F=1/2) \rightarrow d\mu(F=3/2)$ transitions may determine the kinetics and the $dd\mu$ molecule formation mechanism and, hence, the μCF yield and the ratio between nuclear reaction channels (see Section 4).

In mixtures containing a relatively low concentration of heavy hydrogen isotopes, the muonic atom of the heavy isotope formed in the isotopic exchange process (see Section 1) scatters mainly on the nuclei of the light one. It is interesting to note that for the processes

$$d\mu+p \rightarrow d\mu+p, \tag{8a}$$

$$t\mu+p \rightarrow t\mu+p, \tag{8b}$$

there should be the Ramsauer effect observed at energies ~1 eV because the phase of the main S-wave vanishes. Owing to this effect, the $d\mu$-atom formed in process (1a) before forming the $pd\mu$-molecule in which the nuclear reaction proceeds, experiences, at a deuterium concentration less than 0.1% a noticeable range in liquid H_2. L. Alvarez et al. noticed the existence of a gap between the

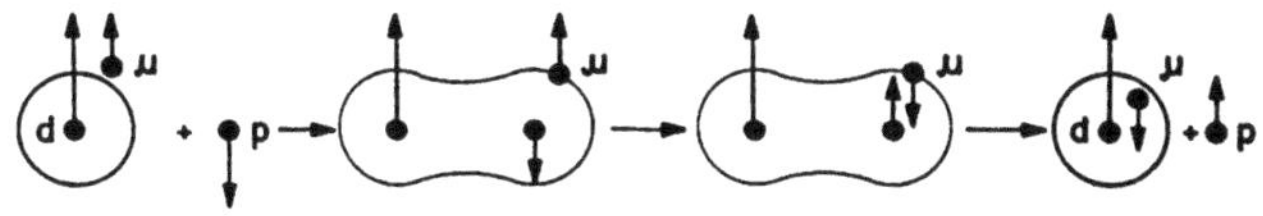

Fig. 3: The dμ(F=3/2)+p → dμ(F=1/2)+p transition mechanism.

point where the moun stopped and that of nuclear reaction - induced muon re-injection, which helped interpret them. At concentrations of ~2% no gaps were observed in these experiments because the scattering process became pronounced. The probability of the dμ(F=3/2)→dμ(F=1/2) transition in collision (8a) is an interesting, though still outstanding, theoretical problem. The early data on this transition were obtained in the CERN experiments [9] and ($\lambda_p^{\mu d} = 4.7\cdot10^6\ s^{-1}$) was measured directly in the SIN experiments [14]. It might seem at first sight that the probability of such a transition should be negligibly small compared to that of process (6b), since it is related to the muon spin flip due to relativistic effects. However, the specific nature of muonic hydrogen atoms may, in my opinion, manifest itself even in this case. With a dμ approaching a proton at a distance of the order of the muonic atom size, there should be a virtual muon-proton transition, even though the energy of the dμ is insufficient for the dμ+p→pμ+d process to proceed. If in this case the muon and proton spins are opposite in direction, then due to hyperfine interaction the rate of muon spin flip in the pμ atoms is $w=\Delta E_{h.f.s}/h{\sim}3\cdot10^{14}\ s^{-1}$. Since at velocities $v(d\mu){\sim}10^5$ cm/s the collision time is $\tau{\sim}5\cdot10^{-16}$ s, the probability of muon spin flip in each collision may, according to the experiment, be about 10^{-1}-10^{-2} (see fig. 3). These qualitative considerations must apparently be verified by calculations. If these are correct, the tμ(F=1)→tμ(F=0) transitions are to be expected in tμ+p → tμ+p collisions.

3. Transition of Muons from Muonic Hydrogen Atoms to Nuclei of Heavy Elements

At a low admixture of heavy nuclei, transition of muons to heavy nuclei occurs mainly from the (1s)-state atoms of hydrogen. For nuclei with Z>3 the transition rate is fairly large, $\sim10^{11}\ s^{-1}$, because of an intersection, or quasi-intersection, of terms in the pμ+Z and Zμ+p system. Owing to this term intersection, transitions of muons take place on the excited levels of muonic atoms with the principle quantum number n≥Z. The studies carried out during recent years by H. Schneuwly revealed quite interesting effects which remain to be explained. Constant addition of low concentrations of elements with Z>3 into the hydrogen isotope mixture proves sometimes to be helpful for the experimental determination of the constants of μ-atomic and μ-molecular processes. In the Z=2,3 case (He,Li), there is no term intersection and muon transitions to the nuclei of these elements proceed via a specific mechanism discovered by N.P. Popov, via formation of excited muonic molecules, e.g. $(pHe\mu)^{++}$. In hydrogen containing chemical compounds, muon transitions are governed largely by the large muonic molecule model [8]. These problems will be treated in more detail in the contributions by H. Schneuwly and N.P. Popov.

4. Muonic Molecule Formation

4.1 Muonic Molecule Levels

To establish the mechanism and to determine the probabilities of muonic molecule formation one should know their energy levels. Calculation of these levels is a delicate problem because the muon-nuclei mass ratio is insufficiently small and the adiabatic approximation requires considerable refinement. The techniques developed within recent years and applied for the solution to the three-body problem made it possible to determine the muonic molecule energy levels to a high accuracy, ~1 meV (see Table 2). This was particularly important for the weakly-bound rotational-vibrational states of the ddμ and dtμ molecules formed via the resonant mechanism.

Table 2 : Binding Energies $|\varepsilon_{Jv}|$ in eV of Muonic Molecule (Jv) States

(Jv) Molecule	(00)	(01)	(10)	(11)	(20)	(30)
ppμ	253.152	--	107.266	--	--	--
pdμ	221.549	--	97.498	--	--	--
ptμ	213.840	--	99.127	--	--	--
ddμ	325.074	35.844	226.682	1.97482	86.434	--
dtμ	319.140	34.834	232.472	0.66017	101.416	--
ttμ	362.910	83.771	289.142	45.206	172.652	48.813

4.2. Nonresonant Mechanism of Muonic Molecule Formation

For a bound state, muonic molecule, to be formed, the binding energy of the muonic molecule in muonic atom - nucleus collision has to be transferred to a third object. The mechanisms of muonic molecule formation are discriminated according to the object this energy is transferred to. Basically, muonic molecules could be formed via binding energy transfer to the photon emitted. This mechanism was considered by A.D. Sakharov as far back as 1948. However, the probability of energy transfer to the atomic electron is a few orders of magnitude higher. This is explained by the fact that for the majority of muonic molecular states the binding energy is higher than the ionization energy of electrons in hydrogen molecules, $I \cong 15$ eV, with the conversion ratio of the energies transferred being fairly high, $\sim 10^3 \div 10^5$. The most probable is in this case the formation of muonic molecules via the electric dipole transition E1. At low collision energies, i.e. in the S-wave, this causes their formation in the rotational J=1 state. In addition, the electric monopole E0 and quadrupole E2 transitions leading to their formation in the J=0 and J=2 states, respectively, are also possible (see Table 3). The strong difference between the probabilities for the ddμ and dtμ molecule formation (see Tables 1 and 3) is explained by the resonant formation mechanism playing at high temperatures the paramount role for these muonic molecules.

Table 3 : Rates λ_{Jv} [10^6 s^{-1}] of Nonresonant Formation of Muonic Molecules of Hydrogen Isotopes*

(Jv) Molecule	(00)	(01)	(10)	(11)	(20)	(30)	Experiment
ppμ	--		2.2	--	--	--	2.5
pdμ	--		5.9	--	--	--	~6
ptμ	--		6.5	--	--	--	--
ddμ	--	0.026	0.013	--	--	--	0.046
dtμ	--	0.093	$5\cdot10^{-4}$	--	$3\cdot10^{-4}$	--	--
ttμ	--	0.002	0.003	3.0	0.06	--	2

* The values are given at liquid hydrogen density ϕ=1.

4.3. The Mechanism of Resonant ddμ and dtμ Molecule Formation

The mechanism of resonant formation of ddμ and dtμ molecules is caused by the existence of their weakly-bound rotational-vibrational (J=1, v=1) states with energies (ε_{11}) close to the difference in those of vibrational levels of "ordinary" D_2 and DT molecules. For example, when a dμ atom collides with a nucleus of a D_2 molecule, the ddμ molecule formed in the (J=1, v=1) state becomes the "nucleus" of the muonic molecular complex (MMC) with its binding energy $|\varepsilon_{11}|$ transferred for the MMC rotational - vibrational state to be excited, the energy differing from the initial level of D_2 by ΔE_{vf} (see fig. 4). In this case the exact coincidence of the values of the initial and final energies accounts for the kinetic energy ε_0 of the incident dμ atom. Hence, the resonance condition takes the form

$$\varepsilon_0 + |\varepsilon_{11}| = \Delta E_{vf}. \qquad (9)$$

According to the presently available calculations, this condition is fulfilled for transitions into the vibrational MMC [(ddμ), dee] state with v=7. If the dμ atoms are thermalized, the temperature dependence of the ddμ formation rate will be determined by the Maxwell distribution for the resonant energy ε and will be maximum at the T satisfying the condition 3/2 K T = ε. The experimental data are in a good agreement with the calculations of the reaction kinetics (see fig. 5). It should be pointed out that the energy $|\varepsilon_{11}|$ being slightly less than the difference in the energy of the relevant vibrational levels, ΔE_f (~4

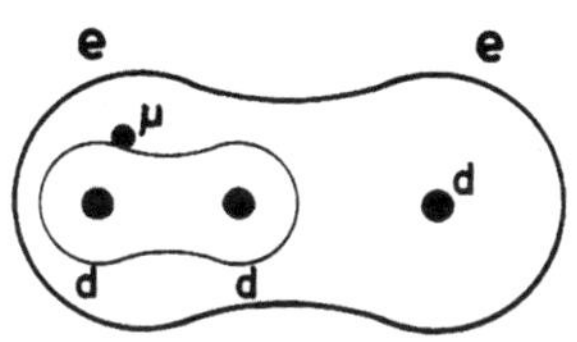

Fig. 4: The muonic molecular complex (MMC) [ddμ, dee].

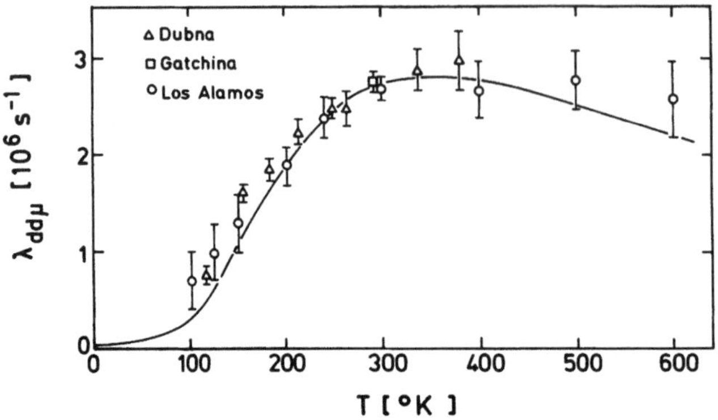

Fig. 5: The temperature dependence $\lambda_{dd\mu}$ (T). A solid line presents the calculation at ε_{11}=1.964 eV, $\lambda_f = 0.41 \cdot 10^9$ s^{-1}. Experimental points: Δ - Dubna, □ - Gatchina, o - Los Alamos (Dubna data are renormalized to the value $\lambda_{dd\mu}$ (T=296 K), measured at Gatchina).

MeV), was the "lucky" circumstance which facilitated the discovery of the resonant mechanism. Otherwise, the resonant condition would have been fulfilled only for the transition to the next vibrational MMC level and achieved in reality only at very high temperatures, at some thousands of degrees, when matter molecules dissociate. The large difference in the rates of resonant and nonresonant ddμ molecule formation (cf. Tables 2 and 3) allowed the physicists, who carried out their splendid experiments at SIN, to understand more deeply the problem of ddμ formation in different h.f.s. states and, on having established the dependence of the reaction on the dμ (F=3/2) → dμ(F=1/2) transition, to evaluate a number of the constants of muonic atoms to a good accuracy.

It might be interesting to consider the case when the value of $|\varepsilon_{11}|$ is somewhat larger than that of $\Delta E_{\upsilon f}$. This is likely to happen when the MMC with the dtμ molecule is formed. For the MMC to be formed, the energy in excess has to be transferred to a third object, e.g. in the process

$$t\mu + D_2 \rightarrow [dt\mu, dee] + D_2'. \tag{10}$$

This assumption is substantiated by the nonlinear density effect established in dtμ molecule formation (see fig. 6). Additional energy transfer mechanisms that might exist in a solid state suggest that **the relevant measurements in solid deuterium with a low admixture of tritium are required.**

The binding energy of the dtμ in the (J-1, ν-1) state being considerably less than that of the ddμ, the MMC with dtμ is formed in the vibrational ν_f=2 state. The overlapping of the wave functions of the initial and final MMC states is in this case much larger than that of the ddμ. Therefore the rate of the resonant dtμ molecule formation turns out to be 2 orders of magnitude higher than that of the ddμ, $\lambda_{dt\mu} \geq 10^8$ s^{-1}.

The experimentally observed difference in the rates of the processes

$$t\mu + [D_2]_{\nu_i=0} \rightarrow [(dt\mu)\, dee]_{\nu_f=2}, \tag{11a}$$

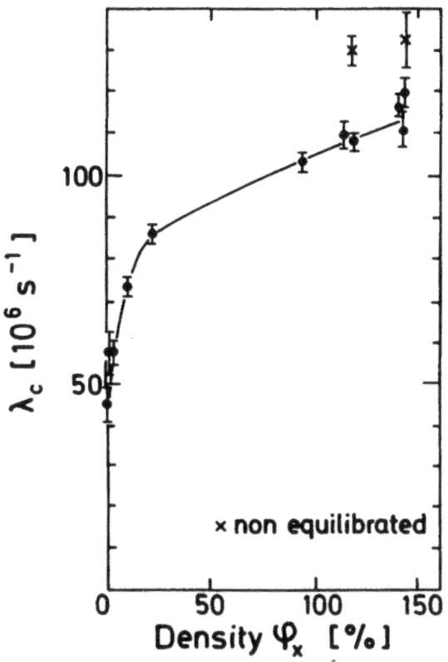

Fig. 6: Cycle rate λ_c vs. gas density ϕ_x at tritium concentration, C_t=40% and T<50 K.

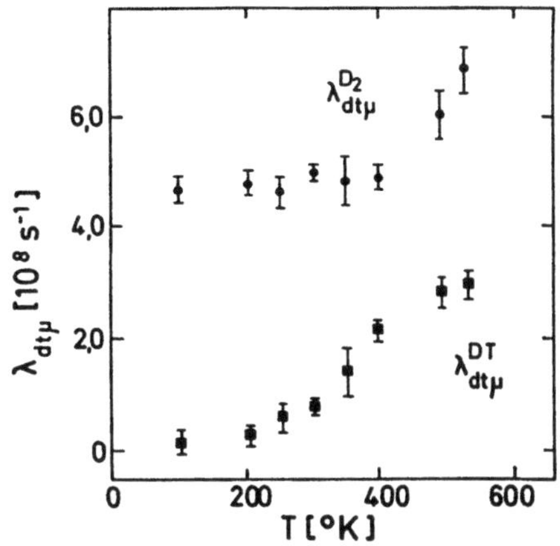

Fig. 7: The dtμ formation rate in D_2 and DT collision.

$$t\mu + [DT]_{v_i=0} \rightarrow [(dt\mu)\ tee]_{v_{f=2}}, \tag{11b}$$

which is due to a difference in the value of the energy of vibrational quanta, ΔE, for different MMC (see fig. 7), is a direct experimental corroboration of the resonant dtμ formation mechanism.

5. Nuclear Fusion Reactions in Muonic Molecules

The nuclei of the muonic molecules formed approach one another at a distance of an order of $2a_\mu$ (~$5 \cdot 10^{11}$ cm) and therefore the following nuclear fusion reactions (via tunnel transitions) become possible:

$$pd\mu \begin{cases} \rightarrow {}^3He + \mu \\ \xrightarrow{\omega} {}^3He\mu + \gamma \end{cases} \quad (Q = 5.4\ \text{MeV}), \tag{12a}$$

$$dd\mu \begin{cases} \rightarrow t + p + \mu & (Q = 4.0\ \text{MeV}) \\ \rightarrow {}^3He + n + \mu & (Q = 3.3\ \text{MeV}), \\ \xrightarrow{\omega} {}^3He\mu + n \end{cases} \tag{12b}$$

$$dt\mu \begin{cases} \rightarrow {}^4He + n + \mu & (Q = 17.6\ \text{MeV}), \\ \xrightarrow{\omega} {}^4He\mu + n \end{cases} \tag{12c}$$

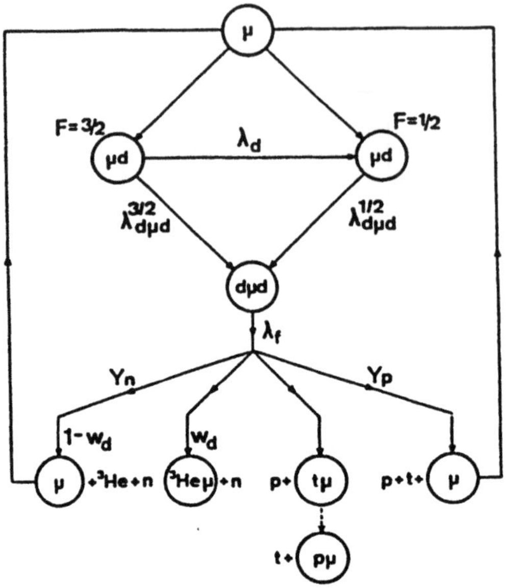

Fig. 8: Simplified scheme of μCF cycle in D_2.

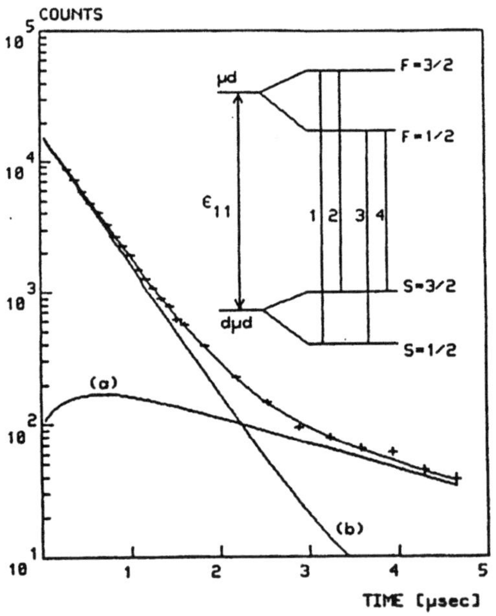

Fig. 9: μCF in D_2. Fusion neutron time spectrum according to molecular formation from the doublet (a) and quartet (b) state and energy levels of μd atoms and dμd molecules.

$$
pt\mu \rightarrow \begin{cases} {}^4He+\mu \\ He\mu+\gamma \\ \xrightarrow{\omega} {}^4He\mu+e^+e^- \end{cases} \qquad (Q=20\ MeV), \qquad (12d)
$$

$$
tt\mu \rightarrow \begin{cases} {}^4He+2n+\mu \\ \xrightarrow{\omega} {}^4He\mu+2n \end{cases} \qquad (Q=11.3\ MeV), \qquad (12e)
$$

The probability of ${}^4He\mu$ muonic atom formation, ω_s, is called a sticking coefficient. Table 1 presents the values of mean process rates (λ_f), those of energies released (Q) and sticking coefficients ω_s.

When making a more consistent consideration, one should bear in mind that the rate of a nuclear reaction and ratio of its channels differ for different h.f.s. states of a muonic molecule.

Therefore they are dependent on the transitions between the h.f.s. muonic atom levels, which precede muonic molecules formation, and also on their formation mechanism. Such effects can be observed in the pdμ and ptμ molecules if one varies the concentration of isotopes in the relevant mixtures (the Wolfenstein-Gerstein effect). Another example of manifestation of the h.f.s. effects is μCF observation in deuterium at temperatures 24-150 K in the experiments of W.H. Breunlich et al. [4]. The scheme of the relevant processes is presented in fig. 8. Studying the time dependence of the μCF yield in the ddμ molecule (fig. 9) one can separate the process of resonant ddμ formation proceeding from the upper h.f.s. level of the dμ (E=3/2) atom and that of ddμ formation from the lower dμ (F=1/2) state. At a low temperature, T<50 K, the latter process is nonresonant due to the E0 and E2 transitions because the thermal energy of the muonic atom proves to be insufficient to satisfy the resonance condition (9) (see fig. 10). In the aforementioned experiments, the time dependence of the μCF yield makes it possible to measure directly the probability of transition (6b) in pure deuterium and also that of transition (8a) in a HD mixture. At a kinetic energy of the dμ atom less than 4 MeV the process of ddμ formation in the dμ (F=3/2) does not satisfy the resonance condition any longer and the reaction yield decreases. It is interesting to note that in this case the ratio of channels of the dd→³He+n and dd→T+p fusion reactions should also vary because with the resonant formation mechanism the deuterium nuclei in the ddμ molecule are found in the S-wave, whereas with the nonresonant one they are in the S-wave.

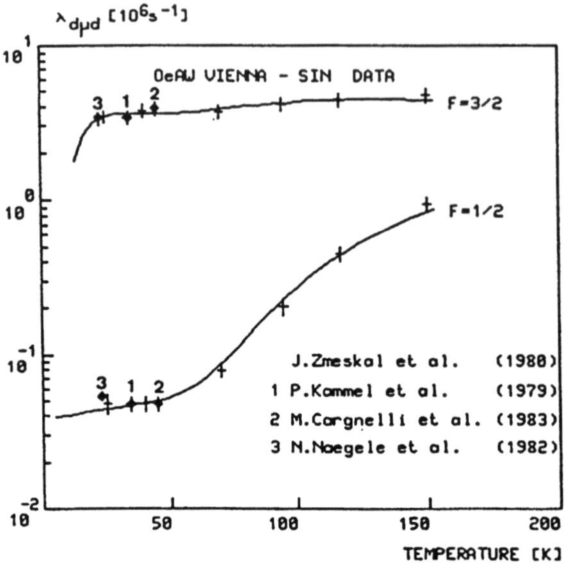

Fig. 10: dμd formation as a function of temperature between 24 and 150 K. Molecular formation from the quartet state is strongly enhanced due to a strong resonance at about 4 MeV. The doublet components (F=1/2) are non-resonant at lowest temperature, and gradually rising due to several resonances at higher temperature (>200 K).

Conclusion

μ-atomic and μ-molecular phenomena and μCF are interesting for a number of problems ranging from the study of weak muonic interactions and the possibility of measuring QED radiation corrections in muonic molecules to the possibility of studying nuclear reactions at low energies, e.g. pd→He^3 and prospects for alternative applications of nuclear energy. Are they of interest for solid state physics?

In my opinion, the study of μCF in solid deuterium with a low admixture of tritium may yield information on the nature of excitation of rotational-vibrational levels in molecules at different pressures and on mechanisms of energy transfer in a crystal lattice.

Unfortunately, similar effects are likely to manifest themselves at very high energies, at which phase transitions in a crystal phase will become possible [16].

Experiments in polarized deuterium (at very low temperatures) might be very interesting.

In any case, the study of the density effect in the muon catalyzed d+t reaction [17] is interesting for solid state physics and μCF.

References

Early papers

[1] F.C. Frank, Nature 160 (1947) 525.
[2] A.D. Sakharov, Report, Lebedev Physics Institute, Moscow, 1948.
[3] Ya. B. Zel'dovich, Doklady Acad. Nauk USSR 95 (1954) 493.
[4] L.W. Alvarez et al., Phys. Rev. 105 (1957) 1127.

Reviews

[5] Ya. B. Zel'dovich and S.S. Gerstein, Usp. Fiz. Nauk 71 (1960) 581 (Sov. Phys.-Uspekhi 3 (1961) 593).
[6] Muon Physics, ed. V. Hughes and C.S. Wu (Academic Press, New York 1975).
[7] L. Bracci and G. Fiorentini, Phys. Rep. 86 (1982) 175.
[8] S.S. Gerstein and L.I. Ponomarev in ref. [6], vol. III, p. 141.
[9] E. Zavattini in ref. [6], vol. II, p. 219.
[10] S.S. Gerstein, Yu.V. Petrov and L.I. Ponomarev. Usp. Fiz. Nauk (1990) in press.
[11] L.I. Menshikov and L.N. Somov, Usp. Fiz. Nauk (1990) in press.
[12] Muon Catalyzed Fusion (μCF), vol. 1,2,3,4. J.C. Baltzer, A.G. Scientific Publisher Company, Basel-Switzerland.
[13] L.I. Ponomarev and G. Fiorentini in ref. [12] (μCF-86), p. 3.
[14] W.H. Breunlich in ref. [12] (μCF-86), p. 29.
[15] H. Schneuwly in ref. [12] (μCF-89), p. 87.
[16] K. Fukushima and F. Iseki in ref. [12] (μCF-86), p. 225.
[17] V. Yu. Petrov and Yu. V. Petrov in ref. [12] (μCF-89), p. 73.

Alternatives in Low Energy Fusion?

C.F. Cerofolini[1] *and A. Foglio Para*[2]

[1]Functional Materials Laboratory, EniChem, I-20097 San Donato MI, Italy
[2]Istituto di Ingegneria Nucleare, Polictecnico di Milano, I-20133 Milano MI, Italy

Though fusion reactions occur almost ubiquitously in the Universe and can be studied quite easily in the laboratory, power production by light-element fusion is an extremely difficult technological task, as confirmed by the difficulties met by the large collaborations engaged in USA, USSR, Japan and Europe to reach the scientific breakeven. Paradoxically, while the difficulties of obtaining power from hydrogen-isotope fusion progressively increase as the breakeven is gradually approached, the number of phenomena where such fusion is claimed to be observed in low-energy experiments is steadily increasing. This work reviews the status of cold-fusion and related phenomena, discusses the 'warm' fusion resulting from the impact of heavy water clusters onto titanium deuteride, and eventually proposes a unifying model explaining not only claims of cold or warm fusion, but also the possible causes of their poor experimental reproducibility.

1 Hot Fusion

Of the hydrogen-isotope fusion reactions,

$$\mathrm{H} + \mathrm{H} \rightarrow \mathrm{D} + \mathrm{e}^{+} + \nu_{\mathrm{e}} \,, \tag{1}$$

$$\mathrm{H} + \mathrm{D} \rightarrow {}^{3}\mathrm{He} + \gamma \,, \tag{2}$$

$$\mathrm{H} + \mathrm{T} \rightarrow {}^{4}\mathrm{He} + \gamma \,, \tag{3}$$

$$\mathrm{D} + \mathrm{D} \rightarrow {}^{3}\mathrm{He} + \mathrm{n} + 3.3\ \mathrm{MeV} \,, \tag{4}$$

$$\mathrm{D} + \mathrm{D} \rightarrow \mathrm{T} + \mathrm{H} + 4\ \mathrm{MeV} \,, \tag{5}$$

$$\mathrm{D} + \mathrm{T} \rightarrow {}^{4}\mathrm{He} + \mathrm{n} + 17.6\ \mathrm{MeV} \,, \tag{6}$$

$$\mathrm{T} + \mathrm{T} \rightarrow {}^{4}\mathrm{He} + 2\mathrm{n} + 10\ \mathrm{MeV} \,, \tag{7}$$

the last four have large cross sections σ already at energies of the order of 20 keV: $\sigma \simeq 3 \times 10^{-4}$ barn for reactions (4) and (5) and $\sigma \simeq 6 \times 10^{-2}$ barn for reaction (6). Their practical exploitation for power production is however difficult because the cross section for Coulomb scattering exceeds by two or more orders of magnitude the cross sections for fusion. This fact implies that all attempts to obtain an energy gain from colliding beams of sufficient energy are hopeless because most of energetic particles are scattered and only few fusions can occur.

Springer Proceedings in Physics, Vol. 59 Exotic Atoms in Condensed Matter
Editors: G. Benedek · H. Schneuwly
© Springer-Verlag Berlin, Heidelberg 1992

However, if the experimental arrangement is such that scattered particles are not lost but rather the whole kinetic energy investing the target is conserved during all Coulomb scatterings, then energy production by fusion becomes feasible. This concept leads to the idea of a high-temperature reacting gas.

In a system with $\mathcal{N}$ particles per unit volume the fusion rate increases as $\mathcal{N}^2$, the constant of proportionality being dependent on temperature T and reactant (H, D or T) composition; if the energy is confined for a time τ, the total energy released in fusion processes varies as $\mathcal{N}^2\tau$, to be compared with the energy required to increase the internal energy of the system to the value $\mathcal{N}u(T)$, where u is the internal energy per particle. If the temperature is high compared to the average interatomic energy, but not enough to ionize the atoms, then $u(T) = \frac{3}{2}k_{\mathrm{B}}T$ (where k_{B} is the Boltzmann constant); if T is sufficiently high to ionize completely the gas and electrons form a non-degenerate gas at the same temperature as the ion gas, then $u = 3k_{\mathrm{B}}T$.

Denoting with $W(T)$ the kinetic constant giving the power produced at unit concentration, the scientific breakeven is reached when

$$W(T)\mathcal{N}^2\tau \geq \mathcal{N}u(T) ,$$

i.e., when

$$\mathcal{N}\tau \geq \mathcal{L}(T) , \tag{8}$$

where the quantity $\mathcal{L}(T) = u(T)/W(T)$ depends on gas temperature and composition. Condition (8) is known as the Lawson criterion. Among reactions (4) to (7), the most convenient one seems to be reaction (6) for which the optimum temperature corresponds to a kinetic energy of 12.5 keV and the quantity $\mathcal{L}(T)$ (which does not vary too much in a wide neighborhood of this energy) is given by $\mathcal{L} \approx 2 \times 10^{14}$ cm^{-3}s.

Two ways have been proposed to confine the impinging kinetic energy inside the target for a time long enough to satisfy the Lawson criterion: the magnetic and inertial confinements, reviewed for instance by Grieger [1] and Johnson [2], respectively. The possible alternatives to these techniques form the mainframe of this paper.

2 Muon-Catalyzed Fusion

The idea of muon-catalyzed fusion was first considered by Frank for H - D fusion and later by Sakharov and Lebedev, Zeldovich, and Jackson for D - D and D - T fusion. The first observation of muon-catalyzed fusion,

$$\mathrm{H} + \mathrm{D} + \mu^- \rightarrow {}^3\mathrm{He} + \gamma + \mu^- ,$$

led Alvarez to the conclusion that power production by this path was technically feasible [3] — as customary when dealing with fusion, partial and preliminary results produced excessive optimism.

The basic idea of muon-catalyzed fusion is very simple. Since the negative muon is essentially a heavy electron, the exotic hydrogenic atoms $(\mathrm{H}^+)\mu^-$, $(\mathrm{D}^+)\mu^-$, $(\mathrm{T}^+)\mu^-$ are formed with binding energy approximately scaled from that of hydrogen E_0 ($E_0 = 13.59$ eV) by the ratio of the muon mass m_μ to the electron mass m_e:

$$E_\mu = E_0(m_\mu/m_e) \simeq 200 \text{ Ry} , \tag{9}$$

and similarly for the radii,

$$a_\mu = a_0(m_e/m_\mu) \simeq 2.5 \times 10^{-11} \text{ cm} , \tag{10}$$

where a_0 is the Bohr radius, $a_0 = 0.53 \times 10^{-8}$ cm.

Owing to the short lifetime τ_μ of the muon ($\tau_\mu \simeq 2.2$ μs in vacuum), all the necessary steps to a fusion reaction must be completed in a time comparable with τ_μ. The steps in this 'race against time' [4] are the following:

- slowing-down of the muon in the mixture of deuterium and tritium;
- formation of a muonic atom, $(D^+)\mu^-$ or $(T^+)\mu^-$; and
- formation of a muonic molecular ion $(D^+T^+)\mu^-$, in which D^+ and T^+ are confined at a distance of the order of 10^{-13} cm.

An alternative process, where the exotic molecule $((D^+T^+)\mu^-)D^+)2e^-$ is formed from the reaction of $(T^+)\mu^-$ with a D_2 molecule and in which the $(D^+T^+)\mu^-$ molecular ion acts as a compound nucleus of charge +1, is greatly enhanced via a resonance mechanism and the fusion probability is correspondingly increased.

From a balance of the energy necessary for obtaining a muon and the energy gained in the D - T fusion, it results that this type of fusion can be of interest for power production only if the muon is able to participate in hundreds of fusion events before its decay. This goal does not seem unrealistic [5] especially in 'hybrid' reactors where the produced neutrons are used to breed fissile matter [3].

3 Cold Fusion

Up to March 23, 1989, cold fusion was the synonym of muon-catalyzed fusion. The almost simultaneous appearance of two papers, by Fleischmann et al. [6] and Jones et al. [7], respectively, has however endowed the definition 'cold fusion' with another meaning.

3.1 Fracto-fusion

The first claims of fusion in condensed matter appeared in the late nineteen twenties (quoted in [8]); 'cold fusion' was recently rediscovered by Deryagin and coworkers [9, 10] at the Institute of Physical Chemistry in Moscow in the fracture following the impact of macroscopic projectiles onto D_2O ice or LiD crystal. Taking into account the accelerating voltages (100 kV or more) which are supposed to be present in the fracture zones of solids, they investigated the emission of radiations following the fracture. A small neutron emission (of the order of 0.1 neutron counts per shot, corresponding to approximately 10 emitted neutrons) was observed and attributed to reaction (4). More recently, Yaroslavskii reported intense neutron emissions during plastic deformation under an applied pressure of solids containing deuterium [11]. A schematic diagram of the fracto-fusion model was presented by Dickinson et al. [12], together with new experimental evidence for the emission of charged particles, photons and radio-frequency signals; almost simultaneously the absence of fracto-fusions in LiD crystals was claimed by Price [13] and Sobotka and Winter [14].

3.2 Electrolytic Fusion

The first announcement from Fleischmann et al. [6] of fusion events in an electrolytic cell, using basically a palladium cathode and a solution of lithium salts in heavy water, followed by the confirmation of Jones et al. [7], who observed fusion events in titanium cathodes in analogous conditions, stimulated immediately a rush to similar experiments in laboratories all over the world. However, it became evident in ashort time that:

- The direct measurements of the developed heat carried out by Fleischmann et al. were highly controversial in their interpretations, owing to possible artifacts. Of course, the major effects claimed by Fleischmann et al. (e.g., melting and evaporation of a palladium cathode while the cell and a part of the fume cupboard housing the experiment were destroyed) cannot be questioned in their experimental evidence, but only in their interpretation. Scarcely convincing were also the nuclear measurements (of gamma radiations, neutrons and tritium activity in the electrolyte), which did not seem to denote experience in the field.
- The neutron measurements by Jones et al. were performed with a sophisticated apparatus constructed and operated by expert people, so that no doubt arose about possible coarse technical artifacts. However, as observed in ref. [15], the net effect was so small that it can be reproduced artificially by a polarized choice of the counting time in the different runs; the reply of Jones and coworkers was however convincing [16].

The work carried out in most laboratories, after disregarding the most controversial experiments, showed that the effect, if any, is indeed tiny and erratic in its appearance [17, 18, 19]: no one was able to reproduce at will the positive experiments by counting protons or neutrons with two or more detectors simultaneously; however, it was impossible to assess that all reported experiments showing an erratical appearance of fusion products were merely due to artifacts.

At present, the major question concerns the results obtained in a few laboratories, such as the reports of repeated positive calorimetric results [20] and of high concentration of tritium activity in the electrolytic solutions [21]. In both cases the expected high fluxes of fusion products (neutrons, protons and gammas) have not been observed [22]

Two major interpretations have been proposed for the supposedly existing electrolytic fusion, either requiring a very narrow D - D confinement or allowing close D - D encounter.

Exotic Molecule. The first model [23], essentially a transcription of what happens in muon - catalyzed fusion, assumes that an exotic D - D molecule is formed with the participation of quasiparticles of TiD_x or PdD_y ($x, y \simeq 1-2$). Indeed, in analogy with eqs. (9) and (10), a heavy quasiparticle forms hydrogenlike configurations with electronic energies and space localizations which are the same as in hydrogen however scaled by the ratio of masses of quasiparticle to electron. It suffices that the quasiparticle has an effective mass $m^* \simeq 5m_e$ to increase the fusion rate from approximately $3 \times 10^{-64}\ s^{-1}$ (calculated for the molecule in standard conditions) to a value of the order of $10^{-23}\ s^{-1}$ which explains the data reported by Jones et al. [7]; an effective mass $m^* \simeq 10m_e$ is able to explain the fusion rate ($\approx 10^{-11}\ s^{-1}$) reported by Fleischmann et al. [6].

However, though a sufficiently large effective mass formally forms very tight hydrogen molecules, the quasiparticle itself is essentially a concept for the quantum mechanical description of the collective motion of the particles forming a macroscopic body and cannot be identified with single constituents of the body.

Screening. The second model considers screening effects. The barrier transparency w for a central potential $U(r)$ is given by

$$w \simeq \exp\left\{-\frac{2}{\hbar}\int_{r_0}^{r_1} \sqrt{2m[U(r)-E]}\,\mathrm{d}r\right\}, \tag{11}$$

where $\hbar$ is the reduced Planck constant, m is the particle mass, and (r_0, r_1) is the classically forbidden region, $(r_0, r_1) = \{r : U(r) - E \geq 0\}$. Limiting the attention to the Coulomb potential,

$$U(r) = \frac{e^2}{r} = \frac{2a_0E_0}{r}, \tag{12}$$

where e is the proton charge and CGS units are used, eq. (11) becomes

$$w \simeq \exp\left\{-\frac{2}{\hbar}\int_{r_0}^{e^2/E} \sqrt{2m\left(\frac{e^2}{r}-E\right)}\,\mathrm{d}r\right\}, \tag{13}$$

where r_0 is determined by the nuclear forces and by the nuclear state. Integral (13) can be calculated in closed form and gives [24]

$$w \simeq \exp\left\{-\frac{2e^2}{\hbar}\sqrt{\frac{2m}{E}}\left[\arccos\sqrt{\frac{Er_0}{e^2}} - \sqrt{\frac{Er_0}{e^2}\left(1-\frac{Er_0}{e^2}\right)}\right]\right\}. \tag{14}$$

In general, the 'nuclear radius' r_0 is of the order of 10^{-13} cm and is much shorter than the size e^2/E which can be explored at thermal energies, so that the limit of eq. (14) for $r_0 \to 0$ is usually considered:

$$w \simeq \exp\left\{-\frac{\pi e^2}{\hbar}\sqrt{\frac{2m}{E}}\right\}, \tag{15}$$

which gives the usual exponential dependence on E.

If an electron gas at density $\mathcal{N}_e$ fills the medium, it screens the Coulomb potential (12) and modifies it to the form

$$U(r) = \frac{e^2}{r} \exp\left(-\frac{r}{a}\right) = \frac{2a_0 E_0}{r} \exp\left(-\frac{r}{a}\right) , \tag{16}$$

where the 'screening length' a is given by

$$a = \sqrt{a_0/4(3/\pi)^{1/3}\mathcal{N}_e^{1/3}} \tag{17}$$

For metals in standard conditions ($\mathcal{N}_e \approx 10^{23}$ cm^{-3}) a is approximately equal to the Bohr radius, $a \simeq 0.5 \times 10^{-8}$ cm.

For $r \ll a$ eq. (16) is reduced to $U(r) \simeq e^2/r(1 - r/a) = e^2/r - e^2/a$ so that, for pure computational reasons, instead of (16) we assume the potential

$$U(r) = e^2 \left(\frac{1}{r} - \frac{1}{a}\right) \tag{18}$$

for $r_0 < r \leq a$, and $U(r) = -U_0$ ($U_0 > 0$) for $r \leq r_0$, and $U(r) = 0$ otherwise. Using approximation (18), eq. (11) becomes

$$w \simeq \exp\left\{-\frac{2}{\hbar}\int_{r_0}^{e^2/(E+e^2/a)} \sqrt{2m\left[\frac{e^2}{r} - \left(E + \frac{e^2}{a}\right)\right]}\,\mathrm{d}r\right\} . \tag{19}$$

The comparison of (19) with (13) shows that the screened problem is translated into the unscreened problem simply by replacing the particle energy E by the 'effective energy' $E' = E + e^2/a = E + 2E_0a_0/a$ — in a way, the effect of screening is to increase the particle energy by an amount $e^2/a = 2E_0a_0/a$.

Though approximation (18) overestimates the screening effect at low kinetic energies and the (overestimated) effective energy due to screening is quite high ($e^2/a \simeq e^2/a_0 = 2E_0 \simeq 27$ eV), this effect seems unable to explain the fusion yield reported by Fleischmann et al. and Jones et al. Indeed, in view of eq. (17), a screening length sufficiently small to explain the reported fusion rates, 2×10^{-3} nm to match Fleischmann et al.'s data and 3×10^{-3} nm for Jones et al.'s data [26], corresponds to a concentration of free charges of 7 orders of magnitude higher than in metals. However, in such extreme conditions (which might occur only in astrophysical environments) the theory of screening in a Fermi gas should be replaced by the theory of screening in a Fermi liquid.

Though *formally* able to explain cold fusion, both the above models require absurd values of known physical parameters in order to explain the reported fusion rates.

4 Warm Fusion

The idea to use cluster or molecular ions for D - D fusion is based on the notion that sufficiently dense collisional cascades form 'hot clouds' and that the cloud temperature can be indefinitely increased by injecting simultaneously in the same microscopic region of the target a number of atoms. This injection can be obtained by implanting clusters [27], molecules [28] or synchronous showers of atomic ions [29]. Local heating produced by cluster or molecular implantation is a consequence of the deterministic space-time superimposition of atomic effects; local heating produced by the implantation of a synchronous shower follows from a deterministic time superimposition and an erratical spatial superimposition. [1] In the case of molecule or cluster impact, the Lawson criterion can be satisfied by implanting sufficiently heavy projectiles at energy higher than 1 GeV into LiD:LiT targets. These considerations seemed to limit the feasibility of fusion by cluster impact to high energy, heavy ion, linear accelerators with suitable source.

The recent surprising observations at Brookhaven by Beuhler et al. [31] have however opened completely new perspectives. These results concern the thermonuclear deuterium-deuterium fusion rate in TiD targets bombarded with $(D_2O)_n$ clusters ($25 \lesssim n \lesssim 1350$) and are collected, in slightly modified representations, in figs. 1 and 2. They are particularly astonishing because of the following reasons:

1. Supposing that deuteron collisions at the surface are responsible for the observed fusions, the quantum theory of tunneling predicts that the fusion yield per impinging deuteron, $y/2n$, increases extremely fast with the deuteron energy ϵ_D,

$$\epsilon_D = \frac{M_D}{nM_{D_2O}} E \simeq 0.1\frac{E}{n} ,$$

   where M_D and M_{D_2O} are the deuteron and heavy water masses, respectively, and E is the total cluster energy. By contrast, the graph at constant E of fig. 1 shows a saturation of $y/2n$ with ϵ_D for $\epsilon_D \gtrsim 10^2$ eV.
2. For assigned cluster composition ($n = 150$) the fusion rate increases (approximately in an exponential way) with the impinging deuteron energy [fig. 1, where the original data have been normalized by imposing that the fusion rates observed in the implantation at variable E (with $n = 150$) and at variable n (with $E = 300$ keV) coincide for the same conditions ($E = 300$ keV and $n = 150$); in practice, the original data at variable E have been reduced by a factor 1.75].
3. The comparison of the two curves of fig. 1 shows that the fusion rate per deuteron does not depend on deuteron energy only but also on cluster composition; this fact is completely inconsistent with a picture of fusions produced by independent collisions of $2n$ impinging deuterons with target atoms.
4. At a cluster energy of 300 keV the total fusion rate is approximately independent of cluster composition (and hence of deuteron energy) for $10^2 < n < 5 \times 10^2$ (see fig. 2).

[1] By contrast, in high current implantation of heavy ions at relativistic energy [30] space and time superimpositions of single events are erratical.

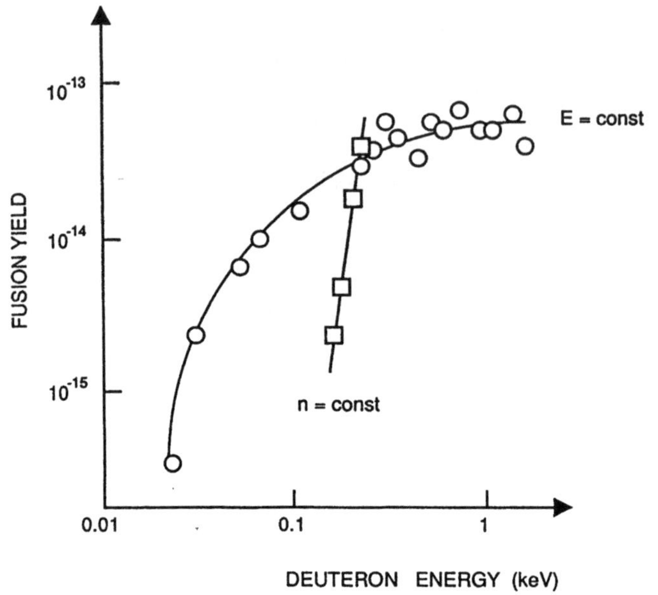

Fig. 1: Fusion yield per impinging deuteron vs deuteron energy for assigned cluster size, $n = 150$, and total energy in the range 225 – 300 keV (□), and for assigned energy, $E = 300$ keV, and cluster size in the interval $25 \lesssim n \lesssim 1350$ (○).

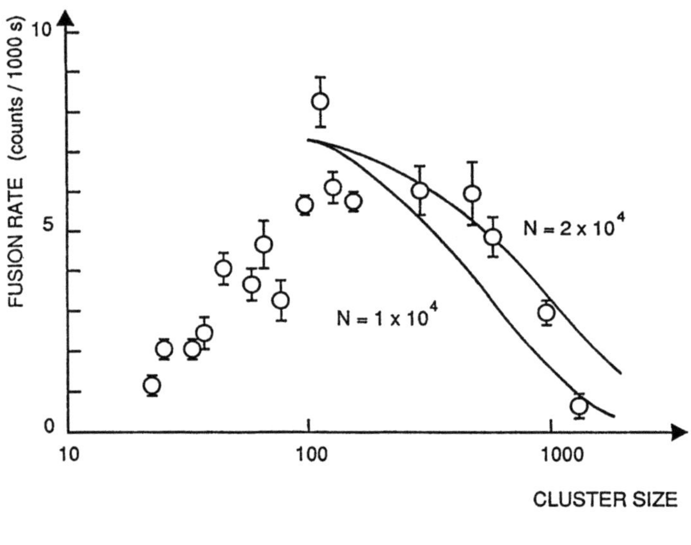

Fig. 2: Fusion rate vs cluster size for a total energy of 300 keV. The interpolation curves are calculated by assuming that the fusion rate is determined by the kinetic energy in the hot cloud and that it scales with energy as shown in fig. 1.

5. The bombardment of $(H_2O)_n$ clusters with the same composition and total energy as $(D_2O)_n$ does not produce appreciable D - D fusion.[2]

The results of Beuhler et al. are so strange as to deserve an accurate analysis in terms of artifacts. Carraro et al., who have recently carried out such an

[2] Very recently the Brookhaven collaboration has also observed D - D fusion in the impact of heavy water clusters onto zirconium deuteride, $ZrD_{1.65}$, and perdeuteropolyethylene, $\leftarrow CD_2 - CD_2 \rightarrow_{n'}$ [32]. In particular it has also been observed that the impact of $(H_2O)_n$ clusters onto $\leftarrow CD_2 - CD_2 \rightarrow_{n'}$ produces approximately 5% of fusions produced by the heavy-water impact.

analysis, have shown that no known model is able to explain them and have considered two possible causes of experimental errors: fragmentation in light ions and poor mass separation [33]. Almost immediately, however, the Brookhaven collaboration has produced a lot of evidence which seems to disregard the hypothesis that the observed fusions can be interpreted in terms whether of fragmentation into light ions or of poor mass separation [32]. Though the analysis of the new Brookhaven results is quite complex, we shall henceforth assume the unrestricted validity of the Brookhaven data.

4.1 The Hot Cloud

Items 3 and 4, taken together, suggest that the deuteron energy ϵ_D does not play any role in the fusion yield at least in the considered energy range, but rather that the total impinging energy E drives the phenomenon. We assume hence the following fundamental hypothesis:

> *Hot Cloud Hypothesis*: Because of the space-time correlation of the original event [the impingement of the $(D_2O)_n$ cluster] if n is large enough, the cascades produced by single atoms superimpose and form a unique collisional cascade so dense to be considered a ***hot cloud*** [35, 36]; as known from low-energy ion implantation [37, 28] the duration τ of this hot cloud and the number N of involved atoms are $\tau \approx 10^{-11}$ s and $N \approx 10^4$, appreciably higher than the analogous quantities estimated by Beuhler et al. ($\tau \approx 10^{-13}$ s and $N \approx 10^3$).

The above value of N is obtained by assuming that energy equipartition in the cascade is possible when the number of atoms involved after k cascade ramifications, 2^k, exceeds the number of atoms contained in a sphere of radius k, $\frac{4}{3}\pi k^3$. The condition $2^k \geq \frac{4}{3}\pi k^3$ identifies a critical branching $\overline{k} \simeq 13.25$ and a cascade size,

$$N = 2^{\overline{k}} = \frac{4}{3}\pi \overline{k}^3 \simeq 10^4 \ , \tag{20}$$

above which the cascade can be seen as a cloud of atoms in fast motion with a local temperature different from that of the target [35, 36]. The estimate (20) is in substantial agreement with the most sophisticated molecular-dynamics calculations which also furnish the value, of the order of 10^{-11} s, for the time required to form the dense collisional cascade [37]. This time can tentatively be assumed of the same order as the duration of the hot cloud, after which the local temperature excess evolves as predicted by the Fourier theory of heat diffusion.

The value of N so determined applies to the dense collisional cascade produced by a single atomic projectile; though a complete theory of the collisional cascade formed by composite projectiles is still missing, the experimental data $y = y(n)$ shown in fig. 2 are easily understood in naive terms of hot cloud. If:

i. the stopping of the impinging deuterium is dominated by nuclear collisions;

ii. the total collisional cascade involves N target deuterium atoms (initially at rest) plus $2n$ impinging deuterium atoms (initially with energy ϵ_D);
iii. heating of the cascade is produced only by impinging deuterium and not by oxygen atoms (see later); and
iv. the process is adiabatic,

then the average kinetic energy $E_{\rm kin}$ in the cascade is given by

$$E_{\rm kin} = \eta \frac{\frac{4}{20}E}{2n + N} , \tag{21}$$

where η ($\eta \leq 1$) is the fraction of the available impinging energy which is not lost in electronic collisions. A precise estimate of this parameter requires both a complete description of the formation of the collisional cascade and of the energy loss mechanisms at low energy. Though none of these topics are accurately understood, a semiquantitative estimate is possible: let ρ be the ratio of the nuclear energy loss to the total energy loss (supposedly constant with the current deuteron energy) and assume that the hot cloud is formed when the ramification χ of the collisional cascade (supposedly occurring via two-body collisions) is such to involve all atoms; in this case the coefficient η is given by $\eta = \rho^{\chi}$. The critical ramification χ is given by the condition $2n \times 2^{\chi} = 2n + N$ [i.e., $\chi = \log_2(1 + N/2n)$], so that $\chi \simeq 2 - 5$ in relation to the cluster size.

The factor 4/20 follows from item 5, which is justified by the following arguments: the implantation of $(H_2O)_n$ clusters is unable to form sufficiently hot clouds because the H-to-D mass ratio does not allow an efficient energy transfer to deuterium atoms of the target [and symmetrically when TiH is bombarded with $(D_2O)_n$ clusters]. This argument requires that oxygen atoms play a negligible role in heating the cloud and rules out the explanation of Echenique et al. [34], essentially based on the hypothesis that *all* the kinetic energy investing the target is released as heat over approximately 10^3 deuterium atoms (without involving titanium and the impinging oxygen).

The above conclusion, i.e. that oxygen does not participate in heating the deuterium collisional cascade, is realistic because oxygen transfers energy mainly to titanium (rather than to deuterium) atoms due to the higher cross section and to a mass ratio closer to unity, while hydrogen and deuterium transfer only a small fraction of their energy to the titanium because of the mass ratio distant from unity. (Likely, the energy released by deuterium to the titanium target compensates exactly the energy transferred from oxygen to deuterium.) Moreover, due to the very different masses of titanium and deuterium, the motions of these atoms are weakly coupled and they quench with different lifetimes.

The kinetic energy $E_{\rm kin}$ does not depend on n only for $2n \ll N$; this inequality applied to the data of fig. 2 gives $N \gg 10^3$. Alternatively, any given value of N combined with the data at variable energy for fixed n allows the function $y = y(n)$ to be calculated. Two calculations, obtained for $N = 1 \times 10^4$ and $N = 2 \times 10^4$, are shown in fig. 2. The resulting function interpolates

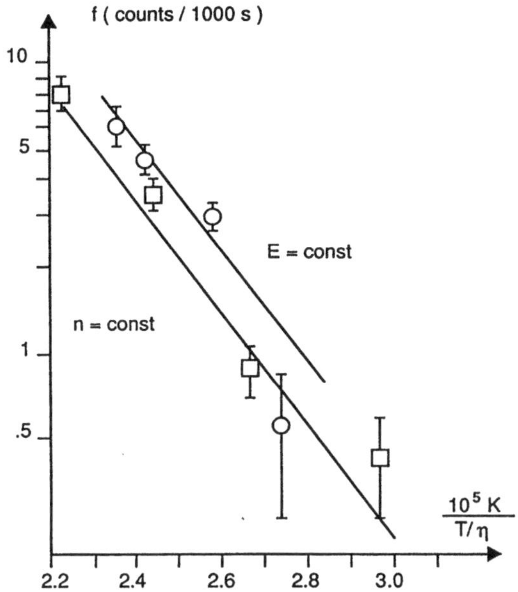

Fig. 3: Arrhenius plots for the data at variable energy (□) and at variable n (○) with $n > 500$; the hot-cloud temperature is calculated assuming $N = 10000$. Symbols and normalization procedure are the same as in fig. 1. The plots give a unique slope, $E^*/\eta = (38.5 \pm 5)$ eV.

satisfactorily, at this level of investigation, all the data for $100 < n < 1350$ — in a way, the hot cloud model removes the contradiction shown in fig. 1.

Assuming the full validity of the hot cloud model [the hot cloud is formed by $2n + N$ deuterons with an average kinetic energy E_{kin} given by eq. (21) and its temperature is $T = \frac{2}{3}E_{\text{kin}}/k_{\text{B}}$] and assuming that the fusion process is a thermally activated process, the data of fusion rate f should obey this relationship

$$f = f_0 \exp\left(-\frac{E^*}{k_{\text{B}}T}\right) , \tag{22}$$

where f_0 is the pre-exponential factor and E^* the activation energy for this process.

Both the data at variable energy and $n = 150$ and the data at $E = 300$ keV with $500 < n < 1350$ can be represented in an Arrhenius plot and give two straight lines with the same slope (fig. 3). Though encouraging, this result is not yet satisfactory because of the following reasons:

- the plots have different pre-exponential factors; and
- most of the data at constant E (namely, all data with $n < 500$) are not described by eq. (22).

However, these anomalies can be removed by considering in greater detail the size and shape of the collisional cascade. As a matter of fact, the projected range R_{p} increases with deuteron energy ϵ_{D} so that at sufficiently high energy R_{p} becomes higher than the diameter $2\overline{k}L$ of the spherical collisional cascade (L is the D - D interatomic distance) and this suggests that N varies as

$$N = N_0\left(1+\frac{R_p}{2\overline{k}L}\right) = N_0\left(1+\frac{\alpha E}{n}\right) \tag{23}$$

where: R_p is assumed to increase with ϵ_D, $R_p = \epsilon_D/(d\epsilon_D/dx)$; $(d\epsilon_D/dx)$ is the deuterium average energy loss (which is dominated by the nuclear energy loss because of assumption i); $N_0 = \frac{4}{3}\pi\overline{k}^{-3} \simeq 10^4$; and $\alpha = R_p/[2\overline{k}L(E/n)] \simeq [20\overline{k}L(d\epsilon_D/dx)]^{-1}$. Inserting eq. (23) into the expression of the kinetic energy in the cloud, eq. (21), we get

$$E_{kin} = \eta\frac{\frac{4}{20}E}{2n + N_0(1+\alpha E/n)}\,. \tag{24}$$

In principle α is determined from stopping power data; however the following considerations:

- for $\epsilon_D \gtrsim 0.5$ keV the electronic energy loss becomes appreciable;
- the formation of the collisional cascade is essentially an end-of-range phenomenon;
- small clusters form sparse collisional cascades ; and
- clusters with high ϵ_D form less dense collisional cascades because the straggling increases with ϵ_D;

reduce the effect of the additive term $\alpha E/n$ and suggest to consider α as a fitting parameter.

Interestingly enough, fitting together the data at constant E and $n > 500$ with the data at constant n and E in the range $225 - 300$ keV indicates $N_0 = 10000$ (± 500), which coincides with the value calculated with the hot cloud model of Cerofolini and Meda [35, 36] and with the numerical analysis of Diaz de la Rubia et al. [37]. The parameter α results in the interval $1.5 - 3.0 \times 10^{-2}$ keV^{-1}; its value cannot be determined better because the term $\alpha E/n$ does not play an important role in the considered range of variation of n. *The same value of N_0, $N_0 = 10000$, and a value of α consistent with the previous estimate, $\alpha = 1.83 \times 10^{-2}$ keV^{-1}, linearize in the Arrhenius plot all data at variable n* (fig. 4).

This surprising result upholds the original hypothesis that the fusion process can be described as a thermally activated phenomenon, though the observed activation energy, $E^*/\eta \simeq (33 \pm 2.5)$ eV, is too low to be associated with the overcoming of the Coulomb barrier hindering D - D fusion. We hence admit that at the energy E^* a *metastable precursor* is formed; the synthesis of this precursor is the rate determining step and the D - D fusion from this precursor takes place with a high probability.

The formation of the precursor is practically feasible only by ion implantation; cluster or molecule implantation seems a particularly economic way to get it [27, 28, 29] because it allows the preparation of dense matter (with density of the order of 10^{23} cm^{-3}) at temperatures in the interval $10^4 - 10^5$ K, that in turn permits the synthesis of compounds whose formation requires activation energies of the order of $10 - 10^2$ eV (in ordinary chemistry reactions take place

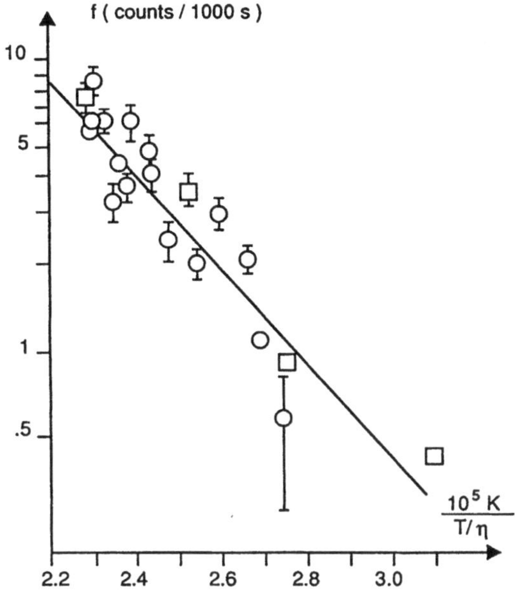

Fig. 4: Arrhenius plot for all data. The hot cloud temperature is calculated assuming $N_0 = 10000$ and $\alpha = 1.83 \times 10^{-2}$ keV^{-1}. The symbols and normalization procedure are the same as in fig. 1. The plot gives a slope $E^*/\eta = (33 \pm 2.5)$ eV.

between 2×10^2 K and 2×10^3 K and the activation energies are in the range $1 - 10$ eV).

4.2 The Activated Precursor

The previous considerations determine the value of the activation energy E^* apart from the parameter η related to the ratio ρ of the nuclear-to-total energy loss $\eta = \rho^\chi$. Taking $\rho = 1$ as done by Carraro et al. [33], one has $\eta = 1$ and $E^* = 33$ eV; taking $\rho = 0.9$ (as done by Cerofolini and co-workers in the description of the amorphization of silicon resulting from the impact of silicon projectiles impinging at similar velocities, of approximately 150 eV/amu [35, 36]) and $\chi \simeq 2 - 5$ in relation to the cluster composition, η ranges between 0.6 and 0.8, and E^* in the interval $20 - 26$ eV. In all cases the activation energy for the precursor formation is of approximately 27 eV, i.e., just twice the ionization energy of the deuterium atom. This indication is a basic clue to formulate a model for the activated precursor.

The precursor can be imagined as a D - D pair at a nuclear distance r simultaneously comparable to, or shorter than, the Bohr radius a_0 and much higher than the range of nuclear forces; a speculation concerning its structure is presented later. Here we intend only to estimate the D - D separation r^* at which the precursor is formed.

Assuming that for $r > r^*$ the potential hindering the formation of the precursor is a truncated Coulomb potential, the D - D separation at which the potential energy saddle is overcome is equal to $r^* \simeq e^2/E^* = 2a_0(E_0/E^*) \simeq a_0$, and the collision leading to the precursor has been assumed to occur when the kinetic energy is higher than E^*.

This estimate neglects an important fact — the screening effect of the orbiting electron. Indeed, according to the Bohr criterion [38], an atom moving in a target is stripped of all electrons with orbital velocities lower than the atom velocity; this criterion implies that the deuteron remains neutral for energies lower than 50 keV, i.e. during the whole process of formation of the collisional cascade. Since in our case the deuterium energy is much lower than this limit, we can confidently assume this second fundamental hypothesis:

> *Neutral Atom Hypothesis*: After the impact onto the target surface, deuterium moves in atomic rather than ionic form, which allows, for any assigned kinetic energy, closer nucleus-nucleus encounter.

According to Bohr [39], a potential of the kind (16) covers most of screening problems of interest in the penetration of atomic particles through matter; in particular it holds with high approximation for the electrostatic field within atoms. In this case, however, the screening length is not given by eq. (17), but can be deduced with the following additional assumptions:

- the atomic screening length is equal to the Bohr radius a_0; and
- the Firsov composition rule [40] holds true;

and is given by $a = a_0/2^{2/3} \simeq 0.63a_0$. The first assumption does not need any discussion; the second assumption, instead, deserves some comments: indeed, the Firsov composition rule is a semi-empirical rule for combining the screening lengths of two atoms; it is based on the Thomas - Fermi model of the atom (and hence in principle it does work well only for heavy atoms), but provides accurate description of both stopping and damage for practically all projectiles in solids [41]; it provides more accurate descriptions than the other semi-empirical rule originally proposed by Bohr [39].

Putting $U(r) = E^*$ in eq. (16) a better estimate is obtained of the separation r^* at which the precursor starts to be formed, $r^* \simeq 0.5a_0$. A similar value is obtained using the Bohr composition rule instead of the Firsov one.

About the precursor structure, the following considerations hold true:

Large Separation. For large D - D separation, where only electrostatic effects are important and the (random) relative spin orientation is mute in the description of the system, the total energy of the system can be calculated in terms of energy of two separate repelling hydrogen atoms:

$$E_{(D+)e^-+(D+)e^-}(r) = \frac{e^2}{r}\exp\left(-\frac{r}{a}\right) - 2E_0 , \tag{25}$$

where the zero-energy configuration has been assumed to be that of two free deuterium nuclei and electrons.

Small Separation. At kinetic energies higher than 27 eV the nuclei approach at distances lower than approximately $0.5a_0$, this nuclear separation being so small that the D - D system can no longer be regarded as formed by two moderately perturbed atoms. In principle, for any given D - D separation the equilibrium electronic configuration can be calculated by solving the stationary Schrödinger equation. It is easily realized that, for D - D separation r much lower than the Bohr radius, the electronic configuration resembles more that of the helium atom rather than that of two separated deuterium atoms or that of an extremely compressed D_2 molecule. The system can therefore be regarded more as a $(D^+D^+)2e^-$ binuclear heliumlike atom than as two deuterium atoms $(D^+)e^- + (D^+)e^-$ repelling each other with potential (16). Though to our knowledge a theory of the *polynuclear atom* does not exist yet, the total energy of the heliumlike atom $(D^+D^+)2e^-$ can be assumed to be estimated by

$$E_{(D^+D^+)2e^-}(r) = \frac{e^2}{r} - E_{el}(r) , \tag{26}$$

where $E_{el}(r)$ is the electronic energy of the system. This energy can be written as

$$E_{el}(r) = [E_{el}(0) - 2E_0]F(r) + 2E_0 , \tag{27}$$

where $E_{el}(0)$ is the electronic energy of the system at null nuclear separation and $F(r)$ is a function of r with the following asymptotic behaviours: $r \to 0 \Rightarrow F(r) \to 1$ and $r \to \infty \Rightarrow F(r) \to 0$.

The energy $E_{el}(0)$ depends on the relative spin orientation; if the spins are antiparallel (para-helium), then $E_{el}(0) = E_1 + E_2$ where E_1 and E_2 are the first and second ionization energies, respectively, of helium ($E_1 = 24.58$ eV and $E_2 = 54.40$ eV); if the spins are parallel (ortho-helium), then $E_{el}(0) \simeq 59$ eV. These situations correspond to two stationary states of the system. Because of the random spin orientation, however, they have null statistical weight and the state of the system is in general a non-stationary linear combination of ortho-helium and para-helium states; correspondingly the electronic energy of the system will be intermediate between that of para-helium and that of ortho-helium. This energy is not constant with time and is coupled with the nuclear energy.

The analysis of para-helium allows the function $F(r)$ to be determined. Indeed, in strict analogy with the theory of the H_2^+ ion (for which an exact solution is known [42]) we may assume

$$E_{el}(r) = (E_1 + E_2 - 2E_0)\exp(-0.46r/a_0) + 2E_0 , \tag{28}$$

which is rigorous for $r = 0$ and matches, in the interval $0.4 - 0.7a_0$, the most accurate calculations available for the H_2 molecule [43] within 1 eV. The comparison of eqs. (27) and (28) determines $F(r)$,

$$F(r) = \exp(-0.46r/a_0) , \tag{29}$$

which is henceforth assumed to hold for all spin configurations.

Equations (26) and (29) are assumed to hold for $r/2 \ll a_m$ (say, $r/2 \leq a_m/2$), where a_m is the radius of the minimum-energy orbital,

$$a_m \simeq a_0\sqrt{E_0/E_2} \simeq 0.5a_0 .$$

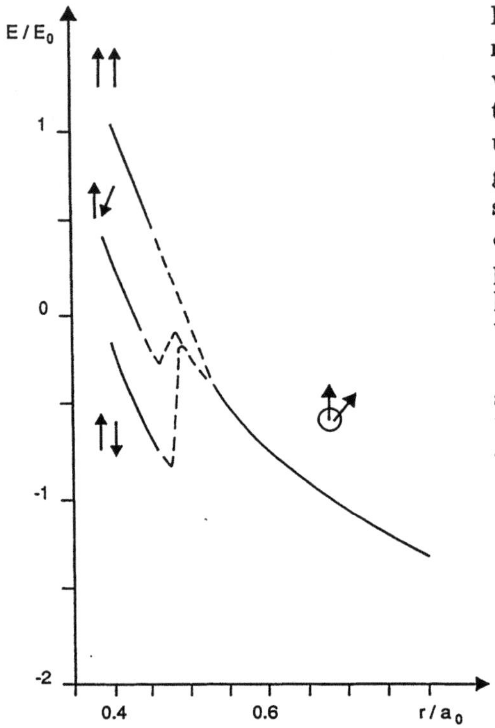

Fig. 5: Total energy vs separation showing that an activation energy is required for the dissociation of the heliumlike atom in two hydrogenlike atoms. The small-separation curve is intermediate between that of the para-heliumlike and ortho-heliumlike atoms [calculated by means of eqs. (26), (27) and (29)]; the large-separation curve is calculated by means of eq. (25). The existence of a region where the binuclear atom is metastable is related to the existence of a separation for which the total energy has a minimum.

Calculating the total energy of the system by means of eqs. (26), (27) and (29) for $r \lesssim 0.5a_0$ and of eq. (25) for $r \gtrsim 0.5a_0$, one immediately realizes that, provided that the relative spin orientation is not too far from the antiparallel one, the energy of the system must have a minimum in the interval $0.4a_0 < r < 0.5a_0$ (fig. 5), so that a metastable configuration with this separation must exist. For $r < 0.4a_0$ the system is unstable with respect to dissociation in all its components (electrons and nuclei).

The D - D collision at sufficiently high energy leads hence to a nuclear configuration which can be stabilized through the formation of a binuclear heliumlike atom. Once this atom is formed, an activation energy is required to bring it into the $(D^+)e^- + (D^+)e^-$ electronic configuration. The binuclear heliumlike atom is metastable and can decay in two ways:

1. dissociation: $(D^+D^+)2e^- \rightarrow (D^+)e^- + (D^+)e^-$; and

2. fusion: $(D^+D^+)2e^- \rightarrow (^4He^{2+})2e^-$;

in turn $(^4He^{2+})2e^-$ is in an excited state and decays mainly through reactions (4) and (5).

The stability of the heliumlike configuration is higher the higher is the activation energy for its dissociation E^*_{diss}; the estimate of this energy requires a knowledge of the spin orientation distribution and a detailed calculation around $r = 0.5a_0$, where neither the heliumlike approximation nor the hydrogenlike one hold true.

5 Conclusions

The current, highly controversial, status of cold-fusion phenomena has been reviewed. The more recent warm fusion produced by the impact of large clusters of heavy water onto TiD target has been discussed in terms of 'new' underlying physics. If this phenomenon is not due to artifacts, it seems to open a new path to power production (as already noted by Beuhler et al. themselves).

Indeed, though the use for power production of fusions induced by large cluster or heavy molecule impact has already been considered, the analysis of refs. [29, 28] (suggesting this possibility for heavy molecules with energy in the giga-electronvolt range) must however be reconsidered in the light of the high fusion yields observed by Beuhler et al. Indeed, the Lawson criterion must presumably be satisfied even in the considered situation but with a density-duration product much lower than the usually accepted value, because of the very low activation energy required to form the activated precursor.

Assuming the validity of the scaling law given by the graph at constant n in fig. 1, the scientific breakeven can be obtained (for appropriate cluster size) for $E \gtrsim 900$ keV. This estimate, however, is an extrapolation for fusion rates exceeding by approximately ten orders of magnitude the ones observed by Beuhler and is not upheld by any theoretical reason.

By contrast, the model developed in this work shows that the pre-exponential factor and activation energy in eq. (22) lead to a fusion rate which, even in the limit $1/T \to 0$, is lower than the one allowing for breakeven by approximately five orders of magnitude. In our model power production by warm fusion is possible only if the titanium host has a negative role in dissociating the metastable precursor (that opens the problem of finding the most appropriate target), if the dissociation of the activated precursor can be prevented (that opens the problem of the role of target temperature), if the intrinsic fusion rate can be increased by changing the reactants [that opens the problem of finding the most appropriate reaction among (2) to (7)], or if the deuterium atoms have spin configurations allowing the heliumlike polynuclear atom to be synthesized with a high-probability (that opens the problem of imposing and maintaining an antiferromagnetic order inside the hot cloud).

Deuterium-rich metastable phases TiD_x or PdD_y with high enthalpy excess (of the order of 1 eV/atom) can explosively revert to the stable phase [44] by releasing deuterium atoms; if these deuterium atoms are weakly coupled with the solid, they may be regarded as forming a gas phase with kinetic energy of the order of 1 eV/atom; moreover, if the activation energy for the synthesis of the metastable precursor is approximately 27 eV, then the observed fusion rate should be of the same order as the one observed by Jones et al. [7] — in a way, the results of Beuhler et al., combined with the concept of hot cloud, furnish us with a possible explanation for cold fusion and the associate complex phenomenology.

Note: Just after the completion of this work, we have read in [32] that the data of fusion yield and rate here reported are underestimated by one order of magnitude. All arguments herein presented continue to hold true, apart from the fact that the distance to the breakeven is reduced by one order of magnitude.

Acknowledgements

We are deeply indebted to Dr. R. Dierckx (EEC Joint Research Center, Ispra), Dr. L. Meda (Functional Materials Laboratory, EniChem, San Donato) and Professor G. Ottaviani (Dipartimento di Fisica, Università di Modena) for helpful discussions on the interpretation of the Brookhaven experiment.

References

[1] G. Grieger, in: B. Brunelli and G.G. Leotta (eds.) *Muon-Catalyzed Fusion and Fusion with Polarized Nuclei*, Plenum Press, New York, NY (1987) p. 251

[2] T.H. Johnson, Proc. IEEE **72**, 548 (1984)

[3] The history of muon-catalyzed fusion is shortly reviewed in:
S. Eliezer, in: B. Brunelli and G.G. Leotta (eds.) *Muon-Catalyzed Fusion and Fusion with Polarized Nuclei*, Plenum Press, New York, NY (1987) p. 19

[4] P. Froelich, Europhys. News **20**, 61 (1989)

[5] S. Jones, in: B. Brunelli and G.G. Leotta (eds.) *Muon-Catalyzed Fusion and Fusion with Polarized Nuclei*, Plenum Press, New York, NY (1987) p. 73

[6] M. Fleischmann, M. Pons and M. Hawkins, J. Electroanal. Chem. **261**, 301 (1989); erratum **263**, 187 (1989)

[7] S.E. Jones, E.P. Palmer, J.B. Czirr, D.L. Decker, G.L. Jensen, J.M. Thorne, S.F Taylor and J. Rafelski, Nature **338**, 737 (1989)

[8] Energy Research Advisory Board, *Cold Fusion Research*, A Report to the U.S. Department of Energy, Nov. 1989

[9] B.V. Deryagin, V.A. Klyuev, A.G. Lipson and Yu.P. Toporov, Coll. J. USSR **48**, 8 (1986)

[10] V.A. Klyuev, A.G. Lipson, Yu.P. Toporov, B.V. Deryagin, V.I. Lusheikov, A.V. Strelkov and E.F. Shabalin, Sov. Tech. Phys. Lett. **12**, 551 (1986)

[11] M.A. Yaroslavskii, Dokl. Akad. Nauk SSSR **307**, 369 (1989)

[12] J. T. Dickinson, L.C. Jensen, S.C. Langford, R.R. Ryan and E. Garcia, J. Mater. Res. **5**, 109 (1990)

[13] P.B. Price, Nature **343**, 542 (1990)

[14] L.G. Sobotka and P. Winter, Nature **343**, 601 (1990)

[15] S. Freedman and D. Krakauer, Nature **343**, 703 (1990)

[16] S.E. Jones, D.L. Decker and H.D. Tolley, Nature **343**, 704 (1990)

[17] J.F. Ziegler, T.H. Zabel, J.J. Cuomo, V.A. Brusic, G.S. Cargill III, E.Y. O'Sullivan and A.D. Marwick, Phys. Rev. Lett. **62**, 2929 (1989)

[18] D.E. Williams, D.J.S. Findlay, D.H. Craston, M.R. Sen, M. Bailey, S. Croft, B.W. Hooton, C.P. Jones, A.R.J. Kucernak, J.A. Mason and R.I. Taylor, Nature **342**, 375 (1989)
[19] G.M. Miskally, M.J. Heben, A. Kumar, R.M. Penner, M.J. Sailor and N.S. Lewis, Science **246**, 793 (1989)
[20] A.J. Appleby, Y.J. Kim, O.J. Oliver and S. Srinivasan, quoted in [8]
[21] N.J.C. Packham, K.L. Wolf, J.C. Wass, R.C. Kainthla and J.O'M. Bockris, J. Electroanal. Chem. **270**, 415 (1989)
[22] M.H. Salamon, M.E. Wrenn, H.E. Bergeson, K.C. Crawford, W.H. Delaney, C.L. Henderson, Y.Q. Li, J.A. Rusho, G.M. Sandquist ans S.M. Seltzer, Nature **344**, 401 (1990)
[23] S.E. Koonin and M. Nauenberg, Nature **339**, 690 (1989)
[24] L.D. Landau and E.M. Lifshitz, *Quantum Mechanics*, Pergamon Press, London (1958) par. 50, problem 2
[25] C. Kittel, *Introduction to Solid State Physics*, 5th Ed., Wiley, New York, NY (1976) ch. 10
[26] G. Benedek and P.F. Bortignon, Nuovo Cimento D **11**, 1227 (1989)
[27] R.J. Beuhler and L. Friedman, Chem. Rev. **86**, 521 (1986)
[28] G.F. Cerofolini, L. Meda and C. Volpones, Mat. Res. Soc. Symp. Proc. **128**, 181 (1989)
[29] G.F. Cerofolini and L. Meda, Italian Patent No. 120683, deposited 30 April 1987, filed 3 May 1989
[30] C. Rubbia, in A. Caruso and E. Sindoni (eds.) *Inertial Confinement Fusion*, Editrice Compositori, Bologna (1988) p. 659
[31] R.J. Beuhler, G. Friedlander and L. Friedman, Phys. Rev. Lett. **63**, 1292 (1989)
[32] R.J. Beuhler, Y.Y. Chu, G. Friedlander, L. Friedman and W. Kunnmann, J. Phys. Chem. **94**, 7665 (1990)
[33] C. Carraro, B.Q. Chen, S. Schramm and S.E. Koonin, Phys. Rev. A **42**, 1379 (1990)
[34] P.M. Echenique, J.R. Manson and R.H. Ritchie, Phys. Rev. Lett. **64**, 1413 (1990)
[35] G.F. Cerofolini and L. Meda, Phys. Rev. B **36**, 5131 (1987)
[36] G.F. Cerofolini, L. Meda and C. Volpones, J. Appl. Phys. **63**, 4911 (1988)
[37] T. Diaz de la Rubia, R.S. Averback, H. Hsieh and R. Benedek, J. Mater. Res. **4**, 579 (1989)
[38] N. Bohr, Phys. Rev. **58**, 654 (1940); Phys. Rev. **59**, 270 (1941)
[39] N. Bohr, Mat. Phys. Medd. Dan. Vid. Selsk. **18** no. 1 (1948)
[40] O.B Firsov, Zh. Eksp. Teor. Fiz. **34**, 447 (1958); English translation: JETP **7**, 308 (1958)
[41] M.T. Robinson and I.M. Torrens, Phys. Rev. B **9**, 5008 (1974)
[42] H. Eyring, J. Walter and G.E. Kimball, *Quantum Chemistry*, Wiley, New York, NY (1944)
[43] W. Kolos and L. Wolniewicz, J. Chem. Phys. **41**, 3663 (1964); J. Chem. Phys. **49**, 404 (1968)
[44] L. Pauling, Nature **339**, 105 (1989)

Part IV

Charge Transfer: Theory and Phenomena

Charge Exchange Processes

N.P. Popov

St. Petersburg Nuclear Physics Institute, Gatchina,
188350 St. Petersburg, Russia

Abstract. Negative muon (pion) transfer from mesic hydrogen to the nucleus of heavier hydrogen isotopes or of elements with Z>1 is investigated. Direct charge exchange as well as the molecular mechanism are considered, the latter going via a quasistationary intermediate molecular state. Both processes are analyzed for the ground state and excited states of the mesic atom.

Kinetics of mesic hydrogen atoms in mixtures of hydrogen isotopes with other elements is quite important for investigating the weak interaction in the nuclear muon capture process (Bertin and Vitale [1]) as well as for the muon catalyzed fusion problem (Ponomarev [2], Petitjean [3], Breunlich et al. [4]).

Isotopic exchange reactions (muon transfer to the heavier isotope nucleus), for example

$$(d\mu)_n + t \rightarrow (t\mu)_n + d, \qquad (1)$$

($n \geq 1$ is the principal quantum number of mesic hydrogen), as well as reactions of muon transfer from mesic hydrogen to nuclei with $Z>1$ ($H \equiv p,d,t$)

$$(H\mu)_n + Z \rightarrow (Z\mu)_{n'} + H \qquad (2)$$

provide, as a rule, fast muon transfer from the initial mesic hydrogen, changing considerably the kinetics of the processes occurring in the mixture. Muonic tritium atoms, produced in the reaction (1), colliding with deuterium atoms, form $dt\mu$ molecules which play the principal role in the muon catalysis. At the same time muon transfer from muonic deuterium or tritium to other nuclei (in particular to helium, which is accumulated in the mixture due to tritium decay and from nuclear syntheses reactions) causes muons to leave the cycle of the muon catalysis. The formation of mesic molecules, as well as muon transfer should be taken into account when extracting the information on weak interaction constants from the reaction of muon capture by the proton. Lasly, the experimental study of reactions (1) and (2) is itself interesting because it allows one to check various theoretical constructions which are used to explain the data. At present being the isotopic exchange reactions (1) are intensely investigated (Bertin and Vitale [1], Ponomarev [2], Petitjean [3], Breunlich et al. [4], Weber et al. [5]). As for muon transfer in reaction (2), a lot of experimental data are now available (e.g., Bertin and Vitale [1]), part of them waits for theoretical explanation (Schneuwly [6]).

As a rule, reactions (1) and (2) are treated in a WKB-approximation, with the exception for n=1 for reaction (1), where the quasiclassical condition

Springer Proceedings in Physics, Vol. 59 Exotic Atoms in Condensed Matter
Editors: G. Benedek · H. Schneuwly © Springer-Verlag Berlin, Heidelberg 1992

$d\lambda/dR \ll 1$ (λ is the wavelength of mesic hydrogen, R is the internuclear distance) is not valid at low collision energies in the effective transition region. In this case the quantal approach in the framework of the adiabatic representation is necessary (Melezhik [7], Kobayashi et al. [8]).

Within the WKB method various approaches could be developed, which lead to different results. Charge exchange of mesic atoms can be considered as so-called second kind collisions, which were treated by Landau [9] and Zener [10]. In this case the quantum treatment of the muonic bound states is combined with the classical description of the motion of colliding nuclei. It is well known that there exists an adiabatic veto upon the probability of the transition considered, which vanishes when the initial and final molecular terms intersect. This means that the transition takes place without change of the internuclear distance. In fact, the transition between muon states occurs at a specific mesoatomic time $\sim 10^{-19}$ s, too short to transfer energy to heavy nuclei.

Gerstein [11] was the first who used this method to calculate the rates of the reaction (2), which occurred to look like $\lambda_{ex} \sim Z\omega \cdot 10^{10}\ s^{-1}$, where ω is the squared transition matrix element. If one takes into account the electron screening of the target nucleus, the simple linear Z dependence of charge exchange rate will change. Fiorentini and Torelli [12] using the Thomas-Fermi model obtained for λ_{ex} a $Z^{2/3}$-dependence. Haff et al. [13] developed the classical theory of processes of charge exchange in which the cross section was determined by the maximal possible impact parameter ρ_{max}, satisfying the condition of vanishing of the Coulomb barrier between the nuclei. This condition determines the internuclear distance R_c, corresponding to the transition, assuming that $R_c > R_1$ (where R_1 is the turning point), i.e. in a classically accessible region. As a result, the fall to the centre takes place for all trajectories with the impact parameter $\rho \leq \rho_{max}$. A similar approach was used by Menshikov and Ponomarev [14] for reaction (1), the nonadiabatic transition probability being considered in a Rosen-Zener-Demkov model [15].

I discuss here the quasiclassical approach by Solovyov and Ovchinnikov [16] based upon the terms behaviour in the complex plane of internuclear distance R.

In WKB approximation the cross section is written as

$$\sigma = \pi \int_0^{\rho_{max}^2} P d\rho^2, \tag{3}$$

where ρ_{max} is a maximal impact parameter, at which there is no turning point between $\mathrm{Re}R_c$ and infinity.

The isotopic exchange reaction (1) is characterized by a series of branching points, which tie the terms corresponding to the initial and final states, the asymptotic splitting of terms being determined by a small resonance defect $\Delta U = (\mu_2 - \mu_1)/(2n^2)$ (μ_1 and μ_2 are reduced masses of the initial and final mesic atoms) $\sim 0.011/n^2$ (mesoatomic energy units).

The transition probability in this case is large, so one should take into account the contribution of all the branching points of the series. The resulting probability of the nonadiabatic quasiresonant transition is (Demkov [15]) as follows

$$P = (2\mathrm{Ch}^2\delta)^{-1}, \tag{4}$$

where the Massey parameter δ is given by

$$\delta = \left| \mathrm{Im} \int_{\mathrm{Re}R_c}^{R_c} (p_R^{(1)} - p_R^{(2)}) dR \right| \tag{5}$$

and ρ_R is a radial momentum

$$p_R^{(i)} = M\upsilon \sqrt{1 - \frac{U^{(i)}(R)}{\varepsilon} - \left(\frac{\rho}{R}\right)^2} . \tag{6}$$

Here $U^{(i)}$ (R) is the term of the two-centre problem with the account of Coulomb repulsion of nuclei for initial (i=1) and final (i=2) states respectively, υ is the relative velocity of atom and mesic atom at $R \to \infty$; $\varepsilon = M\upsilon^2/2$, where M is a reduced mass of the system. The maximal impact parameter ρ_{max} is determined by the condition that $p_R^{(1)}$ be real along the trajectory $R > \mathrm{Re}R_c$. The location R_c of a branching point is determined by the condition (Landau and Lifshitz [17])

$$\Delta U(R) = U^{(1)}(R) - U^{(2)}(R) \sim (R-R_c)^{1/2}. \tag{7}$$

If the Massey parameter is not small (δ~1) or if one has only one branching point instead of a series, the transition probability is (Mott and Massey [18])

$$P = 2e^{-2\delta} (1 - e^{-2\delta}). \tag{8}$$

For δ»1 the expressions (4) and (8) are equivalent.

This approach was used (Kravtsov et al. [19-22]) to calculate the cross sections and rates $\lambda_{ex} = \sigma\upsilon N$ ($N=4.25\cdot 10^{22}$ cm^{-3} is the liquid hydrogen density, LHD) for isotopic exchange reactions (1). The rates λ_{ex} calculated for LHD and ε = 0.04 eV are presented in Table 1. The calculated values of muon transfer

Table 1 : Isotopic exchange rates λ_{ex} (10^{11} s^{-1}) calculated for ε = 0.04 eV and LHD and experimental data on the charge exchange rate from the ground state muonic hydrogen at room temperature.

n	$p\mu \to d\mu$	$p\mu \to t\mu$	$d\mu \to t\mu$
1s	0.17 [23]	0.075 [23]	$2.7\cdot 10^{-3}$ [7]
	0.169±0.026 exp.[1]		$(2.8\pm 0.2)\cdot 10^{-3}$ exp. [24]
2s	17.6	13.7	15.5
3	13.4	10.3	12.0
4	18.3	14.1	16.4
5	22.3	17.1	20.0
6	25.4	19.2	23.0
7	27.5	20.4	25.5

rates decrease with increasing energy ε. Experimental values λ_{ex} for n>1 are not available because their large values are comparable with deexcitation rates of the mesic atom.

One can, however, calculate (and compare with the experiment) the ground state population q_{1s} of the initial mesoatom, which depends on the rates of muon transfer from excited states. These calculations were performed by Menshikov and Ponomarev [25] and by Kravtsov et al. [19,21,26]. The latter results show that better agreement with the experiment is obtained if one assumes that mesic atoms are not thermalized when excited. It should be noted that the information on the q_{1s} value is very important for the problem of the muon catalyzed fusion, because the effective charge exchange rate is

$$\lambda_{ex}^{eff} \cong \lambda_{ex}(1s)/q_{1s}.$$

So, the lower q_{1s}, the greater is the number of cycles per muon. Additional information on the charge exchange from excited states can be obtained from the pion transfer reaction, because the pionic atom cascade does not proceed down to the ground state due to the strong nuclear pion absorption from excited states. One can study the pion transfer reaction $(p\pi)_n + d \rightarrow (d\pi)_n + p$ by observing the gamma quantum from the decay of the neutral pion, produced in the nuclear reaction $\pi^- p \rightarrow n + \pi^\circ$. The comparison of the calculated cross-sections with experimental data shows a good agreement (Weber et al. [5]) and confirms the assumption about nonthermalized excited mesohydrogen (Kravtsov et al. [20]).

Theoretical treatment of the reaction (2) was carried out by Gerstein [11] and Gerstein and Ponomarev [27]. Kravtsov et al. [28,22] obtained the rates of muon transfer from mesohydrogen to helium for excited states with n=2 (metastable 2s-state) and n=3 (see Table 2). Experimental evidence (Bertin et al. [32] and Bubak and Bystritsky [33]) on the large muon transfer rate from mesic hydrogen to helium is in agreement with theoretical predictions.

As for the muon transfer from the ground state mesic hydrogen to nuclei with Z=2 or Z=3, there is no pseudocrossing of terms, corresponding to the initial and final terms in this case (Gershtein [11]).

For this reason the rate of the reaction (2) for Z=2 is suppressed by five orders of magnitude (Matveenko and Ponomarev [34]). However, muon transfer can proceed via an intermediate molecular state (Popov[35]) according to the reaction

$$H\mu + He \rightarrow H\mu He \rightarrow He\mu + H + \gamma. \qquad (9)$$

The spectrum of gamma quanta, emitted by a dissociating mesomolecule, has a maximum at the energy ~7 keV. Such a spectrum was observed by Matsuzaki et al. [36] and by Nagamine et al. [37].

The mesomolecule HμHe is formed in the Auger process

$$H\mu + He \rightarrow [(H\mu He)^{2+}e] + e \qquad (10)$$

due to the polarization of the muonic hydrogen atom in the field of the helium nucleus (quadratic Stark effect). Calculated muon transfer rates, corresponding to the reaction (9), are also given in Table 2 together with experimental data available.

Table 2 : Rates of muon transfer from muonic hydrogen to helium nuclei λ_{ex}^{He}, calculated for $\varepsilon = 0.04$ eV and LHD, and experimental muon transfer rates from the ground state muonic hydrogen.

	λ_{ex}^{He} $(10^8\ s^{-1})$		λ_{ex}^{He} $(10^{11}\ s^{-1})$	
	1s		2s	n=3
	Theory	Experiment		
$p\mu \to {}^3He\mu$	0.53		7.1	78
$p\mu \to {}^4He\mu$	0.35	0.36±0.10 [29]	6.3	75
		0.81±0.08 [30]		
$d\mu \to {}^3He\mu$	1.43	1.27±0.11 [24]	2.8	55
$d\mu \to {}^4He\mu$	3.22	3.68±0.18 [24]	2.2	51
$t\mu \to {}^3He\mu$	8.56	15.0±2.5 [31]	1.7	46
$t\mu \to {}^4He\mu$	1.32		1.2	40

The molecular mechanism for muon transfer from mesic hydrogen to helium can work also for the excited states due to polarization of muonic hydrogen in the field of the helium nucleus as a result of the linear Stark effect (Kravtsov and Popov [38]). In this case, however, it cannot compete with the direct muon transfer (Kravtsov et al. [28]). It should be noted that the formation of the mesomolecule cannot be excluded for any Z and n, however in each case one should be sure the molecular bound state really exists. Molecular charge exchange can be an important channel only if the direct one is suppressed.

Another example of such a reaction is

$$H\mu + Li \to H\mu Li \to Li\mu + H. \tag{11}$$

The main dissociation mechanism for HμLi molecules, resulting in muon transfer to the lithium nucleus, is an Auger process, responsible also for the mesomolecule formation (Kravtsov et al. [39]). The rates of mesomolecular charge exchange of mesic hydrogen on lithium nuclei turns out to depend strongly on the details of the electron cloud and the Auger process. According to calculations, the binding energy of the HμLi molecule is not sufficiently large to emit a $1s^2$ electron from the lithium atom, so only the 2s valence electron can be emitted. The calculation was carried out with the use of Hartree-Fock wave functions for atomic electrons (Kravtsov et al. [39]). The results turned out to be sensitive to very subtle effects, which could hardly be observed in any other process (e.g., photoionization). The account of the intershell correlations (Ivanov et al. [40]), i.e., of virtual excitation of $1s^2$ electrons with the following emission of a 2s electron, has a strong influence on the charge exchange rates.

Table 3 contains the rates of HμLi mesomolecule formation calculated for $\varepsilon = 0.04$ eV and LHD with and without the account of intershell correlations. It is evident, that the investigation of charge exchange of muonic hydrogen in

Table 3 : Rates of muon transfer from muonic hydrogen to lithium nuclei λ_{ex}^{Li} (10^6 s^{-1}) calculated with (a) and without (b) intershell correlations for $\varepsilon = 0.04$ eV and LHD.

	$p\mu \to {}^6Li\mu$	$p\mu \to {}^7Li\mu$	$d\mu \to {}^6Li\mu$	$d\mu \to {}^7Li\mu$	$t\mu \to {}^6Li\mu$	$t\mu \to {}^7Li\mu$
a	0.38	0.28	0.91	0.82	1.75	1.16
b	2.03	0.91	0.018	0.0005	0.12	0.17

lithium allows one to check some fine features of the atomic electron shell. At the same time the problems concerning the molecular structure as well as the condensed state of the target matter have not been considered yet.

One should note in conclusion, that the investigation of the charge exchange of muonic hydrogen is of importance both for elementary particle physics and for applications. In spite of a 25 year long period of research, the questions considered need further investigation, both theoretical and experimental.

References

1. A. Bertin, A. Vitale, Fifty Years of Weak-Interaction Physics. (Eds. A. Bertin, R.A. Ricci, A.Vitale. Bologna: Italian Physical Society. 1984) p. 130.
2. L.I. Ponomarev. Muon Catalyzed Fusion 3, 629 (1988).
3. C. Petitjean, Fusion Engineering and Design 11, 255 (1989).
4. W.H. Breunlich, P. Kammel, J.S. Cohen, M. Leon. Annu. Rev. Nucl. Part. Sci. 39, 311 (1989).
5. P. Weber, D.S. Armstrong, D.F. Measday, A.J. Noble, S. Stanislaus, M.R. Harston, K.A. Aniol, D. Horvath. Phys. Rev. A (in press).
6. H. Schneuwly. Muon Catalyzed Fusion 4, 87 (1989).
7. V.S. Melezhik. Muon Catalyzed Fuzsion 2, 117 (1988).
8. K.Kobayashi, T. Ishihara, N. Toshima. Muon Catalyzed Fusion 2, 191 (1988).
9. L. Landau, Phys. Z. Sow. 1, 88 (1932); 2, 46 (1932).
10. C. Zener. Proc. Roy. Soc. A137, 696 (1932)
11. S.S. Gershtein. Sov. Phys. JETP 16, 501 (1962).
12. G. Fiorentini, G. Torelli, Nuovo Cim. 36, 317 (1976).
13. P.K. Haff, E. Rodrigo, T.A. Tombrello. Ann. Phys. 104, 363 (1977).
14. L.I. Menshikov, L.I. Ponomarev. Z. Phys. D: At. Mol. and Clusters 2, 1 (1986).
15. N. Rosen, C. Zener. Phys. Rev. 40, 502 (1932);
    Yu. N. Demkov. Sov. Phys. JETP 46, 1126 (1964).
16. E.A. Solovyov. Zh. Eskp. Teor. Fiz. 81, 1681 (1981); 90, 1165 (1986).
    S.Yu. Ovchinnikov, E.A. Solovyov. Zh. Eksp. Teor. Fiz. 90, 921 (1986); 91, 477 (1986).
17. L.D. Landau, F.M. Lifshitz. Quantum Mechanics (Moscow, Nauka, 1974).

18. N. Mott, S. Massey. Theory of Atomic Collisions (Oxford University Press, London-New York (1965)).
19. A.V. Kravtsov, A.I. Mikhailov, S.Yu. Ovchinnikov, N.P. Popov, Muon Catalyzed Fusion 2, 183 (1988).
20. A.V. Kravtsov, A.Yu. Mayorov, A.I. Mikhailov, S.Yu. Ovchinnikov, N.P. Popov, V.M. Suvorov, A.I. Shchetkovsky. Muon Catalyzed Fusion 2, 199 (1988).
21. A.V. Kravtsov, A.I. Mikhailov, N.P. Popov. Phys. Lett. A132, 124 (1988).
22. A. Gula, A. Kravtsov, A. Mikhailov, Z. Oziewicz, N. Popov. Muon Catalyzed Fusion 4, 217 (1989).
23. L.I. Ponomarev. 6th Int. Conf. on Atomic Physics Proc. Aug. 17-22, 1978, Zinatne Riga (Plenum Press, New York and London, 1979), p. 182.
24. A.A. Vorobyov. Muon Catalyzed Fusion 2, 17 (1988).
25. L. Men'shikov, L. Ponomarev. Pis'ma Zh. Eksp. Teor. Fiz. 39, 542 (1984).
26. A. Gula, A. Kravtsov, N. Popov. Report INT 237/PS, Cracow, 1989.
27. S.S. Gerstein, L.I. Ponomarev. Muon Physics (Eds. V. Hughes, C.S. Wu, New York. Academic Press, 1975) v.3, p. 141.
28. A.V. Kravtsov, A.I. Mikhailov, N.P. Popov. Zh. Eksp. Teor. Fiz. 96, 437 (1989).
29. V.M . Bystritsky, V.P. Dzhelepov, V.I. Petrukhin, A.I. Rudenko, V.M. Suvorov, V.V.Filchenkov, N.N.Khovanskii, B.A.Khomenko, Sov. Phys. JETP, 57, 728 (1983).
30. R. Jacot-Guillarmod, F. Bienz, M. Boschung, C. Piller, L.A. Schaller, L. Schellenberg, H. Schneuwly, W. Reichart, G. Torelli. Phys. Rev. A38, 6151 (1988).
31. A.J. Caffrey, A.N. Anderson, C. De W. Van Siclen, K.D. Watts, J.N. Bredbury, P.A.M. Gram, M. Leon, H.R. Maltrud, M.A. Paciotti, S.S. Jones. Muon Catalyzed Fusion 1, 53 (1987).
32. A. Bertin, M. Bruschi, M. Capponi, J.D. Davies, S. De Castro, I. Massa, M. Piccinini, M. Poli, N. Semprini-Cesari, A. Trombini, A. Vitale, A. Zoccoli. In Muon Catalyzed Fusion, Sanibel Island, FL., 1988, AIP Conf. Proc. No. 181, Edited by S.E. Jones, J. Rafelski, H.J. Monkhorst (AIP, New York, 1989) p. 161.
33. M. Bubak, V.M. Bystritsky. Preprint JINR E1-86-107, Dubna, 1986.
34. A.V. Matveenko, L.I. Ponomarev. Zh. Eksp. Teor. Fiz. 63. 48 (1973).
35. N.P. Popov. Muon Catalyzed Fusion 2, 207 (1988).
36. T. Matsuzaki, K. Ishida, K. Nagamine, Y. Hirata, R. Kadono. Muon Catalyzed Fusion 2, 217 (1988).
37. K. Nagamine, T. Matsuzaki, K. Ishida, Y. Hirata, W. Watanabe, Y. Miyake, R. Kadono. In Muon Catalyzed Fusion, Sanibel Island, FL, 1988, AIP Conf. Proc. No. 181, edited by S.E. Jones, J. Rafelski, H.J. Monkhorst (AIP, New York, 1989) p. 23.
38. A.V. Kravtsov, N.P. Popov. Z. Phys. D: At. Mol. and Clusters 6, 61 (1987).
39. A.V. Kravtsov, N.P. Popov, G.E. Solyakin. Yad. Fiz. 35, 1498 (1982).
40. V.K. Invanov, A.V. Kravtsov , A.I. Mikhailov, N.P. Popov Z. Phys. D: At. Mol. and Clusters, 7, 349 (1988).

Coulomb Deexcitation of Muonic Hydrogen

W. Czaplinski[1], *A. Gula*[1], *A. Kravtsov*[2], *A. Mikhailov*[2], *S. Ovchinnikov*[3], *and N. Popov*[2]

[1]Institute of Physics and Nuclear Techniques, AGH, Cracow, Poland
[2]St. Petersburg Nuclear Physics Institute, Gatchina, 188350 St. Petersburg, USSR
[3]A.F. Ioffe Physico-Technical Institute, 194021 St. Petersburg, USSR

Abstract. The cross sections and rates of the Coulomb deexcitation of muonic hydrogen are calculated taking into account the electron screening effect and without it. It is shown that for highly excited states of muonic hydrogen one should take into account the Coulomb deexcitation process together with other known deexcitation processes.

Introduction

The study of the kinetics of excited muonic hydrogen is of special interest for the muon catalyzed fusion problem, because one of the most important things here is the information about the energy distribution of muonic hydrogen atoms in the ground and excited states [1]. To calculate the energy distribution function $F_n(\varepsilon)$ (where ε is the collision energy and n is the principal quantum number of the muonic atom) one should know the cross section of the processes, responsible for muonic hydrogen deexcitation. Radiative as well as Auger deexcitation processes have been considered in a number of papers [2-4]. As for the Coulomb deexcitation of muonic hydrogen, first considered in [2,5], the situation here is rather ambiguous, because the rates of the process obtained in [6] were several orders of magnitude lower than those calculated in [5].

In this paper we present the results of calculations of Coulomb transition rates obtained by considering the dynamics of the process in the complex plane of the internuclear distances R in a WKB approximation. We confirm the conclusions made in [5] about the important role of the process for large values of principal quantum numbers.

Adiabatic method

In general, to calculate the cross sections of the processes responsible for the kinetics of muonic atoms of hydrogen isotopes, one should solve the multidimensional system of Schroedinger equations, which is quite a complicated problem. At the same time, while studying the excited states of muonic hydrogen, one can use WKB approximation, which is reliable for $n \geq 2$ according to the validity condition $1 \gg \frac{d\lambda}{dR} = \frac{1}{\sqrt{3Mn}}$, where λ is the wavelength of the muonic atom and M is the reduced mass of the system. In this approach the Coulomb transition

$$(p\mu)^*_n + p \rightarrow (p\mu)^*_{n-k} + p' \qquad (1)$$

 © Springer-Verlag Berlin, Heidelberg 1992

is considered to take place at the (complex) distance R_c, the interacting particles moving along the classical trajectory with the impact parameter ρ. The cross section of the process is then

$$\sigma = \pi \int_0^{\rho_{max}^2} P d\rho^2 \tag{2}$$

where the reaction probability

$$P = 2e^{-2\delta}(1-e^{-2\delta}) \tag{3}$$

and a Massey parameter δ is as follows

$$\delta = \left| \operatorname{Im} \int_{\operatorname{Re}R_c}^{R_c} (p_R^{(1)} - p_R^{(2)})\, dR \right| \tag{4}$$

Here p_R is the radial momentum:

$$p_R^{(i)} = Mv\sqrt{1 - \frac{U^{(i)}}{\varepsilon} - (\rho/R)^2} \tag{5}$$

where M is a reduced mass in the reaction (1), R is internuclear distance, v the relative velocity of colliding particles at $R \to \infty$; $\varepsilon = Mv^2/2$ the collision energy; $U^{(i)}$ the molecular term of two centre problem (with the account of the nuclear Coulomb repulsion) for initial (i=1) and final (i=2) states. The maximal impact parameter ρ_{max} is determined by the condition that $p_R^{(i)}$ be real along the trajectory $R \geq \operatorname{Re}R_c$. The crossing points R_c and Massey parameters are calculated in the complex R plane. According to [7], in our case the terms of the two centre problem have only 2 series of crossing points in the complex R plane:

1. S-series, which connects the terms with equal n_2 and different n_1 (n_1 and n_2 - parabolic quantum numbers);
2. T-series, connecting the terms with equal n_1 and different n_2.

The crossing points, belonging to the S-series, are close to the origin, the corresponding cross sections being small due to small values of ρ_{max} (even in the case when $\operatorname{Re}R_c$ is in the classically accessible region). Thus we consider only T-series of crossing points, corresponding to transitions $(n, n_1, n_2, m) \to (n-1, n_1, n_2-1, m)$, where m is the orbital momentum projection on the internuclear axis.

One should note that the accuracy with which we know the terms and crossing points is very important while calculating Massey parameters and cross sections. The simple asymptotic approximation of terms, used in [5] for rather small R, lead to the crossings of terms at small real R, which do not really exist. This is a possible reason for the too large cross sections, obtained in [5].

We use the asymptotic approximation for terms

$$U(R) \cong -\frac{1}{2n^2} + \left[\frac{3}{2}\, n\Delta - \frac{n^2}{2R}\,(6\Delta^2 - n^2+1)\right] E(R) -$$

$$-\lambda^3 e^{-2\beta}\frac{n^2}{2}\,(9\Delta^2 + n^2 - m^2 + 3), \tag{6}$$

where $\Delta = n_1 - n_2$; $\lambda = a_\mu/a_e$; $\beta = 2R/a_e$. Here a_e, a_μ is a Bohr radius of the atom and muonic atom respectively, E(R) is the atomic electric field

$$E(R) = \frac{1}{R^2}(1 + 2\beta + 2\beta^2)\exp(-2\beta).$$

The screening effect is introduced here by the exponential factor and by the third term. In the absence of screening $\lambda \to 0$ and $\beta \to 0$ in (6). In this case this expression describes the real term within several percent accuracy, as a rule.

In the vicinity of the crossing point (while calculating the Massey parameter δ) it is reasonable to use the following approximation for terms

$$\tilde{U}(R)_{1,2} = \frac{u_1+u_2}{2} \pm \sqrt{\frac{\mathrm{Im}R_c - \mathrm{Im}R}{\mathrm{Im}R_c}}\cdot\frac{u_1-u_2}{2}$$

The coordinates at the crossing points were calculated numerically according to [8]. The resulting cross section for muonic atom with principal quantum number n is then

$$\sigma(n) = \sum_m \frac{2-\delta_{0m}}{n^2}\cdot\frac{1}{2}\,[\sigma_s(n,m) + \sigma_a(n,m)] \tag{7}$$

where $\sigma_{s,a}$ is the cross section corresponding to the symmetric or antisymmetric state in the orbital quantum number ℓ (in the united atom classification). In fact the antisymmetric states contribution is very small because the corresponding crossing points have $\mathrm{Im}R_c$ twice as large as compared with the symmetric case, which leads to much greater Massey parameters and, consequently, to lower cross sections.

The resulting cross sections σ(n) for Coulomb deexcitation of $(p\mu)_n^*$ atoms are given in Fig. 1 versus collision energy ε. As seen the energy dependence is similar to that given in [5], though the cross section values are much lower, especially for large n. The main reason for this is that the crossing points we obtain have much larger $\mathrm{Re}R_c$ and consequently much larger Massey parameters.

Fig. 2 shows calculated values of the rates $\lambda_n = \sigma(n)vN\phi$ ($N = 4.25\cdot10^{22}$ cm^{-3} - liquid hydrogen density) for $\phi = 1.27\cdot10^{-3}$ (≅ 1 atm) with and without electron screening for collision energies of 0.1 and 10 eV. Corresponding Auger deexcitation rates [5] are also given. The effect of screening is evident, especially for small ε and large n. The crossing of the curves λ_n calculated at ε = 0.1 and 10 eV is due to the plateau in energy dependence of σ(n > 10) (see Fig. 1).

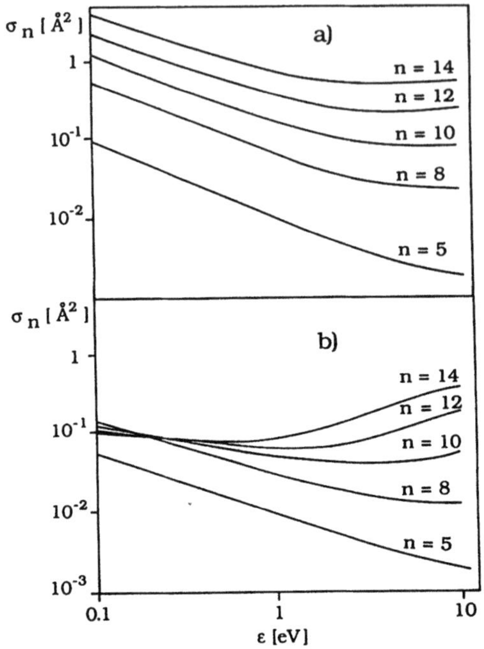

Fig. 1: Coulomb deexcitation cross sections $\sigma_n(\varepsilon)$ calculated without electron screening (a) and with it (b).

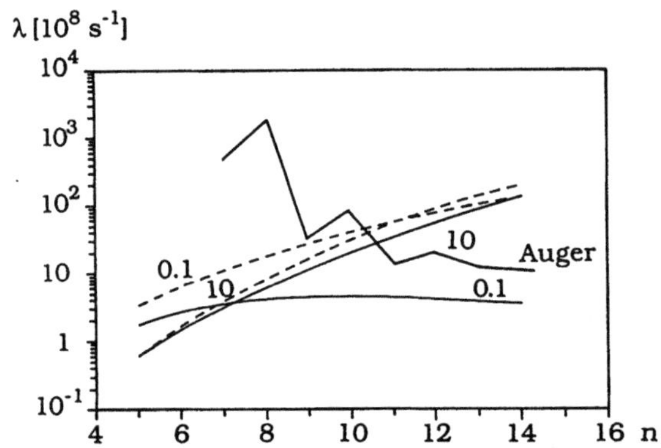

Fig. 2: Coulomb deexcitation rates λ_n calculated for $\varepsilon = 0.1$ and 10 eV with electron screening (solid line) and without it (dashed line) for target density $\phi = 1.27 \cdot 10^{-3}$ of liquid hydrogen density (1 atm).

Sometimes rather simple approximations allow one to trace the calculation scheme and to obtain reasonable results. Let us simplify the Massey parameter (4) as follows [8]: $\delta = Q/\bar{v}_R$, where $Q = \Delta U(\mathrm{Re}R_c) \cdot \mathrm{Im}R_c$.

For large $n \gg 1$ the crossing points of T-series are far from the origin, so one can use the separated atoms approximation, writing then

$$\Delta U(\mathrm{Re}R_c) \cong \Delta U(\infty) = \frac{1}{2(n-1)^2} - \frac{1}{2n^2} = \frac{n - \frac{1}{2}}{n^2(n-1)^2} \qquad (8)$$

As for Im R_c, one can use the approximate expression for the T-series, obtained in [8]:

$$Rc \cong 6n'(n' - \chi) + 4n' ; \qquad (n' = n-1) \tag{9}$$

where

$$\chi = n_1 + \frac{m + 1}{2} = \frac{n + \Delta}{2} \tag{10}$$

Then

$$Q = \frac{4}{n(n-1)} \tag{11}$$

Averaging the radial velocity over the initial and final terms (at $R = ReR_c$) one has

$$\bar{v}_R = \frac{v}{2}\left(\sqrt{a_1^2 - (\rho/ReR_c)^2} + \sqrt{a_2^2 - (\rho/ReR_c)^2}\right) \tag{12}$$

where $$a_i^2 = 1 - u^{(i)}/\varepsilon \tag{13}$$

As $a_2 \gg a_1$, the maximal impact parameter ρ_{max} is determined by v_1, that is $\rho_{max} = a_1 ReR_c$. For the same reason one can neglect the ρ-dependence in the velocity, in the second term, and write

$$\bar{v}_R = \frac{v}{2}\left(a_1\sqrt{1 - \left(\frac{\rho}{a_1 ReRc}\right)^2} + a_2\right) = \frac{va_2}{2}\left(1 + a\sqrt{1 - \frac{\rho^2}{b}}\right)$$

where $a \equiv a_1/a_2$ and $b = (a_1 ReR_c)^2$. As the values of Massey parameters δ for reaction (1) are rather large, one can transform (3) to

$$P \cong 2\exp(-2\delta)$$

so the cross section will read as follows

$$\sigma = 2\pi \int_0^{\rho_{max}^2} \exp\left(-\frac{2Q}{\bar{v}_R}\right) d\rho^2 = 2\int_0^b \exp\left(-\frac{\xi}{1+a\sqrt{1 - \frac{\rho^2}{b}}}\right) d\rho^2 \tag{14}$$

where $\xi = 4Q/(va_2)$. Introducing $x = 1/(1+a\sqrt{1-\rho^2/b})$, one obtains

$$\sigma = \frac{4\pi b}{a^2}\int_v^1 \frac{1-x}{x^3}\exp(-\xi x)dx = \frac{4\pi b}{a^2} f(v,\xi) \tag{15}$$

where $b/a^2 = a_2^2(\mathrm{Re}R_c)^2$ and $f(\nu,\xi) \equiv \int\limits_{\nu}^{1} \frac{1-x}{x^3} \exp(-\xi x)dx$

where $\nu = \frac{1}{1+a} = \frac{a_2}{a_1+a_2} < 1$

So, finally

$$\sigma = 4\pi a_2^2 \tilde{R}_c^2 f(\nu,\xi) \tag{16}$$

where $\tilde{R}_c \equiv \mathrm{Re}R_c$ and a_2 is defined in (13). Function $f(\nu,\xi)$ can be expressed through the exponential integral

$$E_1(x) = \int\limits_{x}^{\infty} e^{-t}t^{-1}dt,$$

tabulated in [9], as follows

$$f(\nu,\xi) = \frac{\xi(\xi+2)}{2} \left[E_1(\nu\xi) - E_1(\xi)\right] + \frac{1+\xi}{2}\exp(-\xi) - \left(1+\frac{\xi}{2} - \frac{1}{2\nu}\right)\frac{\exp(-\nu\xi)}{\nu}.$$

As before, one should sum the cross section over all attracting terms, i.e.

$$\sigma = \sum_{\Delta<0} \frac{n-|\Delta|}{n^2} \sigma(n,\Delta) \tag{17}$$

This expression is equivalent to (7), the cross sections $\sigma(n,\Delta)$ being determined by (16). One can see that Δ-dependence of the cross section (16) is determined only by the value $\tilde{R}_c$, because

$$a_2^2 = 1 - \frac{u^{(2)}}{\varepsilon} \cong 1 + \frac{\Delta u(\infty)}{\varepsilon} = 1 + \frac{n - \frac{1}{2}}{\varepsilon n^2(n-1)^2} \tag{18}$$

does not depend on Δ, as well as $f(\nu,\xi)$. At the same time, as follows from (9) and (10),

$$\tilde{R}_c = 3(n-1)(n-2-\Delta)$$

The resulting total cross section is

$$\sigma = 33\pi n^2(n-1)(n-3)\left(1 + \frac{1}{\varepsilon(n-1)^2 n}\right) f(\nu,\xi) \tag{19}$$

Calculating the Coulomb deexcitation cross section for 13 $\rightarrow$ 12 transition and collision energy $\varepsilon = 1$ eV one obtains

$$\sigma = 0.08 Å^2$$

More accurate calculations described above give the same value. This result may be compared with $\sigma = 1.5 Å^2$, obtained in [5]. At the same time as it was pointed out above, the cross sections obtained in [5], are overestimated because of too small ReR_c used.

Conclusion

As follows from our calculations, the rates of the Coulomb deexcitation show a strong n dependence, very similar to that obtained in [5]. Comparing our rates with the results of ref. [6] one can see that the main contribution to the Coulomb deexcitation cross sections comes from radial transitions ($\Delta m = 0$) and not from Coriolis ones with $|\Delta m| = 1$ [6]. The rates obtained are rather high, which means that this process (at least for large n) should be taken into account together with other deexcitation processes (e.g. Auger deexcitation and target molecule dissociation).

References

1. J.B. Kraiman, G. Chen, P.P. Guss, R.T. Siegel, W.F. Vulcan, R.E. Welsch, W.H. Breunlich, M. Carganelli, P. Kammel, J. Marton, J. Zmeskal, F.J. Hartmann, C. Petitjean, A. Zehnder, J.J. Reidy, H.L. Wolverton, Contribution to μCF 89 (Oxford, 11-13 Sept. 1989), RAL-90-022, p. 47.

2. M. Leon, H.A. Bethe, Phys. Rev. 127 (1962) 636.

3. A.P. Bukhvostov, N.P. Popov, Sov. Phys. JETP 55 (1982) 13.

4. V.E. Markushin, Zh. Eksp. Teor. Fiz. 80 (1981) 35.

5. L. Bracci, G. Fiorentini, Nuovo Cim. 43A (1978) 9.

6. L.I. Men'shikov, Muon Cat. Fusion 2 (1988) 173.

7. E.A. Solovyov, Zh. Eksp. Teor. Fiz. 81 (1981) 1681; 90 (1986) 1165; S. Yu. Ovchinnikov, E.A. Soloviov , Zh. Eksp. Teor. 90 (1986) 921; 91 (1986) 477.

8. D.I. Abramov, S.Yu. Ovchinnikov, E.A. Solov'ev Preprint PTI 1310, 1988, (in Russian).

9. Handbook of Mathematical Functions Eds., M. Abramowitz, I.A. Stegun, National Bureau of Standards, Applied Mathematics Series 55, 1964.

Charge Transfer from Muonic Hydrogen to Other Elements

R. Jacot-Guillarmod[1] *and H. Schneuwly*[2]

[1]Department of Physics, University of British Colombia, Vancouver, Canada
[2]Institut de Physique de l'Université, CH-1700 Fribourg, Switzerland

Abstract. A muonic hydrogen atom can, in collisions, transfer its muon to any other element. By measuring the time distribution of muonic x-rays resulting from muons transferred to the element Z from the μp ground state, one determines the lifetime of μp_{1s} atoms in H_2 + Z gas mixtures. Surprising effects are observed which still wait for explanation.

The muonic hydrogen atom is about 200 times smaller than a normal hydrogen atom. In addition, it is a neutral system and can, therefore, easily penetrate through the electron clouds of other atoms almost like a neutron. In the ground state, the binding energy of about 2.6 keV is very large compared to binding energies of outer electron shells. In the vicinity of an atomic nucleus Z, the muonic hydrogen atom is polarized, and the muon can transfer to an atomic orbit *n* around Z, where the binding energy is greater than 2.6 keV. From this orbit *n*, the muon cascades down to the 1s state by emitting promptly muonic x-rays of characteristic energies and intensities, which are used to identify a muon transfer event. Because of the large binding energy in the ground state, the transfer mechanism is not influenced by the chemical bonding of elements to which the muon is transferred. In atoms with Z < 20, the orbit to which the muon is transferred is geometrically far inside the electronic K-shell, such that the muon transfer is analogous to charge transfer from a hydrogen atom to a bare ion.

A negative muon, stopped in a mixture of natural hydrogen with another gas of atomic number Z, forms a muonic hydrogen or a muonic Z atom in an excited state. In the deexcitation process, muonic x-rays are emitted which appear promptly with regard to the stopping muon. After deexcitation of the $(\mu p)^*$ system (in about 10^{-10} s at 10 bar), which occurs essentially by collisions with hydrogen molecules, if the concentration of the admixed gas is small, the μp atom in the ground state can disappear by muon decay with an associated rate λ_0, by formation of a pμp molecule (λ_{pp}), or by transferring the muon to deuterium (λ_d) or to the Z element (λ_{pZ}). The total disappearance rate, λ, of the μp atom in the ground state is then :

$$\lambda = \lambda_0 + \lambda_{pp} + \lambda_d + \lambda_{pZ} . \qquad (1)$$

The transfer of the muon from the $(\mu p)_{1s}$ atom occurs to much lower *n* levels compared to direct capture in Z, and the associated angular momentum distribution is also very different, such that the muonic x-ray intensity pattern, resulting from transferred muons, is characteristic for the charge transfer process [1,2]. These muonic x-rays, emitted after muon transfer, are delayed

 © Springer-Verlag Berlin, Heidelberg 1992

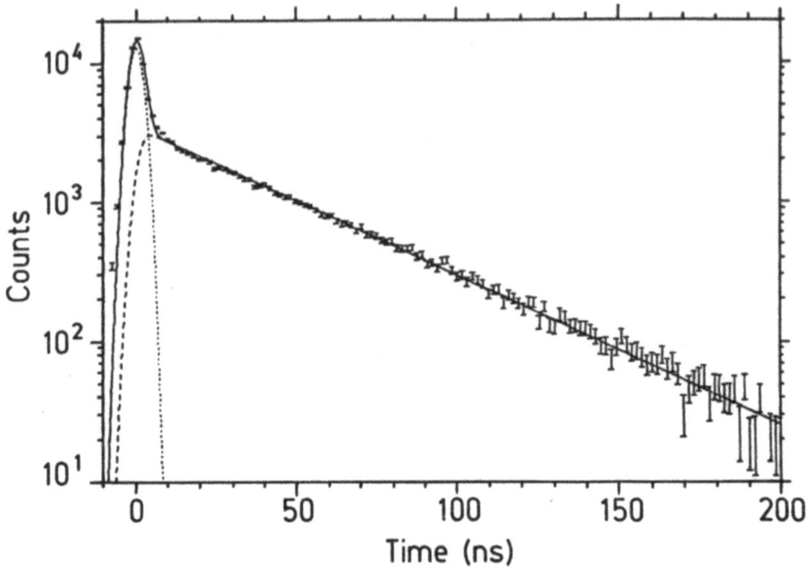

Fig. 1 : Time distribution of the muonic Ar(2p-1s) events measured in a H_2 + 2%Ar mixture at 14.9 bar. Dotted line: fitted function for direct capture. Dashed line: fitted function for transfer from hydrogen. Solid line: total fitted function.

relative to a muon stop in the gas, and their time distribution is expected to be an exponential function with the disappearance rate λ as characteristic time constant. $\tau = 1/\lambda$ is then the lifetime of the μp atom in the ground state.

Already twenty years ago, lifetimes of muonic hydrogen and muonic deuterium atoms have been measured in hydrogen and deuterium gases, where monoatomic gases were admixed [3-5]. Such measurements have recently been repeated in H_2 + Ar mixtures at various pressures and concentrations [1,6]. The measured time distributions of the muonic argon x-rays resulting from transfer could always be fitted using a single exponential function. An example of such a time spectrum is shown in Fig. 1.

In order to compare measurements performed under different experimental conditions, one normalizes the various rates to the atomic density of liquid hydrogen ρ_0. The normalized transfer rate Λ_{pZ} is then :

$$\Lambda_{pZ} = \frac{\rho_0}{N_Z} \lambda_{pZ} \qquad (2)$$

where N_Z is the atomic density of the element Z. The rates λ_0, Λ_{pp} and Λ_d being known, one determines from the measured lifetimes, using eq. (1) and (2), the normalized transfer rate Λ_{pZ}.

Recent measurements of muon transfer to argon, performed under very different conditions of pressure and concentration, yielded the same normalized (or reduced) transfer rate Λ_{pAr} (Table 1). From these results and the measured muonic x-ray intensity patterns in argon, one deduced that the muonic hydrogen atom, from which the muon is transferred, is in the ground state

Table 1 : Comparison of recently measured lifetimes of the muonic hydrogen atom and transfer rates to argon in H_2 + Ar gas mixtures at different pressures and concentrations [1,6].

Pressure [bar]	$C(Ar/H_2)$ $[\cdot 10^{-4}]$	N_{Ar} $[\cdot 10^{17}\ cm^{-3}]$	Lifetime τ $[\cdot 10^{-9}\ s]$	Λ_{pAr} $[\cdot 10^{11}\ s^{-1}]$
100	6	15.0	167	1.42(16)
140	20	68.0	39.1	1.46(5)
9.6	41	9.6	257	1.48(8)
13.3	200	65.0	40.1	1.43(5)

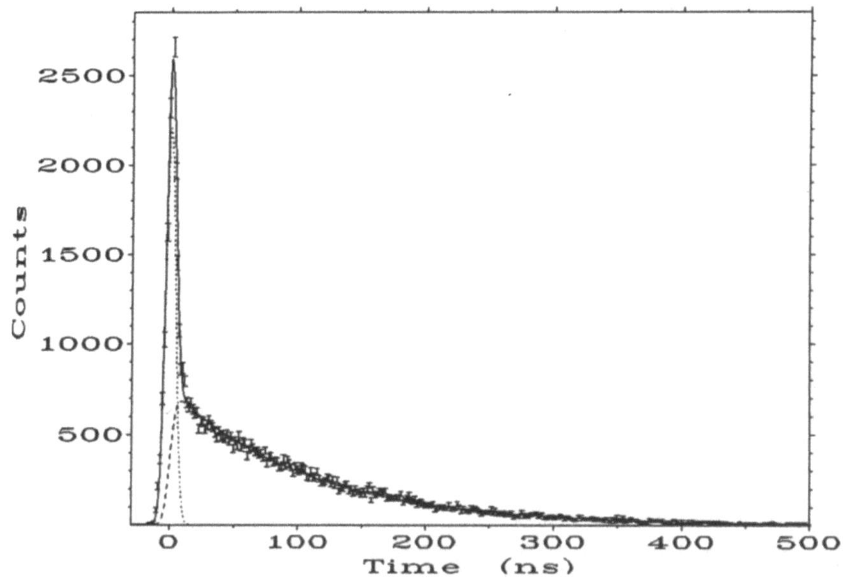

Fig. 2 : Measured and fitted time spectrum of the muonic 2p-1s transition in sulphur of a H_2 + 0.4%SO_2 gas mixture at 15 bar (linear scale).

(singlet) and thermalized less than 20 ns after the muon has entered the target and stopped in the gas.

The lifetime of the muonic hydrogen atom $(\mu p)_{1s}$ in a gas mixture H_2 + SO_2 is expected to be equal to the lifetime in a hypothetical triple gas mixture H_2 + S + O_2, with identical concentrations. Indeed, the muon is transferred in both sulphur and oxygen to levels with binding energies greater than 2.6 keV, and the orbits of these levels lie geometrically inside the electronic K-shell, such that the chemical bond cannot influence the transfer process. From the analysis of the time spectra of the muonic sulphur x-rays (Fig. 2), one determines in a H_2 + 0.4%SO_2 gas mixture at about 15 bar a lifetime of the $(\mu p)_{1s}$

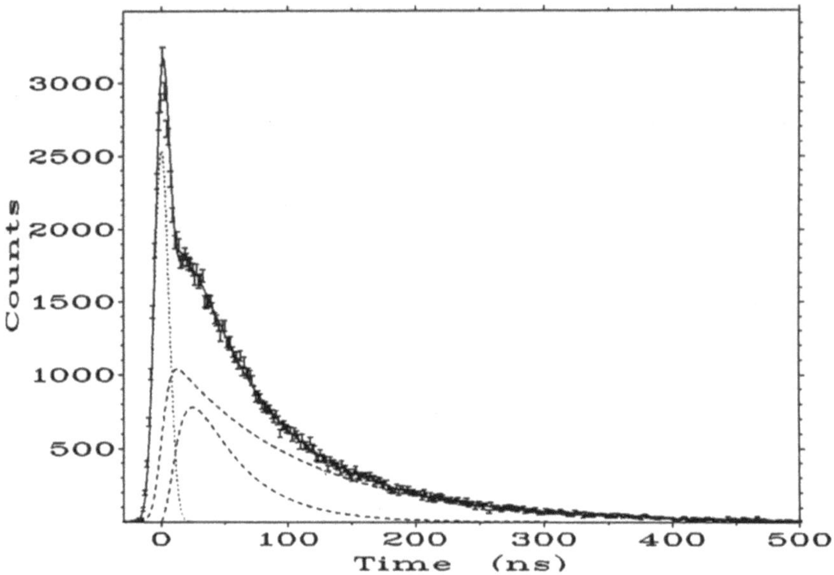

Fig. 3 : Measured and fitted time spectrum of the muonic 2p-1s transition in oxygen of a H_2 + 0.4%SO_2 gas mixture at 15 bar (linear scale).

atom in the ground state of about 110 ns [7], which corresponds to theoretical estimates [8,9]. The reduced transfer rate $\Lambda(SO_2)$, measured in this first mixture, agrees well with the rates measured in two other mixtures, H_2 + 0.2%SO_2 at 10 and at 15 bar [10].

Whereas the time distributions of the muonic sulphur x-rays, appearing delayed with respect to a muon stopping in the gas mixture, have a structure of a single exponential, the time spectra of the corresponding muonic oxygen x-rays exhibit a surprising double-exponential behaviour (Fig. 3). In addition to the prompt peak, these oxygen time spectra show, slightly delayed, another statistically significant "bump" of unknown origin. All four measured x-ray transitions in oxygen, namely O(2-1), O(3-1), O(4-1) and O(5-1), show the same time structures. The fit of these oxygen time spectra gives two "lifetimes" for the muonic hydrogen atom. The longer one agrees with the lifetime determined from the sulphur x-ray time spectra. The shorter one is the same in all four oxygen transitions with a mean value of 41 ns.

The intensity patterns of the delayed muonic sulphur x-rays correspond to what is expected for muons transferred from the ground state of μp atoms [10,11]. The intensity structures of both time components of the muonic oxygen x-rays agree very well with those predicted for transferred muons [10]. The intensity patterns, measured in pure gaseous SO_2, are very different, and the measured x-ray events in sulphur as well as in oxygen appear exclusively as prompt events [12].

These observations lead to speculations about the existence of two kinds of muonic hydrogen atoms in the ground state [13]. Those which transfer their muon to both sulphur and oxygen, have been called "white" muonic hydrogen atoms. The other ones, which transfer surprisingly their muon only to oxygen, have been called "black" μp atoms.

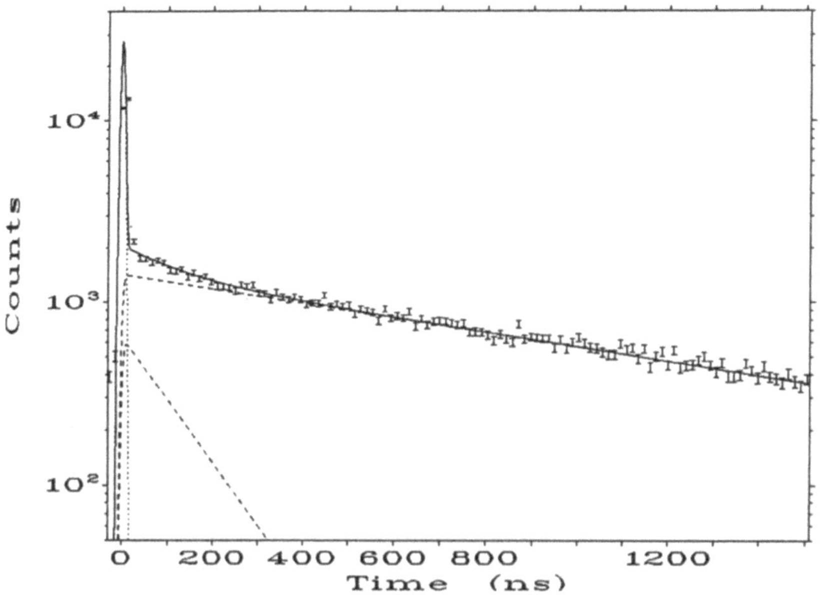

Fig. 4 : Time distribution of muonic neon 2p-1s x-rays events, measured in a binary gas mixture H_2 + 0.7%Ne. The continuous line represents the fit to the data, the broken and dotted lines the contributions of the components.

The surprising results obtained in $H_2 + SO_2$, stimulated new measurements of muon transfer to the monoatomic neon. Such a transfer measurement was already performed twenty three years ago [3] at CERN, using a triple gas mixture H_2 + Ne + Xe. The new measurements [14,15] have been performed in a binary gas mixture H_2 + Ne and in a triple gas mixture H_2 + Ne + Ar.

The time distributions of all the muonic neon x-rays of the Lyman series, measured in the binary mixture H_2 + 0.7%Ne, had the same shape as the Ne(2-1) time spectrum shown in Fig. 4. The main delayed component corresponds to a lifetime of about 1.1 μs. On the basis of the muon transfer rate determined by Alberigi Quaranta et al. [3], this lifetime should have been smaller by almost one order of magnitude. However, the measured time spectra have, in addition, a second time component. This situation is comparable to the one observed in the time spectra of the muonic oxygen x-rays in the $H_2 + SO_2$ gas mixtures. Again, the x-ray intensity patterns of both time components agree with the prediction for muon transfer from hydrogen in the ground state. The fitted decay times in the muonic Ne(n-1) time spectra are given in Table 2.

From these mean decay constants, one determines the reduced transfer rates Λ_{pNe}. In natural hydrogen, part of the muons are transferred to deuterium with a rate $\Lambda_d = 1.7 \cdot 10^{10}\ s^{-1}$. Part of these muonic deuterium atoms transfer their muon to neon with a rate $\Lambda_{dNe} = 1.42 \cdot 10^{11}\ s^{-1}$. In the determination of the reduced transfer rate from protium to neon, Λ_{pNe}, measured in natural hydrogen, one has to correct for the transfer from deuterium [1].

The normalized transfer rate Λ_{pNe}, deduced from the main component of the neon time spectra is $\Lambda_{pNe} = 0.062(4) \cdot 10^{11}\ s^{-1}$, which is almost three times

Table 2 : Fitted decay times in the muonic neon Ne(n-1) time spectra measured in the H_2 + 0.7%Ne gas mixture

Transitions	τ_1 [ns]	τ_2 [ns]
Ne(2-1)	127 (24)	1087 (28)
Ne(3-1)	144 (60)	1111 (100)
Ne(4-1) to Ne (6-1)	156 (53)	1089 (50)
mean values	133 (21)	1089 (24)

Table 3 : Comparison of normalized muon transfer rates Λ_{pNe} with indication of the methods employed : a) intensity method; b) muonic x-ray time structure method (for explanations, see text and ref. [14,15]).

Ref.	Method	Mixture	Λ_{pNe} [$\cdot 10^{11}$ s^{-1}]
[3]	a	H_2 + Ne + Xe	1.16 (28)
[14] + [15]	b	H_2 + Ne	1.15 (20)
[14] + [15]	b	H_2 + Ne	0.062(4)
[14] + [15]	b	H_2 + Ne + Ar	0.056(48)
[14] + [15]	a	H_2 + Ne + Ar	0.062(5)

smaller than the transfer rate to deuterium. From the second time component, which has an intensity of about 4-5% of the main one, a normalized transfer rate of $\Lambda_{pNe} = 1.15(20)\cdot 10^{11}$ s^{-1} is obtained. This last one is, surprisingly, exactly equal to the one measured by Alberigi Quaranta et al. [3], namely $\Lambda_{pNe} = 1.16(28)\cdot 10^{11}$ s^{-1}, twenty years ago.

Alberigi Quaranta et al. employed the intensity method in a triple gas mixture to determine the transfer rate to neon. The transfer ratio $\Lambda_{pNe}/\Lambda_{pXe}$ is equal to the ratio (per atom) of the muons transferred to neon to those transferred to xenon, A_t(Ne/Xe). This ratio is obtained from the analysis of the intensities of the muonic Lyman series. If one of the transfer rates is known, the other can be calculated.

The second transfer measurement to neon, recently performed, was in a triple gas mixture with argon instead of xenon [14,15]. The analysis of the data, using the time structure method and the intensity method, give muon transfer rates, Λ_{pNe}, in agreement with the one obtained from the main time component in the binary H_2 + Ne mixture (Table 3). No second component is observed.

The double decay time, observed in the muonic neon x-rays in the binary H_2 + Ne mixture, is particularly puzzling, although such double exponentials have been measured up to now in all investigated H_2 + SO_2 mixtures. Indeed,

Table 4 : Comparisons of measured reduced muon transfer rates Λ_{pAr} from μp atoms in the ground state to argon.

Year	Ref.	pressure [bar]	$C(Ar/H_2)$ $[\cdot 10^{-4}]$	N(Ar) $[\cdot 10^{17} cm^{-3}]$	Λ_{pAr} $[\cdot 10^{11}\ s^{-1}]$	method
1965	[16]	45	1	1.1	1.20(19)	e
1969	[5]	10	12	2.9	1.46(14)	e
1988	[6]	100	6	15.0	1.42(16)	x
1988	[6]	140	20	68.0	1.46(5)	x
1988	[1]	9.6	41	9.61	1.47(8)	x
1988	[1]	13.3	200	65.0	1.41(5)	x
1967	[3]	26	33	21.0	3.48(60)	x
1971	[17]	900	20	440.0	3.28(78)	x
1982	[18]	2	10	0.5	3.61(220)	x
1982	[18]	3	20	1.5	3.55(112)	x
1982	[18]	3	10	0.7	3.77(136)	x
1982	[18]	4	10	1.0	3.81(167)	x
1980	[19]	600	2	29.0	9.8(15)	x

from the shorter, 4%-component, one deduces a normalized transfer rate Λ_{pNe} which is exactly equal to the normalized transfer rate, observed twenty years earlier at CERN. To the main component corresponds a twenty times smaller transfer rate Λ_{pNe}, which is well reproduced in the triple gas mixture H_2 + Ne + Ar whether one employs for the analysis the method of the time structure or that of the intensity ratio.

As early as 1969,Placci et al. [5] noticed that their normalized transfer rates (they normalized their transfer rates to the molecular density of liquid hydrogen) to xenon disagreed with the corresponding rate measured by Basiladze et al. [16]. However, they did not make any comment on the discrepancy between their new value for the transfer rate to argon and their preceding one [3], where the difference was more than a factor of two. Two different methods had been used to determine these transfer rates, and it was first conjectured that the difference might be related to the methods employed [18]. However, since then, many other muon transfer rates have been measured and the agreement between them did not improve (Table 4).

The present situation is such than one can distinguish three classes of muon transfer rates from the μp ground state to argon [13]. The ratio of the mean values of the successive classes is a factor of about two and a half. No correlation between these classes and particular experimental conditions could up to now be discovered, such that we are confronted with the same problem as in the transfer rates to neon.

Up to now, no other hypothesis than the speculation about the existence of different μp atoms in the ground state [7,13] explains the observed time distributions and the intensity patterns of the muonic x-rays measured in H_2 + Z gas mixtures.

References

1. R. Jacot-Guillarmod, F. Bienz, M. Boschung, C. Piller, L.A. Schaller, L. Schellenberg, H. Schneuwly, W. Reichart and G. Torelli, Phys. Rev. **A38**, 6151 (1988).
2. R. Jacot-Guillarmod, F. Bienz, M. Boschung, C. Piller, L.A. Schaller, L. Schellenberg, H. Schneuwly and D. Siradovic, Phys. Rev. **A37**, 3795 (1988).
3. A. Alberigi Quaranta, A. Bertin, G. Matone, F. Palmonari, A. Placci, P. Dalpiaz, G. Torelli and E. Zavattini, Nuovo Cimento **B47**, 92 (1967).
4. A. Placci, E. Zavattini, A. Bertin and A. Vitale, Nuovo Cimento **A52**, 1274 (1967).
5. A. Placci, E. Zavattini, A. Bertin and A. Vitale, Nuovo Cimento **A64**, 1053 (1969).
6. F. Bienz, P. Bergem, M. Boschung, R. Jacot-Guillarmod, G. Piller, W. Reichart, L.A. Schaller, L. Schellenberg, H. Schneuwly and G. Torelli, J. Phys. B: At. Mol. Opt. Phys. **B21**, 2725 (1988)
7. H. Schneuwly, R. Jacot-Guillarmod, F. Mulhauser, P. Oberson, C. Piller and L. Schellenberg, Phys. Lett. **A132**, 335 (1988)
8. S.S. Gerstein, Zh. Eksp. Teor. Fiz. **43**, 706 (1962) [Sov. Phys.-JETP **16**, 501 (1963)]
9. G. Fiorentini and G. Torelli, Nuovo Cimento **A36**, 317 (1976)
10. F. Mulhauser, R. Jacot-Guillarmod, C. Piller, L.A. Schaller, L. Schellenberg and H. Schneuwly, *in* E.M. Cascade and Chemistry of Exotic Atoms, edited by L.M. Simons, D. Horváth and G. Torelli, Plenum Press(1990)
11. G. Holzwarth and H.-J. Pfeiffer, Z. Physik **A272**, 311 (1975)
12. F. Mulhauser, H. Schneuwly, R. Jacot-Guillarmod, C. Piller, L.A. Schaller and L. Schellenberg, Muon Catalyzed Fusion **4**, 365 (1991)
13. H. Schneuwly, Muon Catalyzed Fusion **4**, 87 (1989)
14. R. Jacot-Guillarmod, F. Mulhauser, C. Piller, L.A. Schaller, L. Schellenberg and H. Schneuwly, *in* E.M. Cascade and Chemistry of Exotic Atoms, edited by L.M. Simons, D. Horváth and G. Torelli, Plenum Press (1990)
15. R. Jacot-Guillarmod, F. Mulhauser, C. Piller and H. Schneuwly, Phys. Rev. Lett. **65**, 709 (1990)
16. S.G. Basiladze, P.F. Ermolov and K.O. Oganesyan, Zh. Eksp. Teor. Fiz. **49**, 1042 (1965) [Sov. Phys.-JETP **22**, 725 (1966)]
17. G. Backenstoss, H. Daniel, K. Jentzsch, H. Koch, H.P. Povel, F. Schmeissner, K. Springer and R.L. Stearns, Phys. Lett. **B36**, 422 (1971), cited in Ref. 18
18. E. Iacopini, G. Carboni, G. Torelli and V. Trobbiani, Nuovo Cimento **A67**, 201 (1982)
19. H. Daniel, H.-J. Pfeiffer, P. Stoeckel, T. von Egidy and H.P. Povel, Nucl. Phys. **A254**, 409 (1980)

Capture and Transfer of Pions in Hydrogenous Materials

D.S. Armstrong

Virginia Polytechnic Institute and State University, Blacksburg, VA 24061, USA
and
TRIUMF, 4004 Wesbrook Mall, Vancouver, B.C., Canada V6T 2A3

Pionic hydrogen is a short-lived exotic hydrogen isotope in which a negative pion replaces the atomic electron. The formation and subsequent interactions of pionic hydrogen are discussed, with emphasis on the process of pion transfer. Recent results using the pion charge-exchange reaction (π^-, π^0) obtained at TRIUMF are reviewed.

1. Introduction

The study of the processes involved when a negative pion stops in matter has been a topic of some interest for the past 30 years. Detecting and understanding these processes is a truly interdisciplinary effort, involving features of atomic, molecular and nuclear physics as well as elements of organic and physical chemistry. Of particular interest is the situation when the pion stops in a material containing hydrogen. In this case a pionic hydrogen atom can be formed, which is then able to wander throughout the material and can end up transferring the pion to a neighbouring atom. A series of experiments have been performed in the last few years at TRIUMF to elucidate the mechanisms governing the formation and transfer of pionic hydrogen. The classic review of the field is that of Ponomarev [1]; more recent work has been reviewed by Schneuwly [2] and Horváth [3].

Negatively charged particles such as pions, muons, kaons etc. are believed to be very similar in the manner in which they slow down and stop in matter. At sufficiently low kinetic energies, they will be captured into molecular or atomic orbitals, typically of high excitation. Being different than the atomic electrons, they do not feel the Pauli exclusion principle, and will cascade down to lower orbitals by radiative transitions or via the emission of Auger electrons. It is only in these lower orbitals where the difference between the different particles manifests itself. For strongly interacting particles like the π^- the interaction is no longer dominantly electromagnetic, and the probability of absorption by the nucleus becomes appreciable. Thus, while it is not usually possible to directly apply knowledge gained from pionic atoms to, for example, muonic atoms, the similarities are enough to prove useful. In fact, as discussed below, the study of pion transfer sheds light on aspects of muonic transfer that are not otherwise readily accessible, in particular, transfer from excited states of the exotic atom. This is of considerable potential interest in the realm of muon-catalyzed fusion. Generally speaking, data on pion capture and transfer provide a valuable "testing ground" for models of similar muonic processes; a successful model should be able to account for the behaviour of pions, muons, kaons etc.

The typical exotic atom is an order of magnitude smaller (10^{-11}-10^{-10} cm) than the corresponding normal atom, due to the reduced mass effect. One might therefore expect that molecular and chemical-bond effects could largely be ignored in studying the atomic and nuclear capture process, since the mesonic orbitals are well within those of the valence electrons. In fact, it turns out that chemical effects are rather important in pionic and muonic systems. This has led to the hope of using these exotic atoms as a probe of the electronic structure of molecules. In particular, pionic hydrogen appears to be a useful tool for studying the bonding of hydrogen in a molecule, and perhaps also the nature of the hydrogen bond between molecules.

 © Springer-Verlag Berlin, Heidelberg 1992

Studies of atomic and molecular effects for stopped pions and muons are of interest for other reasons as well. Transfer effects need to be understood and accounted for in various classes of nuclear and particle physics experiments, especially those using hydrogen targets (e.g. fundamental weak interaction measurements such as muon capture in hydrogen, where muonic transfer processes are a major concern [4]). Finally, the ultimate fate of a pion stopping in a chemical system is important for bio-medical purposes, especially for application to pion radiotherapy which is being developed at TRIUMF and PSI.

Of the pionic atoms, pionic hydrogen is unique in two ways. Like other exotic hydrogen atoms (μ^-p, K^-p), it is electrically neutral. This, along with its small size, enables it to penetrate freely the electron shells of neighbouring atoms. This allows the π^-p to de-excite by giving energy to electrons of another atom; this is known as the external Auger effect. It also allows the pions to be transferred to atoms of higher nuclear charge Z, *i.e.* a reaction of the form

$$\pi^-p + Z \rightarrow \pi^-Z + p \tag{1}$$

can take place. This reaction is irreversible if $Z > 1$, due to the increased binding energy of the π^- in the π^-Z atom. The similarity of reaction (1) to the muon transfer reaction

$$\mu^-p + Z \rightarrow \mu^-Z + p \tag{2}$$

is obvious. There is, however, an important difference. Unlike the μ^- which reaches the ground state before undergoing nuclear absorption, the π^- can be absorbed from nS orbits where $n \leq 5$. The fact that the π^-p system is so ephemeral relative to the comparatively long-lived μ^-p atom means that transfer becomes more difficult to observe. For similar transfer rates, the fraction of π^-'s that will transfer is much smaller than for μ^-'s, simply because the competing π^- nuclear absorption rate is so large. The capture and transfer processes are all typically much more rapid than the pion decay rate (the free pion lifetime is 26×10^{-9} s), and so pion decay can be ignored. (There may be exceptions to this; a recent measurement at KEK found that during the atomic cascade process, 2% of negative kaons stopped in liquid helium were "trapped" in metastable states with an overall lifetime of about 40 ns [5]. It would be very interesting to verify if a similar effect exists for π^-'s).

The second unique aspect of pionic hydrogen is the nature of its demise. If the π^- is not transferred to another atom, it will eventually be absorbed by the proton. This nuclear capture proceeds through one of two branches [6]:

$$\begin{aligned} \pi^- + p &\rightarrow \gamma + n \qquad (39.3\%) \\ \pi^- + p &\rightarrow \pi^0 + n \qquad (60.7\%)\,. \end{aligned} \tag{3}$$

The observation of a π^0 from the charge-exchange reaction is a very clean signature of nuclear capture on hydrogen. For other nuclei the production of π^0's from π^-'s captured at rest is either kinematically forbidden, or strongly suppressed ($\leq 10^{-5}$ branching ratio [7]). Pion absorption for these nuclei leads instead to emission of energetic neutrons, charged particles and occasionally high-energy ($\sim$ 100 MeV) γ rays. The two exceptions to the rule are ^{3}He and the deuteron, with branching ratios for the charge-exchange reaction at rest of $(12.8 \pm 1.2)\%$ [9,10] and $(1.45 \pm 0.19) \times 10^{-4}$ [8] respectively. The potential experimental problem of background from in-flight (π^-, π^0) reactions can be eliminated by using an incident pion beam of low energy (< 25 MeV). This experimental advantage of being able to select π^-'s that have been absorbed by hydrogen is a very useful tool, and is another reason that pionic hydrogen is an attractive system to study.

Molecular and chemical bond effects in pion capture can be quite large. As an example, the capture probability on hydrogen in hydrazine (N_2H_4) is 30 times less than in a gaseous mixture of $N_2 + 2H_2$ [11]. These effects are usually understood in terms of the model of large mesic molecules, introduced by Ponomarev [1]. In this picture, the π^- (or μ^-) initially captures into a molecular orbital by ejecting an Auger electron. These molecular orbitals are similar to those of the ejected electron, hence the sensitivity to electronic wavefunctions and chemical bonds. Imanishi *et al.* [12] have recently extended this model to account for Auger effects in the subsequent de-excitation.

Despite a considerable body of data on pion capture in various hydrogenous materials and mixtures, there are still a considerable number of unanswered questions. In condensed systems, there is even controversy about the basic mechanism of pion transfer. Part of the difficulty lies in the fact that the available models typically have a sufficient number of undetermined parameters to fit a large range of data. Also contributing to the problem is the fact that there are relatively few quantities that are experimentally accessible; typically these are pionic x-ray yields and the charge-exchange probability. One is therefore obliged to select carefully systems that can shed light on particular aspects of the problem. The high-quality low-momentum π^- beams available at TRIUMF, along with an excellent detection system for π^0's produced at rest, have enabled us to improve on some of the older measurements and to begin to answer some of the outstanding questions in the field.

2. Gaseous Mixtures

Perhaps the simplest possible system to consider is a gaseous mixture of hydrogen with other elements. The pion transfer reaction has been extensively studied, largely by the Dubna group, for mixtures of hydrogen with noble gases (He, Ne, Ar, Kr, Xe) [13,14], and simple molecules such as D_2, CO, N_2, and CO_2 [14,15,16]. We have chosen to study the $H_2 + D_2$ system [17], which is particularly interesting because of the relation to similar processes in muon-catalyzed fusion.

The pion transfer is studied by measuring the variation of the π^0 yield as the relative concentration of H_2 and D_2 is varied. Figure 1 shows the experimental setup, which is essentially the same for all our measurements. The production of π^0's is signalled by the decay $\pi^0 \rightarrow \gamma\gamma$ (98.8% branching ratio). The two γ's are produced back-to-back and with equal energies in the rest frame of the π^0; however the π^0 has a small kinetic energy of 2.9 MeV. Therefore the photons are Doppler-shifted in the lab frame and so have a small distribution about 180° and energies of between 55 MeV and 83 MeV. The γ's are detected by two large NaI(Tl) scintillators (TINA ϕ 46 cm $\times$51 cm; MINA ϕ 36 cm $\times$36 cm) at 180° to each other. Plastic scintillators (S_5,S_6) in front of TINA and MINA enable us to reject signals due to charged particles.

The incident π^- beam was at a momentum of 78 MeV/c (kinetic energy of 20 MeV) with a typical flux of 10^5 s^{-1}. The arrival and stopping of an incident π^- is signalled by a series of thin plastic scintillators S_1-S_4 (see Fig. 1). S_3 was made of a very thin (0.5 mm) deuterated plastic $(CD)_n$ to reduce the number π^0's produced from pions stopping in the scintillator.

The essentially continuous time structure of the TRIUMF beam (99% duty factor) is an advantage here. The time between the arrival of the π^- in the target, and its eventual nuclear

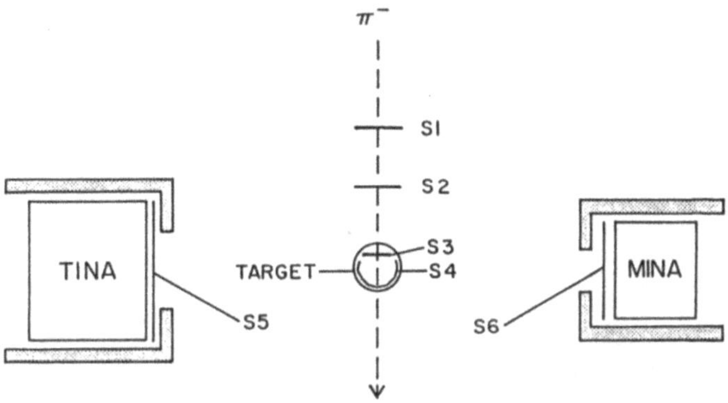

Figure 1: Experimental layout. TINA and MINA are the two large NaI(Tl) gamma-ray detectors. S_1-S_6 are plastic scintillators.

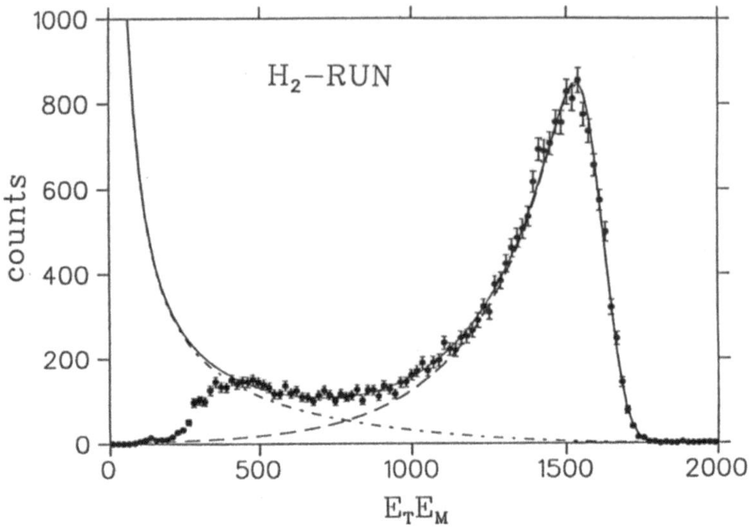

Figure 2: Spectrum of the product of the γ-ray energies $E_T E_M$ from a hydrogen target. The dashed line is a fit to the π^0 peak, and the dot-dashed line is the fitted background.

capture is $< 10^{-9}$ s; the π^0 decays practically instantaneously (10^{-16} s), and so the two γ's must be detected in time coincidence with the arrival of the π^-. This requirement allows the suppression of various backgrounds, notably that from energetic neutrons.

The kinematics of the π^0 decay require that

$$E_T E_M = \frac{M_{\pi^0}^2}{4\sin^2(\psi/2)} , \tag{4}$$

where E_T and E_M are the γ energies deposited in the two detectors, M_{π^0} is the π^0 mass and ψ is the angle between the two γ rays. Therefore the π^0 events will appear as a peak in a histogram of $E_T E_M$, as illustrated in Fig. 2. The width of the peak is primarily due to the energy resolutions of the NaI detectors which were each $\sim 7\%$ at 100 MeV. The number of π^0 events is easily extracted by fitting to a function representing the peak superimposed on a smooth background.

Figure 3 shows the relative π^0 production as the concentration of deuterium (C_d) is increased. If there were no transfer, then the data should fall along the straight line indicated, with the reasonable assumption that the molecular stopping probability is the same for H_2 and D_2. Clearly, significant transfer from the hydrogen to the deuterium does take place. For example, in a 50:50 mixture fewer than half the pions are absorbed on the hydrogen.

To further interpret the data, we adopt a phenomenological approach, inspired by the work of Petrukhin and Prokoshkin [15]. We assume that nuclear capture occurs only in collision, due to Stark mixing. In this approach, the probability of pion capture on hydrogen is viewed as being dependent on a competition between collisionally-induced capture and collisionally-induced transfer. In the simplest case, where the inverse transfer reaction $\pi^- d + p \to \pi^- p + d$ is negligible, one has that the capture probability on hydrogen can be written

$$W_H = \frac{C_p}{C_p + C_d} \frac{\beta_{pp} C_p + \beta_{pd} C_d}{\beta_{pp} C_p + \beta_{pd} C_d + \lambda_{pd} C_d} . \tag{5}$$

Here we have defined β_{pp} as the probability of nuclear capture on the proton in a $\pi^- p + p$ collision, β_{pd} as the probability of nuclear capture on the proton in a $\pi^- p + d$ collision, and

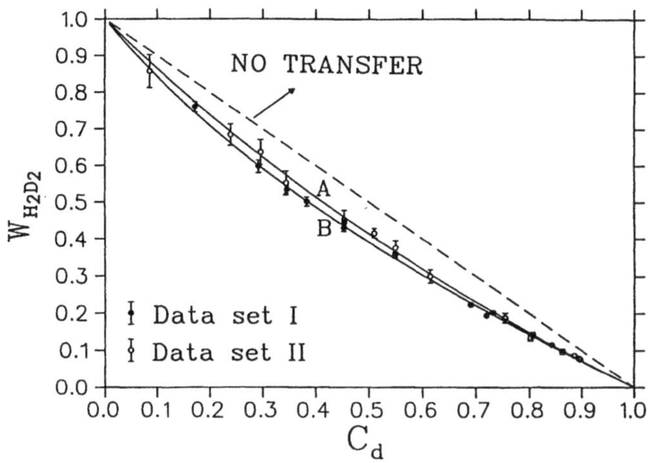

Figure 3: Relative π^0 production versus deuterium concentration for H_2+ D_2 mixtures. The dashed line is the expectation if transfer did not occur. The curves marked A and B are the fits to Eq. (6) for the two data sets.

λ_{pd} as the probability of transfer from p to d in a $\pi^- p + d$ collision. C_p and C_d are the relative concentrations of hydrogen and deuterium respectively. The model can be expressed in terms of only two parameters $\kappa = \beta_{pd}/\beta_{pp}$ and $\Lambda = \lambda_{pd}/\beta_{pp}$. Defining the concentration ratio $C = C_d/C_p$, one has

$$W_H = \frac{1 + \kappa C}{(1 + C)(1 + \kappa C + \Lambda C)} \ . \tag{6}$$

Similarly, the net transfer probability Q can be written

$$Q = \frac{\beta_{pd} C_d}{\beta_{pp} C_p + \beta_{pd} C_d + \lambda_{pd} C_d} = \frac{\Lambda C}{1 + \kappa C + \Lambda C} \ , \tag{7}$$

from which it is easy to see that the transfer probability is simply related to W_H:

$$W_H = \frac{1 - Q}{1 + C} \ . \tag{8}$$

This transfer probability is plotted as a function of C in Fig. 4. The data were obtained during two separate running periods which were analyzed independently to determine our systematic errors. The results of the two fits are shown, as well as a common fit to all our data. Table 1 displays the parameters extracted from the latter fit, and compares them to those obtained by Petrukhin and Prokoshkin [15]. The agreement is reasonable, although our data point to a somewhat larger asymptotic value for the transfer. Two other measurements for W_H are available [18,19]; both are more precise than our data, but both were only taken at one value of the concentration. Inclusion of these two data ("World fit" in Table 1) changes the extracted values for Λ and κ somewhat. This reflects the fact that these two parameters are fairly highly correlated in the fit. The values for κ/Λ and $Q(C \to \infty)$ are essentially unchanged however. The result $\kappa/\Lambda > 1$ indicates that in $\pi^- p + d$ collisions, the probability for capture is higher than that for transfer. Since $\kappa \sim 1$, the probability of pion capture in hydrogen is similar for $\pi^- p + p$ collisions and $\pi^- p + d$ collisions.

In an ancillary experiment, we have searched for a pressure dependence of the transfer, and found none, in agreement with previous observations [18]. In fact, the asymptotic transfer is in good agreement with the value $Q = (28 \pm 13)\%$ at $C = 0.84$ measured at liquid hydrogen

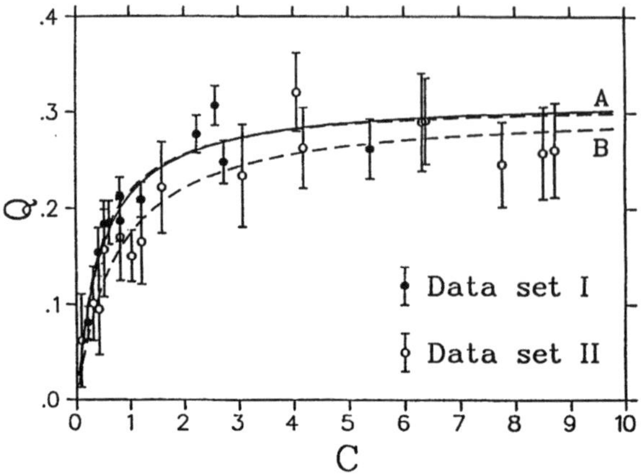

Figure 4: Transfer probability Q versus relative concentration for H_2+ D_2 mixtures. The dashed curves are the fits to Eq. (7) for the two data sets, and the solid curve is the common fit, which yielded the parameters listed in Table 1.

Table 1: The parameters of Eq. (7), extracted from the present data, compared to those obtained by Petrukhin and Prokoshkin [15], and those obtained by a fit to our data as well as the data from Aniol *et al.* [18] and Kravtsov *et al.* [19] ("World fit").

Parameter	Present result	Petrukhin and Prokoshkin [15]	World fit
Λ	0.65 ± 0.07	0.4 ± 0.1	0.45 ± 0.04
κ	1.40 ± 0.22	1.3 ± 0.4	0.93 ± 0.14
κ/Λ	2.2 ± 0.4	3.3 ± 1.3	2.1 ± 0.4
$Q(C \to \infty)$	$(32 \pm 3)\%$	$(23 \pm 4)\%$	$(33 \pm 3)\%$

densities [20]. As emphasized by Horváth [21], this is important because any non-collisional processes would induce a virtual density (pressure) dependence for W_H. The absence of such a dependence is strong evidence that collisional processes dominate the capture and transfer as assumed in our analysis.

3. Alcohols

Since the nuclear capture rate of a π^- on a hydrogen atom is the net result of atomic capture followed by any transfer processes (if they occur), it is often difficult to disentangle the two effects. The analysis of data on pion charge-exchange probabilities to extract transfer rates, often depends on assumptions about the initial pion capture on the different atoms within the molecule. In a complex molecule, the pion charge-exchange probability is an average over the chemically different hydrogen atoms in the molecule. Therefore, systematic studies of charge exchange in closely related chemical species could prove desirable. The hope is to determine an empirical scale of *relative* capture rates on atoms in specific chemical environments. Such a scale could then be useful in isolating effects due to transfer reactions.

With this idea in mind, we proceeded to measure pion charge-exchange probabilities for a series of alcohols. To analyze the results, we postulate that the pion capture on a given

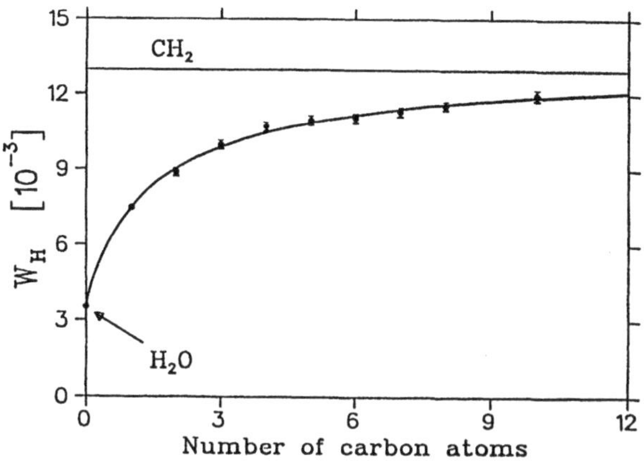

Figure 5: Capture probability in hydrogen, W_H, of alcohols and water. The solid curve is a spline fit through the values calculated using Eq. (9).

hydrogen atom in the molecule depends on its position in the molecule. We define the probability for capture on the hydrogen in the OH group as R_H^O, that for a hydrogen attached to the neighbouring carbon atom as R_H^α (in the chemist's language, this is a hydrogen in the α position with respect to the OH group), and that for all other hydrogens as R_H^β. R_C and R_O are the pion capture rates on carbon and oxygen respectively. The simple model is that W_H is given by

$$W_H = \frac{n_H^\alpha R_H^\alpha + R_H^O + n_H^\beta R_H^\beta}{n_H^\alpha R_H^\alpha + R_H^O + n_H^\beta R_H^\beta + n_C R_C + R_O}, \tag{9}$$

where the n's are the appropriate number of atoms in the particular molecule. It is important to note that the R factors parameterize both the initial selection of the atom by the pion and any transfer. Figure 5 depicts the experimental results for the primary alcohols, plotted as a function of n_C, the number of carbon atoms in the molecule. Also included is the W_H value for water. The data clearly show a monotonic increase in the charge-exchange probabilities with n_C which is easily reproduced by the model. The W_H value for polyethylene is indicated on the figure, and seems to be a plausible asymptotic limit for the fit, as might be expected.

The experimental values are given in Table 2, along with the W_H values calculated using the fit. Also listed are the values for deuterated methanols (to be discussed below) as well as several other alcohols. The fitted capture ratios are given in Table 3. The value for R_H^α/R_C is slightly lower than R_H^β/R_C. This is consistent with the lower electron density expected around hydrogen atoms in the methylene group adjacent to the hydroxyl group, due to the strongly electronegative nature of the oxygen atom. Since the atomic capture process is expected to be dominated by Auger transitions, the capture rate on an atom is expected to depend on the electron density around that atom. Similarly, the value for R_H^O/R_C is much less than either R_H^α/R_C or R_H^β/R_C. This is consistent with the large reduction in the electron density around a hydrogen atom attached directly to the very electronegative oxygen atom. The location of the hydroxyl group does not appear to affect the charge-exchange probabilities significantly (compare the results for Butan-1-ol and Butan-2-ol). The calculation for 2-Methyl-propan-2-ol, which has a very different structure than the primary alcohols, also reproduces the data well. However, the calculation does rather poorly for cyclohexanol. Whether this is related to its ring structure or is a problem with the measurement is not yet clear. We have applied a similar model to the existing data on alkanes [22,23] and found no effects due to ring structures [24].

The results for the deuterated methanols CD_3OH and CH_3OD are particularly interesting. The sum of the capture probabilities in CD_3OH and CH_3OD is considerably less than

Table 2: Pion capture probabilities in hydrogen for alcohols. In the last column the values calculated using Eq. (9) with the parameters given in Table 3 are presented. The fit was determined using the W_H values for water and the primary alcohols only (the first 10 entries).

Molecule	Formula	W_H (10^{-3})	Calculated (10^{-3})
Water	H_2O	3.52 ± 0.07	3.52 ± 0.05
Methanol	CH_3OH	7.47 ± 0.09	7.47 ± 0.10
Ethanol	CH_3CH_2OH	8.88 ± 0.16	9.04 ± 0.08
Propan-1-ol	$CH_3(CH_2)_2OH$	9.98 ± 0.17	9.90 ± 0.07
Butan-1-ol	$CH_3(CH_2)_3OH$	10.70 ± 0.17	10.47 ± 0.08
Pentan-1-ol	$CH_3(CH_2)_4OH$	10.95 ± 0.18	10.87 ± 0.08
Hexan-1-ol	$CH_3(CH_2)_5OH$	11.03 ± 0.18	11.17 ± 0.08
Heptan-1-ol	$CH_3(CH_2)_6OH$	11.29 ± 0.19	11.41 ± 0.08
Octan-1-ol	$CH_3(CH_2)_7OH$	11.54 ± 0.17	11.59 ± 0.08
Decan-1-ol	$CH_3(CH_2)_9OH$	11.96 ± 0.22	11.87 ± 0.08
Deuterated Methanol	CD_3OH	0.845 ± 0.031	1.17 ± 0.02
Deuterated Methanol	CH_3OD	5.55 ± 0.11	6.31 ± 0.10
Butan-2-ol	$CH_3CH_2CH(OH)CH_3$	10.85 ± 0.17	10.54 ± 0.08
2-Methyl-propan-2-ol	$(CH_3)_3COH$	10.66 ± 0.18	10.61 ± 0.09
Cyclohexanol	$(CH_2)_5CHOH$	8.23 ± 0.18	9.57 ± 0.07

Table 3: Best fit capture ratios determined from the primary alcohols.

Capture Ratio	Fitted value
R_H^β/R_C	$(6.72 \pm 0.05) \times 10^{-3}$
R_H^α/R_C	$(6.28 \pm 0.08) \times 10^{-3}$
R_H^O/R_C	$(3.48 \pm 0.07) \times 10^{-3}$
R_O/R_C	1.97 ± 0.02

the probability for CH_3OH. This is also seen by comparing the individual W_H values of the deuterated methanols with the predictions of our model (c.f. Table 2). This is clearly evidence of transfer. However, we can correct for the effect of transfer using the $\pi^- p + d \rightarrow \pi^- d + p$ transfer parameters determined from our H_2/D_2 measurements, or perhaps more applicably, the H_2O/D_2O data discussed below. After this correction the W_H values are 1.33×10^{-3} and 6.24×10^{-3} for CD_3OH and CH_3OD respectively, in quite reasonable agreement with the predictions of our model.

Finally, we have attempted to apply the model parameters deduced from the alcohols to existing data on a very different system, a series of sugar isomers. Earlier work at TRIUMF measured charge-exchange probabilities W_H for several hexose and pentose isomers of sugars, and found evidence of an inverse correlation between W_H and the melting point [25,26]. It was suggested that this correlation was indicative of hydrogen bonding causing a higher melting point and reducing the electron density around the hydrogen atoms and hence reducing the pion capture probability. The present model applied to the sugar data overestimates the W_H values by about a factor of two. This could be considered as additional evidence of the effect of hydrogen bonding on pion capture.

4. Dibromoalkanes

Charge-exchange probabilities were also measured for the first five primary dibromoalkanes (chemicals of the form $Br(CH_2)_nBr$). The results are shown in Fig. 6; the curve is simply a guide to the eye. The W_H values increase dramatically as the chain length increases. This again shows the important effect of a very electronegative group (the Br atoms in this case) on the probability of the π^- capturing on a hydrogen atom. As the chain lengthens, less of the molecule is affected by the distortions in the electron density due to the Br atoms on the ends. Unlike the alcohols series, the curve shows no signs of flattening out, probably indicating that the Br atoms have considerable influence throughout the molecule, even in the case of dibromopentane ($Br(CH_2)_5Br$). Notice that the W_H value for dibromopentane is still far below the value for $(CH_2)_n$ (12.9×10^{-3}). Unfortunately, it appears that there is not sufficient data with only 5 species investigated to apply a model similar to the one used for the alcohols (Eq. (9)) and get meaningful results.

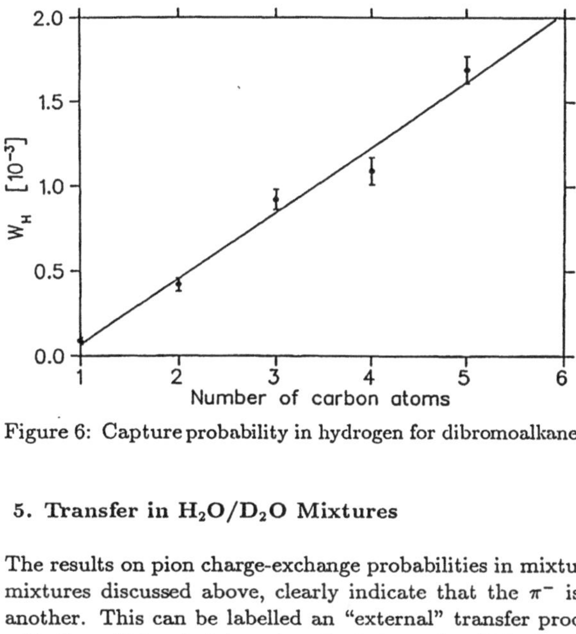

Figure 6: Capture probability in hydrogen for dibromoalkanes. The line is a guide to the eye only.

5. Transfer in H_2O/D_2O Mixtures

The results on pion charge-exchange probabilities in mixtures of gases, for example the H_2/D_2 mixtures discussed above, clearly indicate that the π^- is transferred from one molecule to another. This can be labelled an "external" transfer process. The data are quite consistent with the collisional picture described above, in which the small, neutral π^-p atom breaks away from the molecule and is transferred in a subsequent collision, *i.e.* $\pi^- p + Z \rightarrow \pi^- Z + p$. In more complicated molecules than H_2, the collisional transfer could conceivably also proceed via an "internal" route, that is, the π^-p could move to other atoms in the same molecule.

Such an alternative mechanism for pion transfer in complex molecules has been proposed by D.F. Jackson and co-workers [27,28,29]. In this model, transfer occurs only to the neighbouring atom to which the hydrogen is bonded. The transfer occurs by a barrier penetration method : the π^- tunnels along the Z-H bond. This mechanism is clearly quite different in nature than collisional transfer; it is purely an "internal" transfer process, and so it cannot explain the data on pion and muon transfer in gas mixtures. However, a wide range of both pion charge-exchange and pionic x-ray data on hydrogenous molecules can be fit with this approach [27,30].

Clearly it is important to separate the two proposed mechanisms for pion transfer in condensed systems. An obvious question is whether there is any evidence of a need to invoke an "external" transfer process in condensed systems? To answer this, one approach is to adopt a method similar to the measurements on gas mixtures. The idea is to investigate the sup-

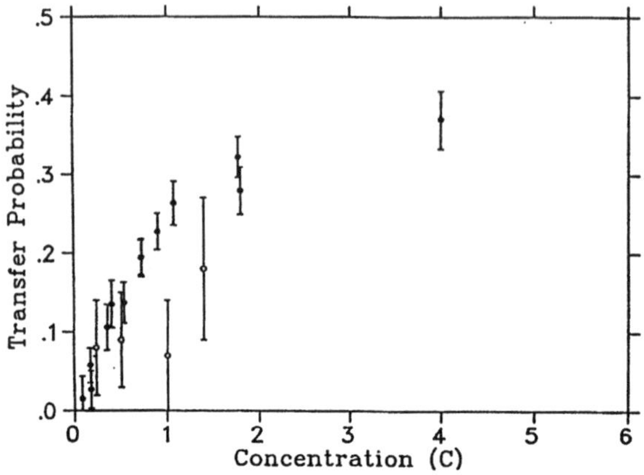

Figure 7: Transfer probability Q as a function of relative concentration for mixtures of liquid H_2O+D_2O. The open circles are the data of Petrukhin *et al.* [31] and the closed circles are the present data.

pression of W_H as one varies the relative concentration of two liquids, only one of which is hydrogenous.

An appealing possibility, obviously analogous to the H_2/D_2 gas mixtures, is to investigate mixtures of liquid H_2O and D_2O. Under the assumption that the initial π^- stopping probability is the same for the D_2O molecule as for the H_2O, the observation of a non-linearity in the concentration dependence of W_H would be evidence for "external" transfer. This measurement was first attempted by Petrukhin *et al.* [31], and while the data were sketchy, they indicated a non-linearity similar to that seen in the gas mixtures. We have repeated the experiment with higher precision and over a broader range of concentrations, and confirm the effect [32]. Figure 7 illustrates our results and compares them with the previous data. The behaviour is similar to that observed in the gas mixtures, but not identical - in particular, the asymptotic value of the transfer is $52 \pm 10\%$ compared to $32 \pm 3\%$ for the gases, using the same phenomenological model to fit the data. Unfortunately, the result is not unambiguous. At high concentrations of D_2O, the hydrogen would be found primarily in a HOD molecule. Therefore the concentration dependence could be interpreted in terms of molecular effects, such as the deuteron preferentially carrying away the π^- in the breakup of the π^-HOD molecule. The lingering possibility also remains that the initial pion stopping probability on the H_2O molecule depends upon the D_2O concentration (such effects have been observed in some gas mixtures [33,34]). If this were the case, W_H would have a non-linear concentration dependence even without any pion transfer.

6. Organic Mixtures

In addition to the H_2O/D_2O system, Petrukhin *et al.* [31] investigated two other systems, mixtures of $C_{10}H_{22}$ and CCl_4 as well as mixtures of C_6H_6 and CCl_4. Again, non-linearities in the concentration dependence of W_H were observed, qualitatively similar to the behaviour seen in the H_2O/D_2O case. However, two assumptions had to be made in the analysis. The first, as discussed above for the H_2O/D_2O mixtures, was that the molecular stopping probabilities do not depend on the concentration. The second assumption was that these stopping probabilities scale with the pion stopping power. Molecular stopping occurs at much lower energies than those at which stopping powers are measured, and situations have been observed where the molecular capture does not scale with the stopping power [9]. Thus the results could not be considered as unequivocal evidence of "external" transfer.

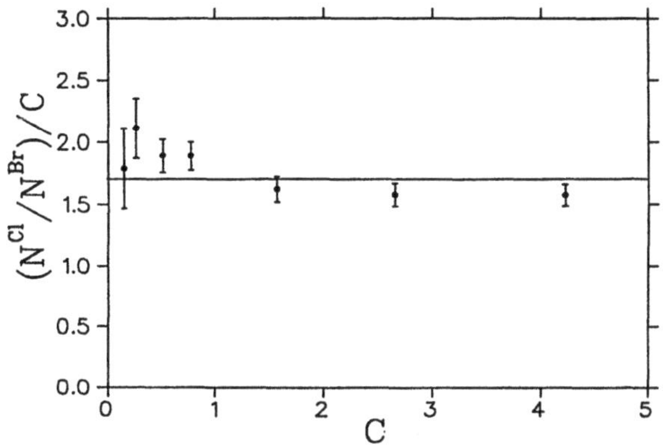

Figure 8: The ratio of the number of pionic x-rays from Cl and Br divided by the relative concentration C, vs. C, for mixtures of bromodecane and carbon tetrachloride. The line is a guide to the eye.

We have recently adopted a similar approach, with one important modification [35]. We used mixtures of bromodecane ($C_{10}H_{21}Br$) and carbon tetrachloride CCl_4. The selection of these two materials allows us to "tag" the π^- stops on each molecule using pionic x-rays. Stopping on the bromodecane is signalled by the 223.5 keV Br $4f \to 3d$ x-ray and stopping on the CCl_4 is signalled by the 150.7 keV Cl $3d \to 2p$ x-ray. The x-rays were detected using an intrinsic germanium detector. Simultaneous measurement of the charge-exchange probability and the pionic x-rays allows us to avoid the assumptions discussed above, and to examine the transfer process unambiguously.

If the molecular stopping distributions are independent of concentration, then the ratio $(N^{\rm Cl}_{3d\to 2p} \;/\; N^{\rm Br}_{4f\to 3d})/C$, should be independent of concentration. Here $N^{\rm Cl}_{3d\to 2p}$ is the number of Cl x-rays detected, $N^{\rm Br}_{4f\to 3d}$ is the number of Br x-rays detected and C is the relative concentration. Figure 8 shows the results for this ratio plotted against the concentration; no significant dependence is seen. This enables us to rule out variations in the molecular stopping probability as an explanation for effects seen in the charge-exchange yield. It also allows us to correct the charge-exchange data for the relative stopping probabilities in the two types of molecules in the mixture, without having to make any assumptions. With Eq. (3), the number of π^0 events detected is given simply by

$$N_{\pi^0} = 0.607\, \epsilon_{\pi^0} P_\alpha P_H (1 - T_T) \;, \tag{10}$$

where ϵ_{π^0} is the π^0 detection efficiency, P_α is the probability of a π^- stopping on the hydrogenous molecule (bromodecane), P_H is the probability that a π^- stopping on the bromodecane will go to a hydrogen atomic orbital and T_T is the total transfer probability to other (higher-Z) atoms. Similarly, the number of bromine $4f \to 3d$ x-ray transitions observed is given by

$$N^{Br}_{4f\to 3d} = \epsilon_{\rm Br} P_\alpha P_{\rm Br} P_{4f\to 3d} \;, \tag{11}$$

where $\epsilon_{\rm Br}$ is the detection efficiency for the Br x-ray, $P_{\rm Br}$ is the probability, given a π^- stopping on the bromodecane molecule, of the π^- going to a bromine atomic orbital and $P_{4f\to 3d}$ is the probability of the $4f \to 3d$ transition in the cascade in Br. By considering the ratio of $N_{\pi^0}/N^{\rm Br}_{4f\to 3d}$, the dependence on the molecular stopping probability P_α cancels out.

The transfer probability T_T can be decomposed into three terms:

$$T_T = T_i + T_\alpha + T_\beta \;. \tag{12}$$

Here T_i, the "internal" transfer term, is the probability for the π^- to transfer to different atoms in the same molecule (*i.e.* the bromodecane). Note that this could represent the barrier penetration mechanism of Jackson *et al.* , the collisional process, a combination of the two, or

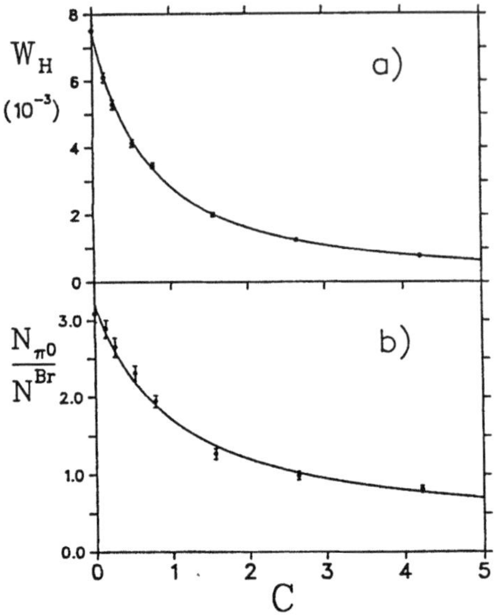

Figure 9: a) The pion capture probability in hydrogen W_H vs. the relative concentration C, and b) the ratio of the number of π^0's to the number of Br x-rays, for mixtures of bromodecane and carbon tetrachloride. The lines are to guide the eye.

possibly other unspecified mechanisms. The terms T_α and T_β represent "external" transfer, to atoms in different bromodecane molecules (T_α) or to atoms in a CCl_4 molecule (T_β). The internal transfer probability T_i should be independent of the molecular ratio $C = n_\beta/n_\alpha$, whereas T_α and T_β are expected to depend (in different ways) on the molecular ratio. In the simplest approximation, the external transfer probability to a molecule of a particular kind is proportional to the mole fraction for that molecule in the mixture. Therefore we have $T_\alpha = \frac{A_\alpha n_\alpha}{n_\alpha + n_\beta}$ and $T_\beta = \frac{A_\beta n_\beta}{n_\alpha + n_\beta}$, where A_α and A_β are transfer constants for α (bromodecane) and β (carbon tetrachloride) molecules respectively. We make no assumption here as to the mechanism for external transfer. Together with (10)-(12) this gives

$$\frac{N_{\pi^0}}{N^{\rm Br}_{4f\to 3d}} = \epsilon\left[1 - T_i - \frac{A_\alpha}{1+C} - \frac{A_\beta}{1+(1/C)}\right] , \tag{13}$$

where the factor ϵ is defined by

$$\epsilon = 0.607\frac{\epsilon_{\pi^0} P_H}{\epsilon_{\rm Br} P_{\rm Br} P_{4f\to 3d}} , \tag{14}$$

and is independent of C. The only concentration dependent terms are those involving the two external transfer terms A_α and A_β. Thus any concentration dependence observed in $N_{\pi^0}/N^{\rm Br}_{4f\to 3d}$ is evidence that these terms must be non-zero.

The concentration dependences of W_H and $N_{\pi^0}/N^{\rm Br}_{4f\to 3d}$ are shown in Fig. 9. Figure 9(b) clearly shows that a significant amount of external transfer is taking place. Unfortunately, there are too many parameters in the model to extract the transfer rates A_β, A_α and T_i from the data independently [35]. Despite this, two conclusions are inescapable: the external transfer terms A_α and A_β cannot both be zero, and $A_\alpha < A_\beta$. The second conclusion comes directly from the fact that the ratio $N_{\pi^0}/N^{\rm Br}_{4f\to 3d}$ decreases as C increases, and implies that the CCl_4 "scavenger" is more efficient at accepting the transferred pions than the bromodecane. We have repeated the measurement using a different molecule, trifluoro-trichloro-ethane (CF_3CCl_3) instead of CCl_4. The results are qualitatively the same. The conclusion is clear:

no model describing pion transfer in condensed systems can be considered complete unless it encompasses "external" transfer.

Acknowledgements

I would like to thank my colleagues who have participated in the pion transfer experiments at TRIUMF, including K.A. Aniol, M. Harston, D. Horváth, A.J. Noble, S. Stanislaus, D. Vetterli and P. Weber. Special thanks are due to D.F. Measday for introducing me to the topic, and guiding the group with verve and insight.

References

[1] L.I. Ponomarev, Annu. Rev. Nucl. Sci. **23**, 395 (1973).

[2] H. Schneuwly, in *Exotic Atoms, First course of the International School of Physics of Exotic Atoms* (Erice, 1977), eds. G. Fiorentini, G. Torelli (INFN, Pisa, 1977)p. 255.

[3] D. Horváth, Radiochim. Acta A **28**, 241 (1981).

[4] G. Bardin, J. Duclos, A. Magnon, J. Martino, A. Richter, E. Zavattini, A. Bertin, M. Piccinini, A. Vitale, and D. Measday, Nucl. Phys. **A352**, 365 (1981).

[5] T. Yamazaki, M. Aoki, M. Iwasaki, R.S. Hayano, T. Ishikawa, H. Outa, E. Takeda, H. Tamura, and A. Sakaguchi, Phys. Rev. Lett. **63**, 1590 (1989).

[6] J. Spuller, D. Berghofer, M.D Hasinoff, R. MacDonald, D.F. Measday, M. Salomon, T. Suzuki, J.-M. Poutissou, R. Poutissou, and J.K.P Lee, Phys. Lett. **67B**, 479 (1977).

[7] B. Bassalleck, F. Corriveau, M.D. Hasinoff, T. Marks, D.F. Measday, J.-M. Poutissou, and M. Salomon, Nucl. Phys. **A362**, 445 (1981).

[8] R. Macdonald, D.S. Beder, D. Berghofer, M.D. Hasinoff, D.F. Measday, M. Salomon, J. Spuller, T. Suzuki, R. Poutissou, J.-M. Poutissou, P. Depommier, and J.K.P Lee, Phys. Rev. Lett. **38**,746 (1977).

[9] A.V. Bannikov, B. Lévay, V.I. Petrukhin, V.A. Vasilyev, L.M. Kochenda, A.A. Markov, V.I. Medvedev, G.L. Sokolov, I.I. Strakovsky, and D. Horváth, Nucl. Phys. **A403**, 515 (1983).

[10] G. Backenstoss, M. Izycki, W. Kowald, P. Weber, H.J. Weyer, S. Ljungfelt, U. Mankin, G. Schmidt, and H. Ullrich, Nucl. Phys. **A448**, 567 (1986).

[11] S.S. Gerstein, V.I. Petrukhin, L.I. Ponomarev, and Yu.D. Prokoshkin, Usp. Fiz. Nauk **97**, 3 (1969)[Sov. Phys. Uspekhi **12**, 1 (1970).]

[12] N. Imanishi, S. Miyamoto, Y. Takeuchi, A. Shinohara, H. Kaji, and Y. Yoshimura, Phys. Rev. A **37**, 43 (1988), and references therein.

[13] V.I. Petrukhin, Yu.D. Prokoshkin, and V.M. Suvorov, Zh. Eksp. Teor. Fiz. **55**, 2173 (1968)[Sov. Phys. JETP **28**, 1151 (1969)].

[14] V. I. Petrukhin and V.M. Suvorov, Zh. Eksp. Teor. Fiz. **70**, 1145 (1976)[Sov. Phys. JETP **43**, 595 (1976)].

[15] V.I. Petrukhin and Yu.D. Prokoshkin, Zh. Eksp. Teor. Fiz. **56**, 501 (1969)[Sov. Phys. JETP **29**, 274 (1969)].

[16] N. Imanishi, Y. Takeuchi, K. Toyoda, A. Shinohara, and Y. Yoshimura, Nucl. Instrum. Methods **A261**, 465 (1987).

[17] P. Weber, D.S. Armstrong, D.F. Measday, A.J. Noble, S. Stanislaus, M.R. Harston, K.A. Aniol, and D. Horváth; Phys. Rev. A **41**, 1 (1990).

[18] K.A. Aniol, D.F. Measday, M.D. Hasinoff, H.W. Roser, A. Bagheri, F. Entezami, C.J. Virtue, J.M. Stadlbauer, D. Horváth, M. Salomon, and B.C. Robertson, Phys. Rev. A **28**, 2684 (1983).

[19] A.V. Kravtsov, A.Yu. Mayorov, A.I. Mikhailov, S.Yu. Ovchinnikov, N.P. Popov, V.M. Suvorov, and A.I. Shchetkovsky, Muon Catalyzed Fusion **2**, 199 (1988).

[20] K. Derrick, M. Derrick, J.C. Fetkovich, T.H. Fields, E.G. Pewitt, and G.B. Yodh, Phys. Rev. **151**, 82 (1966).

[21] D. Horváth, Phys. Rev. A **30**, 2123 (1984).

[22] V. I. Petrukhin, V.E. Risin, I.F. Samenkova, and V.M. Suvorov, Zh. Eksp. Teor. Fiz. **69**, 1883 (1975).[Sov. Phys. JETP **42**, 955 (1976)].

[23] Z.V. Krumshtein, V.I. Petrukhin, V.E. Risin, L.M. Smirnova, V.M. Suvorov, and I.A. Yutlandov, Zh. Eksp. Teor. Fiz. **65**, 455 (1973) [Sov.Phys.JETP **38**, 222 (1974)].

[24] M.R. Harston, D.S. Armstrong, D.F. Measday, S. Stanislaus and P. Weber, accepted in Phys. Rev. A.

[25] J.R.H. Smith, A.S. Clough, D.F. Jackson, K.R. Smith, D.F. Measday, F. Entezami, A.J. Noble, S. Stanislaus, C.J. Virtue, M. Salomon and K.A. Aniol, Nucl. Instrum. Methods **A242**, 465 (1986).

[26] D. Horváth, K.A. Aniol, F. Entezami, D.F. Measday, A.J. Noble, S. Stanislaus, C.J. Virtue, A.S. Clough, D.F. Jackson, J.R.H. Smith, and M. Salomon, Phys. Rev. A (in press).

[27] D.F. Jackson, C.A. Lewis, and K. O'Lcary, Phys. Rev. A **25**, 3262 (1982).

[28] D.F. Jackson and C. Tranquille, Phys. Lett. A **91**, 324 (1982).

[29] C. Tranquille and D.F. Jackson, Phys. Rev. A **34**, 742 (1986).

[30] D.F. Jackson and J.R.H. Smith, Phys. Rev. A **34**, 763(1986).

[31] V. I. Petrukhin, V.E. Risin, and V.M. Suvorov; Yad. Fiz. **19**, 626 (1973)[Sov. J. Nucl. Phys. **19**, 317 (1974).

[32] S. Stanislaus, D.F. Measday, D. Vetterli, P. Weber, K.A. Aniol, M.R. Harston, and D.S. Armstrong, Phys. Lett. A **143**, 239 (1990).

[33] P. Ehrhart, F.J. Hartmann, E. Köhler, and H. Daniel, Phys. Rev. A **27**, 575 (1983).

[34] P. Ehrhart, F.J. Hartmann, E. Köhler, and H. Daniel, Z. Phys. A **311**, 259 (1983).

[35] M.R. Harston, D.S. Armstrong, D.F. Measday, S. Stanislaus, and P. Weber, Phys. Rev. Lett. **64**, 1887 (1990).

Part V

Condensed Matter Studies with Positive Muons

Positive Muons as Probes in High-T_c Superconductors

H. Keller

Physik-Institut der Universität Zürich, Schönberggasse 9,
CH-8001 Zürich, Switzerland

Abstract. A short review on some recent muon spin rotation (μSR) investigations on high-T_c superconductors and their related magnetically ordered parent compounds is presented. The possibilities and limitations of the μSR technique for studying the microscopic magnetic properties of these materials are briefly discussed. In the superconducting copper oxides, μSR provides relevant information on the microscopic magnetic flux distribution in the bulk of the sample, allowing a direct determination of the London penetration depth in the mixed state. In addition, μSR is a sensitive tool for investigating the magnetic phase diagram (interplay of magnetism and superconductivity) of high-T_c systems.

1. Introduction

The discovery of high-temperature superconductivity by Bednorz and Müller in 1986 has triggered a world-wide effort on experimental and theoretical investigations on the superconducting copper oxides and closely related materials. A precise knowledge of the magnetic and electronic properties of these novel materials is important in order to search for possible pairing mechanisms of superconductivity and to test theoretical models. At present the nature of the microscopic pairing mechanism and the origin of the high transition temperatures (up to 125 K) are not yet understood. More detailed experiments on high quality and well characterized samples, as well as more theoretical work, are required for getting a better understanding of superconductivity in the copper oxide systems.

Over the past fifteen years the muon spin rotation (μSR) technique has been developed into a powerful and well recognized tool for investigating a large variety of problems in condensed matter physics [1-6]. Although most experiments are performed with positive muons (μ^+), there are also some unique applications where negative muons (μ^-) can be used to investigate specific problems in solid state physics [4,5]. In this paper we focus our attention on μSR experiments in high-T_c systems with positive muons (μ^+SR). Short descriptions of μSR studies in solids involving negative muons (μ^-SR) are given by Nagamine [4] and Maier [5] in this workshop.

In contrast to standard *macroscopic* magnetization experiments, the μSR technique provides a sensitive *microscopic* probe (the positive muon) of local magnetic fields (at the muon site) in solids. In particular, extensive μSR investigations on high-T_c systems have demonstrated that this rather exotic method yields crucial information on the microscopic magnetic properties of the superconducting oxides and their magnetically ordered parent compounds which are not easily obtained with other ex-

 © Springer-Verlag Berlin, Heidelberg 1992

perimental techniques. As a result of these experiments, a large number of papers on this subject has appeared in the literature [6-15]. Since the field is growing so rapidly, only a very small selection of some typical examples of recent μSR experiments in high-T_c systems demonstrating potential and unique applications of this technique is presented.

The organization of the paper is as follows: In Section 2 the basic principles of the μSR technique, insofar as is relevant to the interpretation of the following experiments, are briefly described. Some fundamental properties of the cuprate high-T_c systems in connection with the experiments presented in this paper are discussed in Section 3. A short description of the magnetic flux distribution in the mixed state of an extreme type-II superconductor (isotropic and anisotropic case) and its relation to the London penetration depth is given in Section 4. In Section 5 recent μSR measurements of the London penetration depth (temperature dependence, anisotropy) in various $YBa_2Cu_3O_x$ samples ($x = 6.9$) are presented. In addition, a detailed study of the muon-spin relaxation rate in oxygen deficient $YBa_2Cu_3O_x$ polycrystalline samples ($6.5 < x < 7.0$) is reported. The results are discussed in connection with the 'Uemura plot' [16,17]. Section 6 is devoted to μSR experiments involving the interplay between magnetism and superconductivity in cuprate high-T_c systems. Such experiments are of particular interest, since in several proposed theories of high-T_c superconductivity, magnetic interactions play an important role. As an example, recent experiments on $La_{2-x}Sr_xCuO_4$ and $YBa_2Cu_3O_x$ are discussed. The summary follows in Section 7.

2. Basic Principles of the μSR Technique

Before describing briefly the basic concepts of the μSR method, we first recall some fundamental properties of the positive muon (μ^+), summarized in Table 1. The positive muon is a lepton with a mass of $m_\mu \simeq 207 m_e$ (electron mass m_e), spin $S = 1/2$ and a finite lifetime of 2.2 μs. From a particle physics point of view, the μ^+ may be considered as a heavy positron (e^+). In condensed matter, however, positive muons behave more like light unstable protons ($m_\mu \approx m_p/9$). Due to its positive charge, the muon is repelled by the nuclei in a solid and therefore goes preferentially to interstitial regions in the lattice. In a magnetically ordered solid or superconductor the positive muon serves as a *microscopic* magnetic probe ($\mu_\mu \simeq \mu_B/207$) of the local magnetic field B_μ seen by the muon.

Table 1: Some properties of the positive muon.

Property	Value
Rest mass m_μ	105.658 MeV/c^2
	206.768 m_e
	0.1124 m_p
Charge q	$+e$
Spin S	1/2
Magnetic moment μ_μ	$4.836 \times 10^{-3} \mu_B$
	3.183 μ_p
Gyromagnetic ratio $\gamma_\mu/2\pi$	135.5387 MHz/T
Lifetime τ_μ	2.197 μs

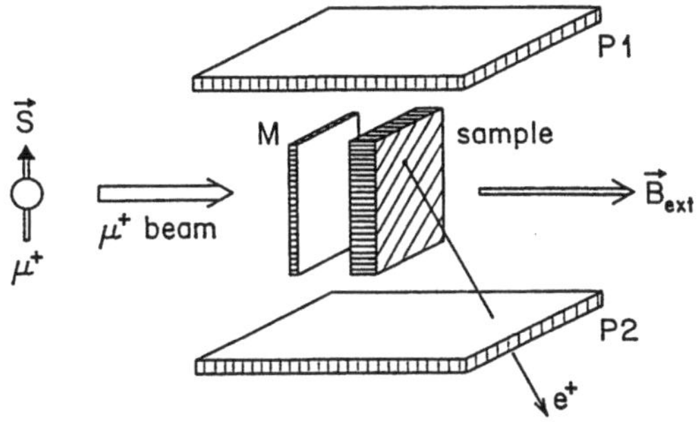

Figure 1: Schematic experimental set-up for a time-differential, transverse field μSR experiment. The muon detector (M) registers the incoming muons (μ^+) with their spins pointing perpendicular to the external magnetic field (B_{ext}). The positron detectors ($P1$ and $P2$) are used to count the decay positrons (e^+) arising from muons stopped in the sample.

The experimental set-up for a standard time-differential, transverse-field μSR experiment is shown schematically in Fig. 1. Nowadays highly polarized muon beams are available which are polarized either parallel or perpendicular to the muon momentum. Here we discuss the situation for low-energy muons ($\approx$ 4 MeV) with their spins pointing perpendicular to the muon momentum (Fig. 1). These so-called spin-rotated 'surface' muons are ideal for μSR studies in superconductors for the following reasons: (i) the muon beam is focussed by the external magnetic field B_{ext} on the sample (low background signal from muons stopping in the cryostat windows and target holder), (ii) relatively small samples (typically 1 cm^2 $\times$ 1 mm) can be used, and (iii) the unwanted background signal arising from beam positrons is small. In a standard μSR experiment, spin-polarized positive muons are implanted into a sample (one at a time), which is situated in an external magnetic field B_{ext} perpendicular to the initial muon-spin direction (parallel to the muon momentum for spin-rotated muons). In the sample, the muons come to rest in a thin layer several hundred μm below the surface. During the short thermalization process ($< 10^{-10}$ s) in the sample the muons are not depolarized and subsequently undergo Larmor precession in the local magnetic field B_μ (at the muon site) with a precession frequency $\omega_\mu = \gamma_\mu B_\mu$, where γ_μ is the gyromagnetic ratio of the muon (see Table 1). The stopped muons decay with a mean lifetime of 2.2 μs by emitting a positron (e^+) and two neutrinos ($\nu_e, \bar{\nu}_\mu$) according to the decay scheme indicated in Fig. 2. Due to parity violation in this decay process, the emission probability of the decay positrons with respect to the muon-spin direction is anisotropic (see Fig. 2) and therefore allows a direct observation of the muon-spin precession by detecting the number of decay positrons as a function of time after a muon has stopped in the sample. Experimentally, this is realized by using an arrangement of several particle detectors as shown schematically in Fig. 1. Before stopping in the sample, a muon passes through a thin muon counter M and starts a clock (t_{start}). The muon stops in the sample and starts to precess in the local magnetic field B_μ seen by the muon. After a certain time (mean lifetime 2.2 μs) the muon decays. When the corresponding decay positron is detected in one of the two positron counters $P1$ or $P2$, the clock is stopped (t_{stop}). The μSR signal is then obtained by counting the number of decay positrons N_{e^+} as a function of the

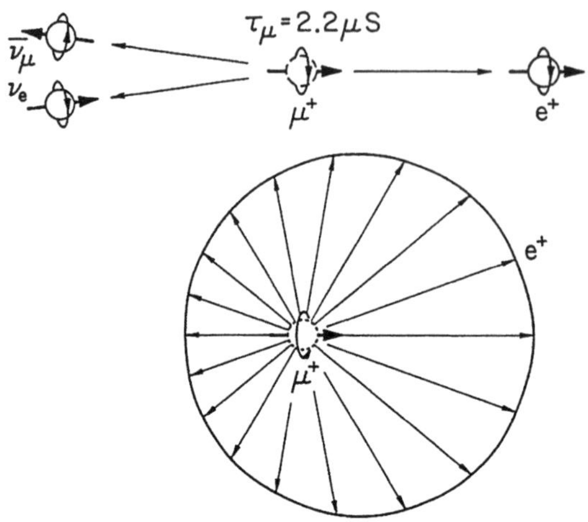

Figure 2: Muon-decay scheme and angular distribution pattern (emission probability) of the decay positrons with respect to the muon-spin direction. Note that the decay positrons are preferentially emitted along the muon-spin direction.

time difference $t = t_{stop} - t_{start}$ and storing these events in a histogramming memory. The resulting μSR time spectrum is given by the following expression [1]:

$$N_{e^+}(\varphi, t) = N_o \exp(-t/\tau_\mu)[1 + A\, R(t) \cos(\omega_\mu t + \varphi)] + b, \tag{1}$$

where N_0 is a normalization constant, τ_μ is the muon lifetime, A is the precession amplitude, $\omega_\mu = \gamma_\mu B_\mu$ is the muon-spin precession frequency, φ is the initial phase, and b is a time-independent background. For the detector geometry shown in Fig. 1 the initial phase is 0 and π for the positron detectors $P1$ and $P2$, respectively. The relaxation function $R(t)$ describes the damping of the precession signal. Note that in magnetic solids (ferromagnets, superconductors) the local field B_μ seen by the muon is in general not identical with the external field B_{ext}. In the special case of a Gaussian distribution of local static fields $p(B_\mu)$, which is of particular importance in this paper (cf. Sec. 5), the relaxation function has the simple form [1]:

$$R(t) = \exp(-\sigma^2 t^2). \tag{2}$$

Here σ is the so-called muon-spin depolarization rate which is related to the second moment (linewidth) of $p(B_\mu)$ according to the expression:

$$\langle \Delta B_\mu^2 \rangle = \langle B_\mu^2 \rangle - \langle B_\mu \rangle^2 = 2\sigma^2/\gamma_\mu^2, \tag{3}$$

where the first moment of $p(B_\mu)$ is simply given by the relation $\langle B_\mu \rangle = \omega_\mu/\gamma_\mu$. Beside the standard time-differential, transverse-field μSR technique described above, other experimental techniques such as zero-field μSR (no external magnetic field) and longitudinal-field μSR (external magnetic field parallel to the initial muon-spin polarization) have been employed in order to study specific aspects of high-T_c systems [6-15]. A detailed description of μSR spectroscopy and its applications in condensed matter physics is given in Ref. [1].

3. Some Properties of Copper Oxide Superconductors

Up to now there are, roughly speaking, four families of copper oxide superconductors: (1) La-Sr-Cu-O with critical temperatures (T_c's) up to 40 K (La-Ba-Cu-O was originally discovered by Bednorz and Müller in 1986), (2) Y-Ba-Cu-O with T_c's up to 93 K, (3) Bi-Ca-Sr-Cu-O with T_c's up to 108 K, and (4) Tl-Ca-Sr-Cu-O with T_c's up to 125 K, the highest value of T_c reached so far. In all these systems the superconducting charge carriers are holes, or vacancies in the valence band. Recently, a new class of cuprate superconductors Nd-Ce-Cu-O with T_c's as high as 24 K was discovered [18]. In contrast to the hole-doped systems, the charge carriers in this new class are electrons.

A common feature of all these layered cuprate superconductors (hole-doped or electron-doped systems) are the copper-oxygen (CuO_2) planes which are responsible for superconductivity in these novel materials. In the insulating parent compound La_2CuO_4 (Nd_2CuO_4) both copper and oxygen are divalent, and the Cu^{2+} spins order antiferromagnetically. When a fraction of La^{3+} (Nd^{3+}) is replaced by Sr^{2+} (Ce^{4+}) in the parent compound La_2CuO_4 (Nd_2CuO_4), holes (electrons) are introduced into the CuO_2-planes. These injected holes (electrons) are mobile and may form pairs under certain conditions giving rise to superconductivity. However, the nature of the microscopic pairing mechanism of the holes (electrons) is still unclear.

The superconducting oxides may be considered as extreme type-II superconductors having an extremely short coherence length ξ (very large upper critical field B_{c_2}) and a rather large magnetic penetration depth λ (very small lower critical field B_{c_1}), i.e. the Ginzburg-Landau parameter $\kappa = \lambda/\xi \gg 1$. As an example, for Y-Ba-Cu-O one finds $\xi \approx 2$ nm, $\lambda \approx 140$ nm, and $\kappa \approx 70$. The short coherence length in these superconductors gives rise to flux pinning at twin boundaries and defects, leading to unusual magnetic behavior in the presence of an external magnetic field [19-23].

Due to their layered structure (weakly coupled CuO_2-planes) the superconducting copper oxides are strongly anisotropic (uniaxial) superconductors (cf. Sect. 4). In these materials the superconducting screening currents are flowing dominantly in the CuO_2-planes (crystallographic ab-planes). Consequently, the magnetic penetration depth is highly anisotropic, and the Abrikosov flux-line lattice is very complex for a general orientation of the external magnetic field with respect to the symmetry axis (crystallographic c-axis) [24-27].

For a more detailed discussion of the magnetic and electronic properties of cuprate high-T_c systems, we refer to the articles of Gough [14], Forgan [15], and Bianconi [28] of this workshop.

4. Magnetic Flux Distribution in the Mixed State and London Penetration Depth

The London penetration depth λ and the coherence length ξ are two fundamental microscopic lengths that characterize the magnetic behavior of a superconductor in an external magnetic field. At the boundary between a normal and superconducting region, the superconducting carrier density can rise from zero in the normal region to its value in the superconducting region only gradually over a distance which is called the coherence length ξ. The magnetic penetration depth λ, on the other hand, is a measure of the distance over which an external magnetic field can penetrate from the normal into the superconducting region (diamagnetic screening). In the clean limit (mean free path $\ell \gg \xi$) the London penetration depth is given by [29]

$$\lambda = \sqrt{m^*/\mu_0 e^2 n_s}, \tag{4}$$

where m^* is the effective mass of the superconducting carriers and n_s is the superconducting carrier density. Since n_s is closely related to the superconducting order parameter Ψ ($n_s = |\Psi|^2$), the temperature dependence of λ provides essential information on the type of pairing mechanism (s-wave or p-wave pairing, weak or strong coupling) [29,30]. For a conventional s-wave superconductor, the temperature dependence of λ is usually described by the *empirical* expression (two-fluid model) [29]

$$\lambda(T) = \lambda(0)[1 - (T/T_c)^4]^{-1/2}. \tag{5}$$

For an extreme type-II superconductor ($\kappa = \lambda/\xi \gg 1$), such as the superconducting copper oxides, this empirical formula represents a good approximation of $\lambda(T)$ for an s-wave superconductor with strong coupling [30]. Moreover, it was argued by Uemura *et al.* [12] that the short coherence length copper oxides are clean superconductors ($\ell/\xi \gg 1$) and therefore one expects the London formula in Eq. (4) to be valid for these materials.

From a μSR experiment, the London penetration depth λ can be determined directly. The positive muon serves as microscopic probe of the magnetic flux distribution in the *bulk* of a superconductor. In the mixed state of a type-II superconductor, an external magnetic field $B_{ext} > B_{c_1}$ (lower critical field) penetrates the superconducting specimen in the form of a regular Abrikosov vortex lattice, each vortex carrying an elementary flux quantum $\Phi_0 = h/2e$. This gives rise to a broad distribution $p(B)$ of local fields (at the muon site). The width (second moment) of $p(B)$ is determined by λ. The magnetic flux distribution $p(B)$ in an extreme type-II superconductor ($\lambda/\xi \gg 1$) may be calculated using the simple London-theory approach [31]. As a result, for an isotropic superconductor the second moment of $p(B)$ for a perfect triangular vortex lattice is given by [31]

$$\langle \Delta B^2 \rangle = 0.00371 \Phi_0^2 \lambda^{-4}. \tag{6}$$

This equation is only valid for high magnetic fields ($B_{ext} > 2B_{c_1}$), where $\langle \Delta B^2 \rangle$ is independent of B_{ext} [31], and was recently confirmed by μSR experiments on Y-Ba-Cu-O [32,33].

In the case of an anisotropic superconductor, such as the layered copper oxide systems, the effective mass m^* in Eq. (4) has to be replaced by a diagonal effective mass tensor m^*_{ik} [24]. For a uniaxial superconductor, the tensor m^*_{ik} has a degenerate eigenvalue m^*_{ab} associated with supercurrents flowing in the ab-planes (CuO_2-planes) and an eigenvalue m^*_c associated with supercurrents flowing along the c-axis (perpendicular to the CuO_2-planes). The London formula given in Eq. (4) then has the more general form [25,34]

$$\lambda_{ab,c} = \sqrt{m^*_{ab,c}/\mu_0 e^2 n_s}, \tag{7}$$

where λ_{ab} (λ_c) is the magnetic penetration depth associated with the effective mass m^*_{ab} (m^*_c). A quantity of interest is the anisotropy ratio defined by

$$\gamma = (m^*_c/m^*_{ab})^{1/2} = \lambda_c/\lambda_{ab}. \tag{8}$$

Theoretical considerations [24-27] show that the magnetic flux distribution $p(B)$ in the mixed state of a uniaxial superconductor for an arbitrary orientation of the external magnetic field B_{ext} with respect to the symmetry axis (crystallographic c-axis) looks very complex. For example, for $B_{ext} \parallel c$ the equilibrium state is a regular triangular vortex lattice as in the isotropic case, whereas for $B_{ext} \perp c$ a distorted vortex lattice consisting of isosceles triangles is formed [27]. This prediction was recently confirmed by Bitter decoration experiments on single-crystal Y-Ba-Cu-O by Dolan *et al.* [35,36]. For the two principal-axis field orientations in a uniaxial superconductor, Barford and Gunn [34] derived the following expressions for the second moment of $p(B)$:

$$\langle \Delta B^2 \rangle_{\parallel} = 0.00371 \Phi_0^2 \lambda_{ab}^{-4} \tag{9}$$

and

$$\langle \Delta B^2 \rangle_{\perp} = 0.00371 \Phi_0^2 (\sqrt{\lambda_{ab}\lambda_c})^{-4} \tag{10}$$

for $B_{ext} \parallel c$ and $B_{ext} \perp c$, respectively. It is interesting to note that for $B_{ext} \perp c$, Eq. (10) implies that the effective penetration depth is the geometric mean of λ_{ab} and λ_c. This is due to the fact that for $B_{ext} \perp c$ the screening currents are flowing along the c-axis (m^*_c) as well as in the ab-planes (m^*_{ab}). A convenient quantity is the ratio

$$\langle \Delta B^2 \rangle_{\parallel} / \langle \Delta B^2 \rangle_{\perp} = (\lambda_c/\lambda_{ab})^2 = m^*_c/m^*_{ab} = \gamma^2, \tag{11}$$

which reflects directly the magnetic anisotropy γ of the system. For an arbitrary angle θ between B_{ext} and the c-axis, the angular dependence of the second moment of $p(B)$ may be described by the more general expression [37]

$$\langle \Delta B^2 \rangle(\theta) = \langle \Delta B^2 \rangle(\theta = 0)[\gamma^{-2} \sin^2\theta + \cos^2\theta]. \tag{12}$$

This expression was derived from an anisotropic London-equation approach [27] and is a good approximation as long as the flux-lattice structure is qualitatively the same at all angles [37]. Note that Eq. (11) is just a special case of Eq. (12) for $\theta = \pi/2$.

One should emphasize, however, that the application of Eqs. (6) and (9-12) requires that a stable and regular vortex lattice actually exists in the sample. Although at present the magnetic ($B-T$) phase diagram of the short coherence length superconducting copper oxides is not completely understood, there is experimental evidence from Bitter decoration experiments [35,36] and recent neutron diffraction experiments [38,15] on Y-Ba-Cu-O that a short-range ordered vortex lattice is indeed formed at low temperatures. Therefore, Eqs. (6) and (9-12) may be used in a first approximation to extract the London penetration depth from the second moment of $p(B)$ measured by μSR at low temperatures.

5. Measurement of the London Penetration Depth

The principal-axis penetration depths λ_{ab} and λ_c and the anisotropy ratio γ can in principle be directly measured by a μSR experiment on a single crystal. However, high quality single crystals which are large enough for μSR experiments (at least $5 \times 5 \times 1$ mm^3) are not readily available. On the other hand, polycrystalline samples can nowadays be prepared in large sizes and of high purity. Therefore, most μSR experiments are performed on polycrystalline specimens. In a ceramic sample an effective penetration depth λ_{eff} is measured which is a powder average over all orientations. However, it was shown by Barford and Gunn [34] that for a strongly anisotropic superconductor ($\gamma \geq 5$) the effective penetration depth λ_{eff} is independent of γ and proportional to λ_{ab}:

$$\lambda_{eff} \simeq 1.23\lambda_{ab}. \tag{13}$$

Thus, by means of a μSR experiment on a polycrystalline sample, it is possible to determine λ_{ab} for a uniaxial superconductor, provided the anisotropy ratio is large ($\gamma \geq 5$). Under the simple assumption that the local magnetic field distribution $p(B)$ may be approximated by a Gaussian distribution [33], the depolarization rate σ may be expressed in terms of λ_{eff} by combining Eqs. (3) and (6):

$$\sigma = 0.0431\gamma_\mu \Phi_0 \lambda_{eff}^{-2}. \tag{14}$$

This implies that the temperature dependence of σ in terms of the two-fluid model [Eq. (5)] may be written as

$$\sigma(T) = \sigma(0)[1 - (T/T_c)^4]. \tag{15}$$

These equations have been widely used to determine λ_{eff} in high-T_c superconductors from μSR experiments [6-17,32,33]. Note that the magnetic flux distribution in a perfect type-II superconductor is in general strongly asymmetric and not Gaussian-like [31]. However, it was shown that for polycrystalline samples under certain conditions a Gaussian distribution represents a reasonable approximation and yields reliable results [33].

As an example, a systematic μSR investigation of the London penetration depth in the mixed state of $YBa_2Cu_3O_x$ as a function of temperature, angle between external magnetic field and c-axis (anisotropy), and oxygen content x is presented. The following experiments

(a) Accurate measurement of the effective London penetration depth λ_{eff} and its temperature dependence in a high quality sintered $YBa_2Cu_3O_x$ sample [x = 6.970(1)] (Refs. [32,33]) prepared at the ETH Zurich [39,40].

(b) Determination of the anisotropy ratio γ in a c-axis-oriented $YBa_2Cu_3O_x$ polycrystalline sample [$x \approx 6.9$] (Refs. [32,11]) prepared at the University of Frankfurt [41].

(c) Determination of the anisotropy ratio γ by measuring the angular dependence of the muon relaxation rate σ on a mosaic of single crystals of $YBa_2Cu_3O_x$ [$x \simeq 6.9$] prepared at the University of Birmingham [37].

(**d**) Systematic study of the muon relaxation rate σ in a series of oxygen deficient sintered $YBa_2Cu_3O_x$ samples $[6.5 < x < 7.0]$ (Refs. [42,43]) prepared at the ETH Zurich [39,40].

performed on various samples will be discussed in detail.

Such μSR experiments must be performed on a slowly field-cooled (FC) sample in a high external magnetic field ($B_{ext} > 2B_{c_1} \approx 100$ mT) in order to make sure that a 'regular' vortex lattice is established in the mixed state of the specimen [10,11,33]. In the FC case the magnetic flux density is nearly constant above and below T_c in a high external field, and flux-pinning effects are in general small. In a recent μSR experiment, the influence of a dc-transport current on the magnetic flux-line distribution in the mixed state of sintered Y-Ba-Cu-O was investigated [44]. It was found that in a FC sample the flux-line distribution at low temperatures is not appreciably affected by a small transport current, indicating that the FC state represents a stable flux-line configuration (vortex state). This supports the assumption of a stable vortex lattice, a necessary condition for evaluating reliable values of the London penetration depth from μSR experiments at low temperatures.

(**a**) Measurements on a high quality sintered $YBa_2Cu_3O_x$ sample [$x = 6.970(1)$] were performed in an external field of 350 mT (FC) [32, 33]. The μSR time-spectra were analyzed assuming a Gaussian distribution of local magnetic fields [Eqs. (1-3)], yielding the depolarization rate σ. In order to make sure that the external field was high enough for Eq. (6) to be valid, a field scan between 5 and 350 mT was performed at 10 K (FC). Indeed, σ was found to be field-independent for $B_{ext} > 150$ mT, as expected for a vortex lattice [Eq. (6)]. Assuming that a 'regular triangular vortex lattice' is established in the sample at 350 mT and that the local magnetic field distribution may be approximated by a Gaussian distribution [33], Eq. (14) can be used to extract the effective penetration depth $\lambda_{eff}(T)$ from the measured depolarization rate $\sigma(T)$. Figure 3 shows the measured temperature dependence of σ after subtraction of a small temperature-independent background contribution ($\sigma_0 \approx 0.1\mu s^{-1}$) from the copper nuclear moments which was determined above T_c. As indicated by the solid line in Fig. 3, $\sigma(T)$ is very well described by the empirical formula of the two-fluid model given in Eq. (15). This means that the observed temperature behavior of σ at 350 mT is consistent with conventional s-wave pairing and suggests evidence for strong coupling [30]. A detailed discussion of the temperature dependence of the penetration depth in Y-Ba-Cu-O measured by μSR and other techniques is given in Refs. [11,33]. For comparison, the temperature dependence of λ_{eff} as obtained from Eq. (14) is displayed in Fig. 4. Again the solid line represents a fit to the empirical expression of the two-fluid model [Eq. (5)]. In order to extract an accurate value of λ_{eff} from the data at $T = 0$ K, a systematic data analysis was performed yielding $\lambda_{eff}(0) = 155(10)$ nm [32,33]. This value is in fair agreement with those previously measured by other μSR groups [6-11,16].

(**b**) The 'in-plane' penetration depth λ_{ab} is proportional to λ_{eff} for an anisotropy ratio $\gamma \geq 5$ as given in Eq. (13). In order to get an estimate of the magnitude of γ in Y-Ba-Cu-O, similar experiments to those described above were performed on a c-axis-oriented polycrystalline sample [32,11]. The temperature dependence of the

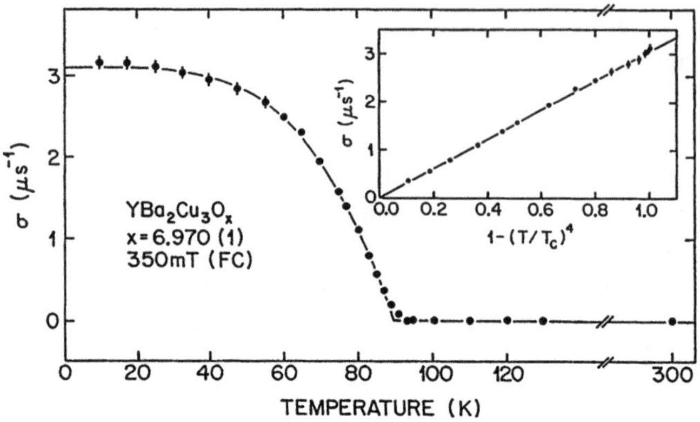

Figure 3: Depolarization rate σ as a function of temperature for a high quality sintered $YBa_2Cu_3O_x$ sample [$x = 6.970(1)$] measured in a field of 350 mT (FC). The solid curve corresponds to a fit using the empirical temperature dependence of the two-fluid model [Eq. (15)]. In the inset, σ versus $1-(T/T_c)^4$ is plotted.

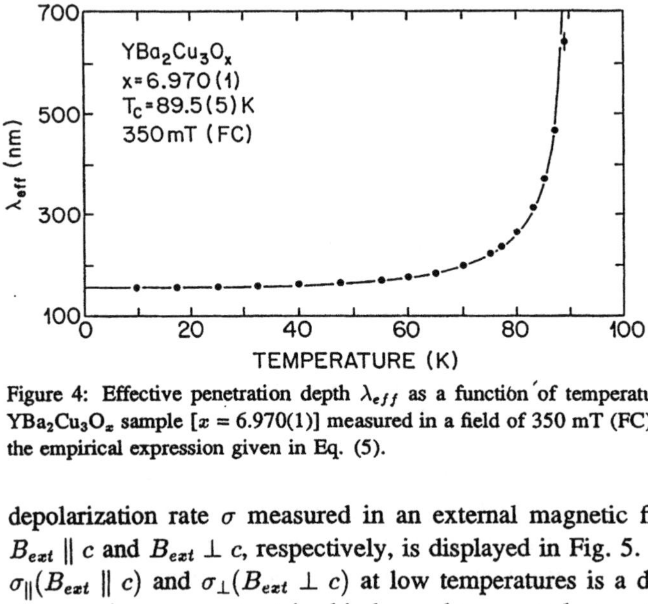

Figure 4: Effective penetration depth λ_{eff} as a function of temperature for a high quality sintered $YBa_2Cu_3O_x$ sample [$x = 6.970(1)$] measured in a field of 350 mT (FC). The solid curve is a fit using the empirical expression given in Eq. (5).

depolarization rate σ measured in an external magnetic field of 350 mT (FC) for $B_{ext} \parallel c$ and $B_{ext} \perp c$, respectively, is displayed in Fig. 5. The different behavior of $\sigma_{\parallel}(B_{ext} \parallel c)$ and $\sigma_{\perp}(B_{ext} \perp c)$ at low temperatures is a direct manifestation of the strong anisotropy present in this layered superconductor. Similar experiments were also reported by Uemura *et al.* [9]. As shown in Fig. 5 the temperature dependence of both $\sigma_{\parallel}$ and $\sigma_{\perp}$ is rather well described by the two-fluid model [Eq. (15)]. Although the quality of the measurements is not sufficient to extract reliable values of the principal-axis penetration depths λ_{ab} and λ_c, nevertheless, they allow an estimate of $\gamma = \lambda_c/\lambda_{ab}$ [32,11]. Combining Eqs. (3) and (11), the anisotropy ratio is given by the simple relation: $\gamma = \sigma_{\parallel}/\sigma_{\perp}$. At low temperatures (10 K) one obtains $\gamma \simeq 5(1)$, in fair agreement with recent Bitter decoration experiments [36] and other experiments summarized in Refs. [11,33].

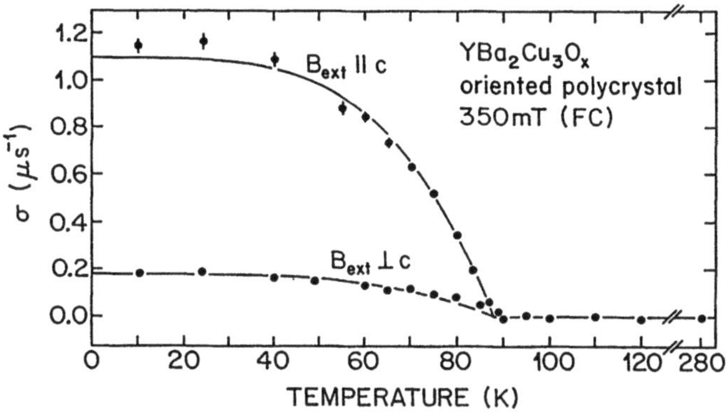

Figure 5: Depolarization rate σ as a function of temperature for a c-axis-oriented polycrystalline $YBa_2Cu_3O_x$ sample measured in 350 mT (FC) for $B_{ext} \parallel c$ and $B_{ext} \perp c$, respectively. The solid lines represent fits to the data using the empirical expression defined in Eq. (15).

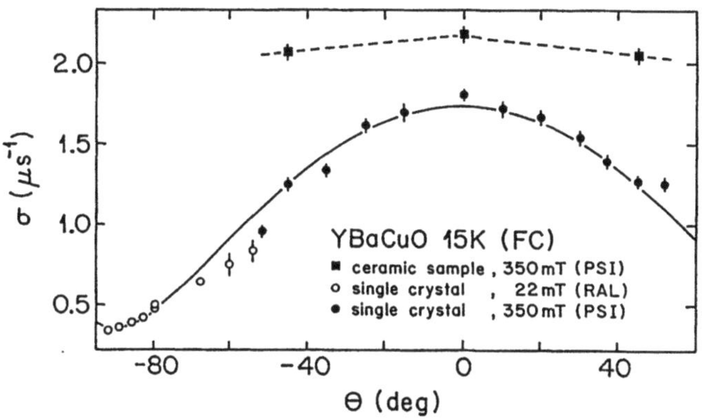

Figure 6: Low-temperature angle dependence of the depolarization rate σ for a mosaic of single crystals of $YBa_2Cu_3O_x$ (circles) and a sintered $YBa_2Cu_3O_x$ sample (squares). The solid line is a fit to the theoretical expression defined in Eq. (12).

(c) In another experiment the depolarization rate σ in the mixed state of Y-Ba-Cu-O was studied as a function of the angle θ between the external magnetic field and the crystallographic c-axis [37]. The sample was a mosaic of carefully aligned single crystals ($x \simeq 6.9$). Figure 6 shows the angular dependence of the depolarization rate σ measured at low temperature. The experiments with B_{ext} nearly parallel to the ab-planes were performed in a field of 22 mT at the Rutherford Appleton Laboratory (RAL), U.K., (open circles), whereas the results for B_{ext} nearly parallel to the c-axis were obtained in a field of 350 mT at the Paul Scherrer Institute (PSI), Switzerland, (closed circles). For each data point (angle θ) the sample was slowly field-cooled (FC) from above T_c in order to establish a stable flux-line configuration at low temperature. A pronounced angular dependence of σ and especially the rapid decrease of σ when B_{ext} is nearly parallel to the ab-planes is evident in Fig. 6. For comparison, the results obtained for an unoriented sintered sample are also included in Fig. 6. The slight angle dependence of σ observed for this sample is most likely due to the texture effect. The solid curve in Fig. 6 is a fit to the theoretical expression in Eq. (12) by using the simple relation $\langle \Delta B^2 \rangle(\theta) \propto \sigma(\theta)^2$. Although there are systematic deviations from the theoretical prediction, the general description of the data is very satisfactory. A preliminary data analysis [37] yields an effective mass ratio $\gamma^2 = m_c^*/m_{ab}^* \approx 25$, which is consistent with $\gamma^2 \simeq 25$ obtained for the c-axis oriented polycrystalline sample (see above).

In summary, two independent μSR experiments on two different Y-Ba-Cu-O samples gave essentially the same value of the anisotropy ratio $\gamma \simeq 5(1)$. Since the condition $\gamma \geq 5$ is fulfilled, Eqs. (13) and (8) can be used to determine λ_{ab} and λ_c from the value of the effective penetration depth $\lambda_{eff}(0) = 155(10)$ nm obtained for the high quality sintered sample and the measured value of $\gamma \simeq 5(1)$. As a result, the following values were found for $T = 0$ K: $\lambda_{ab}(0) = 130(10)$ nm and $\lambda_c(0) \simeq 500 - 800$ nm. The present values of λ_{ab}, λ_c, and γ are consistent with those obtained for different types of samples using other experimental techniques. For comparison, representative lists of recent values are given in Refs. [11,33,43]. At this point we should note that values of the penetration depth measured by μSR should be considered as lower limits. Indeed, any microscopic field inhomogeneity in a sintered sample not caused by the vortex lattice, like flux pinning and demagnetization effects, leads to a misleading broadening of the measured field distribution giving rise to a systematic error in the effective penetration depth λ_{eff} deduced from Eq. (14). However, a misinterpretation of $\langle \Delta B_\mu^2 \rangle$ of say 25 % would according to Eq. (14) affect λ_{eff} by only 7 %, so that the systematic errors in the quoted values are expected to be rather small.

(d) All experiments presented so far were performed on almost fully oxygenated $YBa_2Cu_3O_x$ ($x \approx 6.9 - 7$) samples with $T_c \approx 90$ K. It is well known, however, that the structural and electronic (superconducting) properties of $YBa_2Cu_3O_x$ sensitively depend on the oxygen stoichiometry x [45,39]. For instance, the superconducting transition temperature T_c increases with increasing x, exhibiting two characteristic plateaus at 60 K and 90 K (see Fig. 7). The oxygen content x determines the average occupancy of oxygen sites in the whole system (CuO_2-planes and CuO-chains) and therefore controls the density of the holes in the CuO_2-planes which are responsible for superconductivity in these materials. Recently, it was observed by Cava *et al.* [45] that the 60 K and 90 K plateaus in T_c versus x (see Fig. 7) are associated with plateaus in the effective valence of the copper in the CuO_2-planes. This interesting observation was discussed in terms of a positive charge transfer from the CuO-chains to the CuO_2-planes for $x > 6.45$ when the system becomes superconducting [45,46,39]. In this transfer mechanism, the CuO-chains play the role of a charge reservoir. From this point of view, a detailed study of the influence of the oxygen content x on the behavior of the London penetration depth $\lambda \propto \sqrt{m^*/n_s}$ (zero-temperature value and temperature dependence) is of considerable interest, since it provides important information on the actual carrier density n_s in the CuO_2-planes.

For this reason a series of oxygen deficient sintered $YBa_2Cu_3O_x$ samples ($6.5 < x < 7.0$) were prepared at the ETH Zurich [39,40] for a systematic μSR investigation

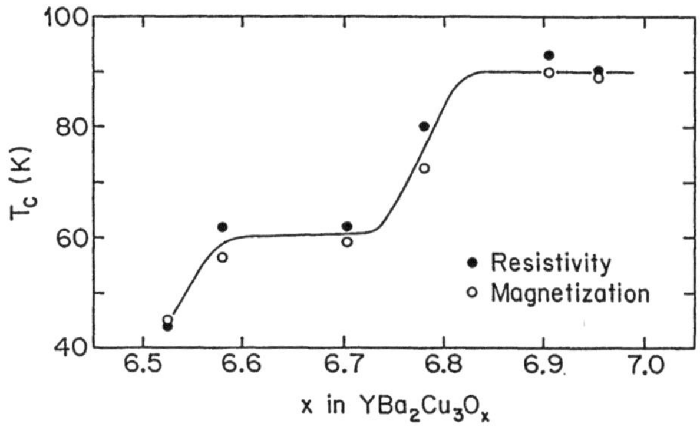

Figure 7: Superconducting transition temperature T_c as a function of oxygen content x for six sintered $YBa_2Cu_3O_x$ samples from the same 'mother batch'. The values of T_c were deduced from resistivity (closed circles) and magnetization measurements (open circles).

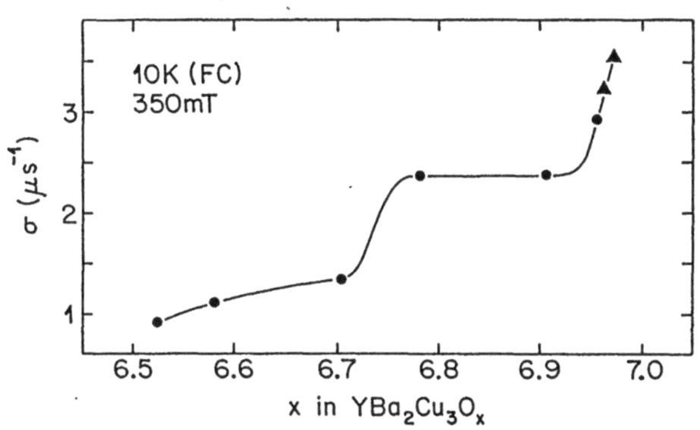

Figure 8: Depolarization rate σ as a function of oxygen content x for eight sintered $YBa_2Cu_3O_x$ samples measured in an external field of 350 mT (FC) at 10 K. The circles are data points obtained for samples from the same 'mother batch'. The triangles are data points obtained from samples of other batches. The solid line is a guide to the eye.

[42,43]. The oxygen content x of each sample was determined by high precision volumetric analysis (accuracy $\Delta x = \pm 0.001$) described elsewhere [40]. The variation of the transition temperature T_c with x for these samples is depicted in Fig. 7. Figure 8 shows the measured μSR depolarization rate σ as a function of x after field-cooling each sample to 10 K in an external magnetic field of 350 mT. Above $x = 6.7$ and 6.9, pronounced step-like increases in $\sigma(x)$ are observed. We recall that for a sintered sample of a uniaxial superconductor with $\gamma \geq 5$ the following relation applies in the clean limit [Eqs. (14), (13) and (7)]:

$$\sigma \propto 1/\lambda_{ab}^2 \propto n_s/m_{ab}^*. \tag{16}$$

This means that σ is proportional to the superconducting carrier density n_s divided by the effective mass m_{ab}^*. From a μSR experiment alone it is not possible to separate the effects of n_s and m_{ab}^* on σ. It was suggested, however, that m_{ab}^* does not vary appreciably within a certain class of CuO_2-based systems, so that σ is mainly

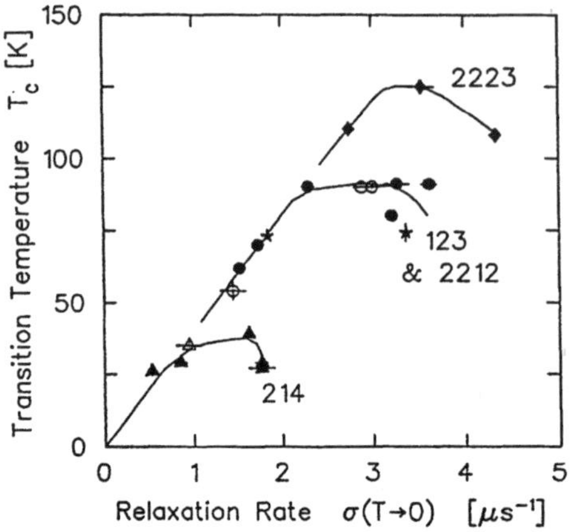

Figure 9: The superconducting transition temperature T_c of various cuprate high-T_c superconductors plotted versus the low-temperature values of the muon-spin relaxation rate $\sigma(T \rightarrow 0)$. Both T_c and σ were determined by μSR experiments (see figure caption in Ref. [17] for details). Figure taken from Uemura *et al.* [17].

controlled by the carrier density n_s (density of holes in the CuO_2-planes) [16,12]. Therefore, the step-like behavior of $\sigma(x)$ in Fig. 8 may be interpreted as a positive charge transfer from the CuO-chains to the 'superconducting' CuO_2-planes [45,46,39] which is a basic ingredient of the theoretical model in Refs. [47,48]. The fact that this charge transfer seems to occur in a discontinuous way around $x = 6.7$ and 6.9 is a remarkable result. In addition, it is very interesting to note that the x-dependence of the bond valence sum around the copper in the CuO_2-plane, derived from structural data by Cava *et al.* [45], exhibits striking similarities to the behavior of $\sigma(x)$. The bond valence sum is a measure of the actual charge associated with the in-plane copper atoms. This finding further supports the assumption that the variation in σ versus x is mainly determined by n_s rather than by m^*_{ab}. In the concentration range $6.5 < x < 6.9$ the general behavior of $T_c(x)$ and $\sigma(x)$ in Figs. 7 and 8 is very similar, suggesting that the step-like changes in both quantities are probably induced by the same mechanism. Above $x \geq 6.9$ a sudden increase of σ of more than 1 μs^{-1} in a rather small range of $\Delta x \simeq 0.06$ is observed, whereas T_c interestingly remains constant. It is evident from this observation that an accurate determination of the oxygen content x for this kind of investigations is very important. Recently, Rusiecki *et al.* [39] discovered a distinct anomaly (change of slope) in the lattice constant c versus x. It is noteworthy that this anomaly appears to coincide with the sudden increase of σ at $x > 6.9$. This suggests that some structural and electronic (superconducting carrier density) properties of $YBa_2Cu_3O_x$ are closely related to each other and are very sensitive to small changes of the oxygen content x [39].

Uemura *et al.* [16,17,6,8,9,12] performed extensive μSR measurements of the depolarization rate σ in more than thirty different specimens of cuprate high-T_c superconductors in order to study systematically the relation between T_c and $\sigma(T \rightarrow 0)$ in these CuO_2-based systems. Figure 9 shows a plot of T_c versus $\sigma(T \rightarrow 0) \propto n_s/m^*_{ab}$ [49] for sixteen different samples of various cuprate high-T_c superconductors [17]. This graph which is now often called 'Uemura plot' shows the following interesting

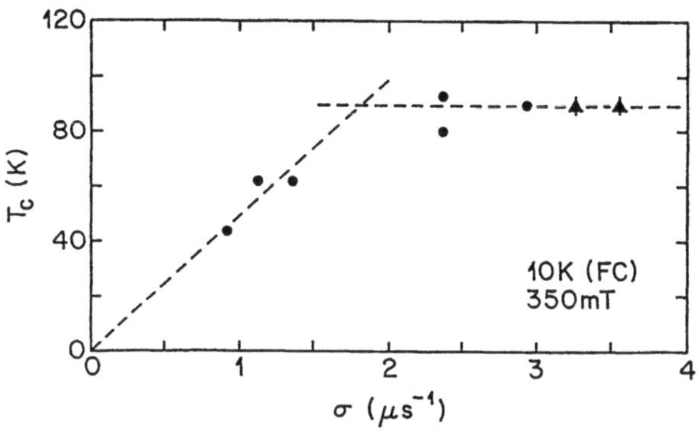

Figure 10: Superconducting transition temperature T_c (obtained from zero resistance) versus low-temperature depolarization rate $\sigma(T = 10\ \mathrm{K})$ for eight sintered $YBa_2Cu_3O_x$ samples (same samples as in Fig. 8). The dashed lines are guides to the eye.

features that seem to be 'universal' for all copper oxide high-T_c systems [17,12]: (i) with increasing carrier doping, T_c initially increases linearly, then saturates, and finally is suppressed for high carrier doping, (ii) this general trend is observed in single layer $La_{2-x}Sr_xCuO_4$ (214), double layer $YBa_2Cu_3O_x$ (123) and $Bi_2Sr_2CaCu_2O_8$ (2212), triple layer $Bi_2Sr_2Ca_2Cu_3O_{10}$ (2223), and many other related systems, and (iii) the initial slope of the linear increase of $T_c(\sigma)$ appears to be the same for all different families of copper oxide systems, however, the saturation of T_c starts at different values of σ for systems with different multiplicities of CuO_2-planes.

For comparison, the corresponding 'Uemura plot' for the eight oxygen deficient $YBa_2Cu_3O_x$ ($6.5 < x < 7.0$) samples described above (see Fig. 8) is displayed in Fig. 10. The behavior of T_c as a function of $\sigma(T = 10\ \mathrm{K})$ shows the same 'universal' tendency as in Fig. 9, and the slope of the linear increase of T_c at low oxygen content x is consistent with the results of Uemura *et al.* [17]. As already discussed above, it seems very likely that the 'universal' behavior of T_c versus $\sigma \propto n_s/m^*_{ab}$ (Figs. 9 and 10) is mainly governed by changes of n_s rather than changes of m^*_{ab} [16,12]. For low doping concentrations this would imply that T_c is proportional to the superconducting carrier density n_s. From this point of view, it is also interesting to compare $T_c(x)$ in Fig. 7 with $T_c(\sigma)$ in Fig. 10 for $YBa_2Cu_3O_x$. We note that the plateau at 60 K in $T_c(x)$ is not present in $T_c(\sigma)$. This is because the depolarization rate σ reflects the actual superconducting carrier density n_s in the CuO_2-planes which predominantly determines T_c, whereas the oxygen content x controls the total number of holes located somewhere in the crystal (CuO_2-planes and CuO-chains). However, the complex relation between the 'real' superconducting carrier density n_s and the oxygen content x, which is sensitively reflected by the quantity $\sigma(x) \propto n_s/m^*_{ab}$ displayed in Fig. 8, is not well understood.

The 'Uemura plot' in Fig. 9 is of particular importance for testing theoretical models, since any successful theory of superconductivity in these unique materials should also provide an explanation of the 'universal' behavior of $T_c(\sigma)$. A few general remarks can be made on the 'Uemura plot' [16,17,12]: (i) the linear relation between T_c and n_s/m^*_{ab} cannot be explained within the framework of weak-coupling BCS theory, (ii) a linear relation, $T_c \propto \varepsilon_F \propto n_s/m^*_{ab}$, is found for a two-dimensional non-interacting electron gas, provided the energy scale of the attractive pairing interaction is comparable to or larger than the Fermi energy ε_F, and (iii) the linear relation $T_c \propto n_s/m^*_{ab}$ is also consistent with theories based on Bose-Einstein condensation, as well as with the recently proposed 'anyon' theory [50]. However, one has to wait for further experimental and theoretical achievements in order to reach a deeper understanding of these interesting empirical results.

6. Magnetism and Superconductivity

A central question regarding high-temperature superconductivity is whether or not magnetic interactions play an essential role in the pairing mechanism as proposed in several theories [7,8]. Some of the pioneering experiments which first suggested a possible relationship between antiferromagnetism and superconductivity in the CuO_2-based systems were μSR studies on La-Sr-Cu-O and Y-Ba-Cu-O [6-10,51,52]. The μSR technique is a very sensitive method for detecting small frozen electronic moments (magnetic order) in zero external magnetic field. However, one should emphasize that because μSR is sensitive only to static local fields on a time scale of the muon lifetime (2.2μs), it generally cannot distinguish between long-range magnetic order *or* freezing of short-range magnetic correlations.

Extensive μSR investigations on various cuprate high-T_c systems were performed in order to search for a possible interplay between magnetism and superconductivity by carefully studying the magnetic phase diagram of these CuO_2-based systems. They include measurements on the hole-doped systems La-Sr-Cu-O, Y-Ba-Cu-O, and related systems, as well as on the electron-doped system Nd-Ce-Cu-O [6-10,12,13,51,52]. For all these systems, the undoped parent compound is an insulator where the Cu^{2+} spins in the CuO_2-network order antiferromagnetically with a rather high Néel temperature between 200 and 400 K. When these systems are doped with carriers, holes (electrons) are very likely added to the CuO_2-planes. Consequently, the ordering temperature is drastically reduced with increasing doping concentration, and finally magnetic order disappears at around the doping concentration where superconductivity starts to occur (see Fig. 11).

For the hole-doped systems, Aharony *et al.* [53] proposed a simple model in which the disorder in the Cu^{2+} spin system is introduced by the holes which are assumed to be mainly located on the oxygen atoms in the CuO_2-planes. The magnetic holes on the oxygen sites ($O^{1-}, S = 1/2$) induce local ferromagnetic exchange between, otherwise antiferromagnetically coupled, adjacent copper spins. As a consequence, with increasing doping concentration the induced holes gradually destroy antiferromagnetic order in the copper spin system due to the competing ferromagnetic and antiferromagnetic exchange interactions ('frustration'). In the insulating phase this model predicts a phase diagram, including an antiferromagnetic and a spin-glass phase, which for instance is consistent with μSR and other experiments on La-Sr-Cu-O [51,52]. However, this simple model cannot explain the appearance of superconductivity in the CuO_2-based systems. For the electron-doped systems a similar picture applies [12]. In this case, however, the doped electrons are assumed to be mainly located on the copper sites, changing magnetic Cu^{2+} ($S = 1/2$) to non-magnetic Cu^{1+} ($S = 0$). This leads to a reduction of the magnetic ordering temperature due to a 'dilution'

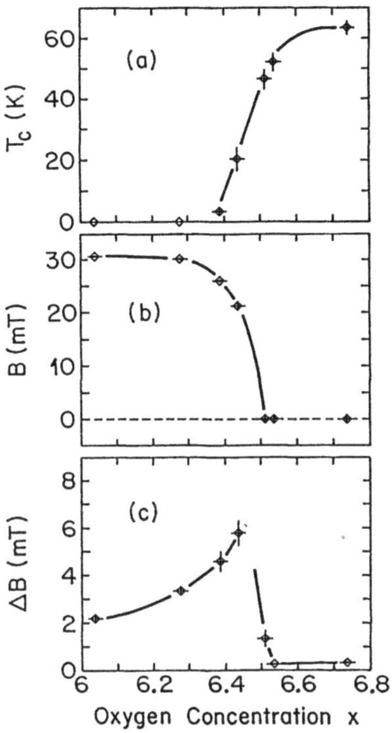

Figure 11: Internal fields in oxygen deficient $YBa_2Cu_3O_x$: The x dependence of (a) the midpoint of the superconducting transition temperature T_c, (b) the average internal field at the muon site B, and (c) the rms deviation ΔB. The curves are meant to guide the eye. Figure taken from Kiefl *et al.* [52].

effect in the copper spin system, rather than due to a 'frustration' effect as in the hole-doped systems [12].

It is an experimental fact that only slight changes in the carrier concentration in the CuO_2-based systems cause a phase transition from a magnetically ordered state to a superconducting state and vice versa (see Fig. 11). Therefore, the fundamental question arises: Is there a coexistence of magnetic order and superconductivity in the cuprate high-T_c systems such that the magnetic and the superconducting order parameters have some overlap? In order to answer this question, extensive μSR investigations were performed [6-10,12,13,51,52]. Here, only two representative examples are presented. Weidinger *et al.* [51] reported zero-field μSR experiments on superconducting $La_{2-x}Sr_xCuO_4$ samples with $0.07 \leq x \leq 0.15$ at low temperatures (35 mK). They observed internal magnetic fields of electronic origin, coexisting with bulk superconductivity in these sintered samples, even for the doping concentration $x = 0.15$ with the highest superconducting transition temperature (midpoint) $T_c = 32$ K. In the superconducting state the internal magnetic fields were found to be an order of magnitude smaller than the local field ($\simeq 40$ mT) in the undoped parent compound La_2CuO_4. Kiefl *et al.* [52] performed similar measurements on $La_{2-x}Sr_xCuO_4$ with $x = 0.15$ in zero applied field at 20 mK. However, in contrast to Weidinger *et al.* [51], they found little evidence for internal fields of electronic origin in this heavily doped superconducting sample. In the same paper, Kiefl *et al.* [52]

present a detailed μSR investigation of internal fields in a series of oxygen deficient $YBa_2Cu_3O_x$ ($6.0 < x < 6.8$) samples below 90 mK. Figure 11 shows the superconducting transition temperature T_c, the measured magnitude of the internal field B, and the rms deviation ΔB as a function of the oxygen concentration x. In the antiferromagnetic phase the internal field B is almost constant in the concentration range $6.0 < x < 6.3$. However, around $x \approx 6.4$ where the onset of superconductivity occurs, a sharp drop of B, together with a sudden increase of the rms deviation ΔB, is observed. Finally, in the heavily doped region ($x \geq 6.54, T_c \geq 52$ K) *no* indication of static or slowly fluctuating electronic moments in these samples was found. This strongly suggests that static magnetic order is most likely not present in these $YBa_2Cu_3O_x$ samples with a high oxygen concentration.

In conclusion, there is little or no evidence that static electronic moments and superconductivity coexist in heavily doped $YBa_2Cu_3O_x$ ($x \geq 6.54$) at very low temperatures, although in some superconducting samples with a low doping concentration static internal fields were observed (see Fig. 11). On the other hand, for heavily doped $La_{2-x}Sr_xCuO_4$ ($x \simeq 0.15$) the situation is less clear. While Kiefl *et al.* [52] find no obvious indication of internal magnetic fields of electronic origin at low temperatures (20 mK), the experiments of Weidinger *et al.* [51] suggest that magnetism coexists with superconductivity at low temperatures (35 mK). In a comment Weidinger *et al.* [54] argued that magnetism and superconductivity affect each other on a microscopic scale and therefore appear to 'coexist' in microscopic spatial regions, even though the spatial distribution is unknown. However, since the coherence length of the copper oxide superconductors is extremely short, variations in the stoichiometry even on a microscopic scale could give rise to magnetism and superconductivity in microscopic regions, although in these regions the magnetic and superconducting order parameters have no actual overlap.

7. Summary

We have demonstrated that μSR is a powerful and unique tool for investigating various *microscopic* magnetic properties of copper oxide superconductors and their magnetically ordered parent compounds.

In the superconducting compounds μSR provides important information on the local magnetic flux distribution $p(B_\mu)$ (at the positive muon) in the *bulk* of the sample (no surface effect). Particularly, in the mixed state of a type-II superconductor, the London penetration depth λ can be determined from the second moment of $p(B_\mu)$ in a direct way. A precise measurement of λ is of fundamental importance, since $\lambda(T)$ essentially reflects the type of the pairing mechanism (s- or p-wave, weak or strong coupling) [29,30]. An investigation of the second moment of $p(B_\mu)$ as a function of the angle between the external magnetic field and the c-axis allows a determination of the effective mass ratio $\gamma^2 = m_c^*/m_{ab}^*$ and the principal-axis penetration depths λ_{ab} and λ_c in these uniaxial layered superconductors. As an example, some recent μSR measurements on various $YBa_2Cu_3O_x$ samples were discussed in detail. The measured values of the principal-axis penetration depths $\lambda_{ab}(0) = 130(10)$ nm and $\lambda_c(0) \simeq 500 - 800$ nm, as well as the value of the anisotropy ratio $\gamma \simeq 5(1)$ for fully oxygenated $YBa_2Cu_3O_x$ are consistent with those obtained from other experimental techniques [11,33,43]. For a high quality sintered $YBa_2Cu_3O_x$ sample [$x = 6.970(1)$] the temperature dependence of the effective London penetration depth λ_{eff} was found to be well described by the empirical two-fluid model. This temperature behavior of λ_{eff} is consistent with conventional s-wave pairing and suggests evidence for strong coupling [30]. Moreover, a systematic investigation of the London penetration depth in $YBa_2Cu_3O_x$ as a function of oxygen content x is of great interest for studying the influence of x on the superconducting carrier density n_s in the CuO_2-planes. First experiments on a series of oxygen deficient samples ($6.5 < x < 7.0$) revealed some remarkable results. In particular, the muon relaxation rate $\sigma(x)$ exhibits a peculiar behavior, with two step-like increases above $x = 6.7$ and $x = 6.9$. Because σ is proportional to n_s/m^*_{ab}, the step-like behavior of σ may be associated with a positive charge transfer from the CuO-chains to the CuO_2-planes [45,46,39] taking place discontinuously with increasing x. Indeed, $\sigma(x)$ shows interesting similarities to the x-dependence of the bond valence sum around the CuO_2-plane copper [45], reflecting the actual charge sitting on that particular copper atom. This observation may have some important consequences for theoretical considerations, since it has been suggested that charge-transfer excitations and phonons play an essential role in high-temperature superconductivity [47,48]. Furthermore, there is experimental evidence that the behavior of the muon-spin relaxation rate $\sigma(x)$ and slight changes in the structural properties (oxygen ordering [45], anomaly in the lattice constant c [39]) of $YBa_2Cu_3O_x$ are strongly correlated.

Uemura *et al.* [16,17] found empirically an interesting 'universal' relation between the superconducting transition temperature T_c and n_s/m^*_{ab} which seems to be observed 'universally' for various families of cuprate high-T_c superconductors ('Uemura plot'). Any reasonable theory of superconductivity in these CuO_2-based materials should be able to give an explanation for this 'universal' behavior.

Due to its high sensitivity to small frozen magnetic moments, zero-field μSR was extensively used to study the magnetic phase diagram of the cuprate high-T_c systems. One fundamental question of interest is whether or not magnetic interactions play an important role in the pairing mechanism in these CuO_2-based systems as suggested in some theories [7,8]. As an example, two representative μSR experiments on superconducting samples of $La_{2-x}Sr_xCuO_4$ and $YBa_2Cu_3O_x$ were presented [51,52]. For low doping concentrations x (low T_c) in both systems, internal magnetic fields of electronic origin were seen at low temperatures ($\approx 20 - 35$ mK). On the other hand, for high doping concentrations x (high T_c), Kiefl *et al.* [52] found little evidence for static electronic moments in $YBa_2Cu_3O_x$ and $La_{2-x}Sr_xCuO_4$, in contrast to Weidinger *et al.* [51] who observed internal magnetic fields in heavily doped $La_{2-x}Sr_xCuO_4$ at low temperatures. An observation of internal magnetic fields of electronic origin in these short coherence length superconductors, however, does not necessarily imply that magnetism and superconductivity actually coexist. Microscopic variations in the stoichiometry on a length scale larger than the coherence length may give rise to internal magnetic fields in *microscopic* regions in the sample in such a way that the magnetic and superconducting order parameters do not really overlap. In order to interpret unambiguously these controversial results one has to wait for further experimental work on high quality single crystals.

Acknowledgments

I would like to thank G. Benedek, H. Schneuwly and G. Torelli for organizing this International Workshop on Exotic Atoms in Condensed Matter, and for the opportunity to contribute. The kind collaboration with W. Assmus, E.M. Forgan, E. Kaldis, J. Kowalewski, Y. Maeno, I. Mangelschots, W. Odermatt, B. Pümpin, C. Rossel, S. Rusiecki, I.M. Savić, J.W. Schneider, H. Simmler and P. Zimmermann, who participated in some of this work, is gratefully acknowledged. I also wish to thank A. Baratoff, A. Bianconi, E.H. Brandt, C.E. Gough, W. Kündig, J. Mannhart, I. Morgenstern, K.A. Müller, A.M. Portis and T. Schneider for many helpful and stimulating discussions and their interest in our work. I am very thankful to Y.I. Uemura and J.H. Brewer for their permission to include Fig. 9 and Fig. 11 in this paper. This work was partly supported by the Swiss National Science Foundation.

References

[1] See, e.g., A. Schenck, *Muon Spin Rotation Spectroscopy: Principles and Applications in Solid State Physics* (Adam Hilger, Bristol 1985).

[2] See, e.g., C. Bucci, this workshop.

[3] See, e.g., S.F.J. Cox, this workshop (two contributions).

[4] See, e.g., K. Nagamine, this workshop (two contributions).

[5] See, e.g., K. Maier, this workshop.

[6] See, e.g., *Proceedings of the 5th International Conference on Muon Spin Rotation, Relaxation and Resonance*, Oxford, U.K., to be published in Hyperfine Interactions.

[7] See, e.g., *Proceedings of the International Conference on High-Temperature Superconductors and Materials and Mechanism of Superconductivity*, Interlaken, Switzerland, Physica C **153-155** (1988).

[8] See, e.g., *Proceedings of the International Conference on Materials and Mechanisms of Superconductivity, High Temperature Superconductors II*, Stanford, U.S.A., Physica C **162-164** (1989).

[9] See, e.g., Y.J. Uemura, B.J. Sternlieb, D.E. Cox, V.J. Emery, A. Moodenbaugh, M. Suenaga, J.H. Brewer, J.F. Carolan, W. Hardy, R. Kadono, J.R. Kempton, R.F. Kiefl, S.R. Kreitzman, G.M. Luke, P. Mulhern, T. Riseman, D.L. Williams, B.X. Yang, W.J. Kossler, X.H. Yu, H. Schone, C.E. Stronach, J. Gopalakrishnan, M.A. Subramanian, A.W. Sleight, H. Hart, K.W. Lay, H. Tagaki, S. Uchida, Y. Hidaka, T. Murakami, S. Etemad, P. Barboux, D. Keane, and D.C. Johnston, J. Phys. (Paris) Colloq **49**, C8-2087 (1988).

[10] See, e.g., H. Keller, IBM J. Res. Develop. **33**, 314 (1989), and references quoted therein.

[11] See, e.g., H. Keller, in *Earlier and Recent Aspects of Superconductivity*, ed. by J.G. Bednorz and K.A. Müller, Springer Series in Solid-State Sciences, Vol. 90 (Springer-Verlag Berlin, Heidelberg 1990), pp. 222-239, and references quoted therein.

[12] See, e.g., Y.J. Uemura, G.M. Luke, B.J. Sternlieb, L.P. Le, J.H. Brewer, R. Kadono, R.F. Kiefl, S.R. Kreitzman, T.M. Riseman, C.L. Seaman, J.J. Neumeier, Y. Dalichaouch, M.B. Maple, G. Saito, and H. Yamochi, in *Proceedings of NATO Advanced Research Workshop on 'Dynamics of Magnetic Fluctuations in High Temperature Superconductors'*, Crete, Greece, 1989 (Plenum, 1990).

[13] See, e.g., A. Schenck, in *Proceedings of the ICMAS 89*, ed. by A. Niku-Lari, IITT-International, Technology Transfer Series, Gournay-sur-Marne (France), 1989, pp. 165-171.

[14] See, e.g., C.E. Gough, this workshop.

[15] See, e.g., E.M. Forgan, this workshop.

[16] Y.J. Uemura, V.J. Emery, A.R. Moodenbaugh, M. Suenaga, D.C. Johnston, A.J. Jacobson, J.T. Lewandowski, J.H. Brewer, R.F. Kiefl, S.R. Kreitzman, G.M. Luke, T. Riseman, C.E. Stronach, W.J. Kossler, J.R. Kempton, X.H. Yu, D. Opie, and H.E. Schone, Phys. Rev. B **38**, 909 (1988).

[17] Y.J. Uemura, G.M. Luke, B.J. Sternlieb, J.H. Brewer, J.F. Carolan, W.N. Hardy, R. Kadono, J.R. Kempton, R.F. Kiefl, S.R. Kreitzman, P. Mulhern, T.M. Riseman, D.Ll. Williams, B.X. Yang, S. Uchida, H. Tagaki, J. Gopalakrishnan, A.W. Sleight, M.A. Subramanian, C.L. Chien, M.Z. Cieplak, Gang Xiao, V.Y. Lee, B.W. Statt, C.E. Stronach, W.J. Kossler, and X.H. Yu, Phys. Rev. Lett. **62**, 2317 (1989).

[18] Y. Tokura, H. Takagi, and S. Uchida, Nature **337**, 345 (1989).

[19] K.A. Müller, M. Takashige, and J.G. Bednorz, Phys. Rev. Lett. **58**, 1143 (1987).

[20] I. Morgenstern, K.A. Müller, and J.G. Bednorz, Z. Phys. B - Condensed Matter **69**, 33 (1987).

[21] H. Keller, B. Pümpin, W. Kündig, W. Odermatt, B.D. Patterson, J.W. Schneider, H. Simmler, S. Connell, K.A. Müller, J.G. Bednorz, K.W. Blazey, I. Morgenstern, C. Rossel, and I.M. Savić, Physica C **153-155**, 71 (1988).

[22] B. Pümpin, H. Keller, W. Kündig, W. Odermatt, B.D. Patterson, J.W. Schneider, H. Simmler, S. Connell, K.A. Müller, J.G. Bednorz, K.W. Blazey, I. Morgenstern, C. Rossel, and I.M. Savić, Z. Phys. B - Condensed Matter **72**, 175 (1988).

[23] C. Rossel, Y. Maeno, and I. Morgenstern, Phys. Rev. Lett. **62**, 681 (1989).

[24] V.G. Kogan, Phys. Rev. B **24**, 1572 (1981).

[25] N. Schopohl and A. Baratoff, Physica C **153-155**, 689 (1988).

[26] L.J. Campbell, M.M. Doria, and V.G. Kogan, Phys. Rev. B **38**, 2439 (1988).

[27] S.L. Thiemann, Z. Radović, and V.G. Kogan, Phys. Rev. B **39**, 11406 (1989).

[28] See, e.g., A. Bianconi, this workshop.

[29] M. Tinkham, *Introduction to Superconductivity* (McGraw-Hill, New York 1975).

[30] J. Rammer, Europhys. Lett. **5**, 77 (1988).

[31] E.H. Brandt, Phys. Rev. B **37**, 2349 (1988).

[32] B. Pümpin, H. Keller, W. Kündig, W. Odermatt, I.M. Savić, J.W. Schneider, H. Simmler, P. Zimmermann, J.G. Bednorz, Y. Maeno, K.A. Müller, C. Rossel, E. Kaldis, S. Rusiecki, W. Assmus, and J. Kowalewski, Physica C **162-164**, 151 (1989).

[33] B. Pümpin, H. Keller, W. Kündig, W. Odermatt, I.M. Savić, J.W. Schneider, H. Simmler, P. Zimmermann, E. Kaldis, S. Rusiecki, Y. Maeno, and C. Rossel, to be published in Phys. Rev. B.

[34] W. Barford and J.M.F. Gunn, Physica C **156**, 515 (1988).

[35] G.J. Dolan, G.V. Chandrashekhar, T.R. Dinger, C. Feild, and F. Holtzberg, Phys. Rev. Lett. **62**, 827 (1989).

[36] G.J. Dolan, F. Holtzberg, C. Feild, and T.R. Dinger, Phys. Rev. Lett. **62**, 2184 (1989).

[37] E.M. Forgan, S.L. Lee, S. Sutton, J.S. Abell, S.F.J. Cox, C.A. Scott, H. Keller, B. Pümpin, J.W. Schneider, H. Simmler, P. Zimmermann, and I.M. Savić, to be published.

[38] E.M. Forgan, D. McK. Paul, H.A. Mook, P.A. Timmins, H. Keller, S. Sutton, and J.S. Abell, Nature **343**, 735 (1990).

[39] S. Rusiecki, B. Bucher, E. Kaldis, E. Jilek, J. Karpinski, C. Rossel, B. Pümpin, H. Keller, W. Kündig, G. van Tendeloo, T. Krekels, and C. Amelinckx, to be published in *Proceedings of the E-MRS Spring Meeting 1990, Symposium A, High-T_c Superconductor Materials*, Strasbourg, 1990.

[40] K. Conder, S. Rusiecki, and E. Kaldis, Materials Research Bulletin **24**, 581 (1989).

[41] J. Kowalewski, D. Nikl, and W. Assmus, Physica C **153-155**, 429 (1988).

[42] B. Pümpin, H. Keller, W. Kündig, I.M. Savić, J.W. Schneider, H. Simmler, P. Zimmermann, E. Kaldis, S. Rusiecki, and C. Rossel, to be published in Hyperfine Interactions.

[43] B. Pümpin, H. Keller, W. Kündig, I.M. Savić, J.W. Schneider, H. Simmler, P. Zimmermann, E. Kaldis, S. Rusiecki, C. Rossel, and E.M. Forgan, to be published in *Proceedings of the E-MRS Spring Meeting 1990, Symposium A, High-T_c Superconductor Materials*, Strasbourg, 1990.

[44] P. Zimmermann, H. Keller, W. Kündig, B. Pümpin, I.M. Savić, J.W. Schneider, H. Simmler, E. Kaldis, and S. Rusiecki, to be published in Hyperfine Interactions.

[45] R.J. Cava, A.W. Hewat, E.A Hewat, B. Batlogg, M. Marezio, K.M. Rabe, J.J. Krajewski, W.F. Peck Jr., and L.W. Rupp Jr., Physica C **165**, 419 (1990).

[46] E. Kaldis, P. Fischer, A.W. Hewat, E.A. Hewat, J. Karpinski, and S. Rusiecki, Physica C **159**, 668 (1989).

[47] A.R. Bishop, R.L. Martin, K.A. Müller, and Z. Tesanovic, Z. Phys. B - Condensed Matter **76**, 17 (1989).

[48] K.A. Müller, Z. Phys. B - Condensed Matter, in press.

[49] Note that σ used in Ref. [17] is per definition a factor $\sqrt{2}$ larger than σ used in this paper [Eqs. (2) and (3)].

[50] R.B. Laughlin, Phys. Rev. Lett. **60**, 2677 (1988); Science **242**, 525 (1988); B.I. Halperin, J. March-Russell, and F. Wilczek, Phys. Rev. B **40**, 8726 (1989).

[51] A. Weidinger, Ch. Niedermayer, A. Golnik, R. Simon, E. Recknagel, J.I. Budnick, B. Chamberland, and C. Baines, Phys. Rev. Lett. **62**, 102 (1989), and references quoted therein.

[52] R.F. Kiefl, J.H. Brewer, J. Carolan, P. Dosanjh, W.N. Hardy, R. Kadono, J.R. Kempton, R. Krahn, P. Schleger, B.X. Yang, Hu Zhou, G.M. Luke, B. Sternlieb, Y.J. Uemura, W.J. Kossler, X.H. Yu, E.J. Ansaldo, H. Takagi, S. Uchida, and C.L. Seaman, Phys. Rev. Lett. **63**, 2136 (1989), and references quoted therein.

[53] A. Aharony, R.J. Birgeneau, A. Coniglio, M.A. Kastner, and H.E. Stanley, Phys. Rev. Lett. **60**, 1330 (1988).

[54] A. Weidinger, Ch. Niedermayer, A. Golnik, R. Simon, E. Recknagel, J.I. Budnick, B. Chamberland, and C. Baines, Comment, Phys. Rev. Lett. **63**, 2539 (1989).

Muonium and Hydrogen Defect Centres in Solids

S.F.J. Cox

ISIS Pulsed Muon Facility, Rutherford Appleton Laboratory,
Chilton, Oxfordshire, OX11 0QX, UK

Hydrogen and its light isotope, muonium, may both be observed in insulating crystals as trapped interstitial atoms. Their difference in mass gives rise to a small dynamic isotope effect in hyperfine coupling constant and a large difference in diffusive behaviour - muonium showing in some lattices a striking minimum in mobility versus temperature.

Muonium may also be detected as isolated paramagnetic centres in the semiconductors whereas information on hydrogen in these materials is restricted to diamagnetic states in association with other impurities. Arguably one of the most significant findings of the μSR technique is the coexistence of two distinct types of muonium centre in many of the tetrahedrally co-ordinated lattices.

In literature spanning more than a decade, the isotropic state was described as "normal" muonium and assigned to the tetrahedral interstitial site, while the nature of the "anomalous" anisotropic state remained a puzzle. Within the last few years the situation has entirely reversed: the anisotropic state has been recognised as the more stable of the two and its electronic structure and (bond-centre) location have been well established, whereas the precise location of the isotropic state and the nature of its metastability have become open questions.

1. VACUUM STATE MUONIUM

As far as atomic, solid state and chemical physics are concerned, the positive muon behaves like a light weight proton. Emphasizing this behaviour, it is able to bind an electron to form a hydrogen-like atom, μ^+e^-. This is known as muonium, and given the chemical symbol Mu. It contrasts with positronium in having an essentially central nucleus - the electron reduced mass is only 0.5% lower than in protium - so that its size and binding energy are very similar to those of ordinary atomic hydrogen and its more familiar heavy isotopes(Table 1).

TABLE 1. Properties of the positive muon, referred to those of the proton, and properties of (vacuum state) muonium, referred to those of protium.

μ^+	Charge	$Q_\mu = 1 = Q_p$
	Spin	$I_\mu = 1/2 = I_p$
	Mass	$m_\mu = 0.11(\simeq 1/9)m_p$
	Magnetic moment	$\mu_\mu = 3.18\ \mu_p$
Mu	Bohr radius	$a = 1.004a_o$
	Ionisation potential	$I = 0.996\ I_H$
	Hyperfine constant	$A = 4.46\text{GHz} \simeq (\mu_\mu/\mu_p)A_H$

Exotic Atoms in Condensed Matter
© Springer-Verlag Berlin, Heidelberg 1992

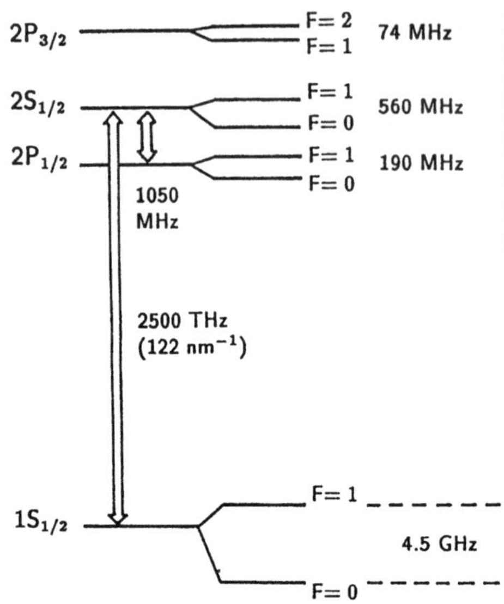

Figure 1. Energy level diagrams for atomic muonium, including hyperfine splittings (not to scale!). The n=1 to n=2 separation (the Rydberg constant) and the n = 2 splitting (the Lamb shift) are shown and refer to the experiments mentioned in Section 1. The diagram is schematically equivalent to that for atomic hydrogen.

Muonium shares with positronium the property of being an all leptonic system, however, and as such lends itself to extremely accurate QED calculations of its spectroscopic constants /1/. In hydrogen, these are limited by uncertainties in the proton size. Measurements of the muonium Rydberg constant and Lamb shift are as yet far from being precise enough to test the theory, although the relevant transitions (Figure 1) have been detected /2,3/. More ambitious measurements are in progress.

In these experiments the muon acquires an electron in passage through a thin target. Fine silica powder yields thermalised muonium which escapes or is ejected into the surrounding vacuum; thin metallic foil gives some admixture of excited states.

2. HYPERFINE CONSTANT

The most precise spectroscopic measurement to date is of the muon-electron hyperfine coupling constant in the ground state. From the frequencies of the high-field transitions in Figure 2, induced by microwave resonance on muonium formed by stopping muons in a low pressure gas, Casperson et al./4/ obtained the value

$$A = 4.463\ 302\ 88\ (16)\ \text{GHz}\ (0.04\text{ppm}).$$

The remainder of this Lecture is concerned with muonium formed by stopping or thermalising muons in bulk solid media. For this purpose such precision is not required; the hyperfine constant of the free atom serves as a reference point and is essentially equal to that of free hydrogen, scaled by the ratio of magnetic moments of the muon and proton.

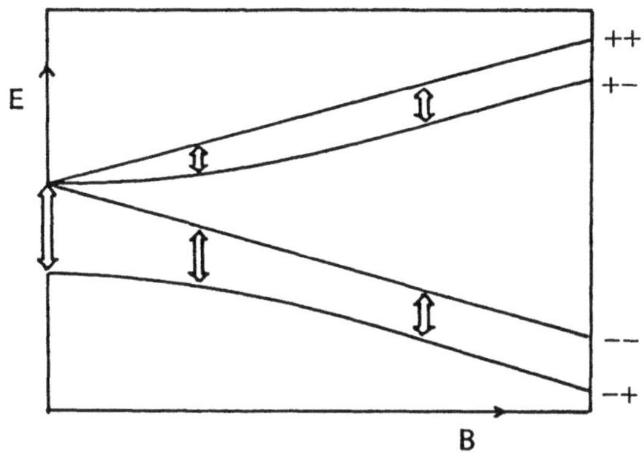

Figure 2. The hyperfine and Zeeman splittings of muonium in its ground state. A schematically equivalent diagram is valid for muonium in solids, in the case that the hyperfine interaction is isotropic and there is no interaction with other nuclear spins. μSR spectroscopy of isotropic muonium centres in solids (Section 3 et seq) makes use of all the transitions indicated, although only the high field transitions may be detectable in the presence of super-hyperfine interactions.

3. MUONIUM IN SOLIDS: SYSTEMATICS OF THE HYPERFINE CONSTANT

Values of hyperfine constant for muonium in a variety of solids are reported in Figure 3, scaled to the free atom value. This represents the data of various different groups, working world wide, though the precision measurements are principally from the Zurich group. Measurements are made from the μSR precession frequencies detected in high field, intermediate field or (most directly) zero field; the transitions involved are indicated in Figure 2 and the techniques described elsewhere /5,6/.

Values close to the free atom value (unity in Figure 3) indicate that the muonium, though trapped at interstitial sites in the crystal lattices, is only weakly perturbed by the host. Values slightly greater than unity must indicate that the hydrogen-like atom is slightly compressed or squeezed in the interstitial cages: spin density is enhanced at the nucleus (the interstitial muon) as though the electronic wave function were forbidden to overlap with surrounding atoms in the most ionic hosts (the alkali fluorides). Slightly lower values suggest instead a dilation of the atom, i.e. overlap of the wavefunction onto the neighbours.

The substantial lowering of hyperfine constant in the semiconductors, to around 50% of the free atom value in silicon and germanium, must correspond to a very considerable overlap or delocalisation in these (covalent) lattices: a molecular orbital is now required to describe the electronic structure of these defect centres (see Section 7).

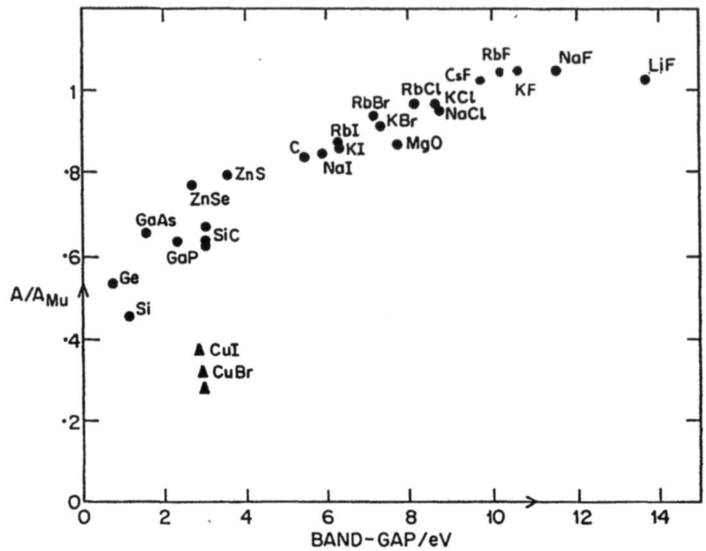

Figure 3. Hyperfine constants for isotropic muonium centres in various crystalline solids, scaled to the free atom value as a measure of "spin density" on the muon. The correlation with band-gap of the host material is taken from /6/, which contains references to the original measurements.

4. NOMENCLATURE

No entirely satisfactory (i.e. monotonic) correlation of hyperfine constant with an electronic property of the host material has been discovered. Ionicity, for instance, cannot be a relevant parameter since such similar couplings are found in diamond (perfectly covalent with tetrahedral co-ordination) and in MgO (highly ionic with octahedral co-ordination). A broad correlation with band gap, illustrated in Figure 3, is found for many of the centres /6/; it is reasonable to suppose that these have a fundamentally similar electronic structure and in the following they are designated Mu' centres (reserving plain Mu for the free atom). The scheme fails to embrace the data for the cuprous halides, which suggests that in these materials the electronic structure of the muonium defect centres is rather different; they are therefore distinguished by the symbol Mu''.

There are no such paramagnetic centres in metals, i.e. no states in which a magnetic moment is centred on the muon or proton. In metals, muonium and hydrogen adopt diamagnetic states. Conduction electrons are attracted to the positive charge of the interstitial muon or proton, screening the Coulomb potential so that a bound state with a particular electron is impossible /7/. In Figure 3, metals could be represented at the origin! The hyperfine or contact interaction which arises from a polarisation of the screening change in magnetic metals, and the very high muon Knight shift in some semimetals which suggests incipient muonium formation, are dealt with elsewhere /6/.

The nature of the Mu' and Mu'' centres is examined in Sections 5-9. Not represented in Figure 3 are the highly anisotropic muonium centres which also exist in certain materials: these are commonly designated Mu* and are the subject of Sections 10–13.

5. COMPARISON WITH HYDROGEN

Hydrogen defect centres are well known in the oxides and halides: these are the traditional hosts for their study by ESR and optical spectroscopy /7/. The systematics of the hyperfine coupling from host to host are similar, but the absolute values of spin density are consistently lower (by about 5%) for muonium than for hydrogen /8/. This can be understood by the greater zero-point energy and vibrational amplitude or spread of the lighter particle. The muon explores more of the interstitial cavity in which it is confined than does the proton. It is on average slightly closer to the neighbouring atoms and forces greater overlap or delocalisation of its electronic wavefunction onto them.

Most interestingly, however, no such hydrogen defect centres are known in the semiconductors. Referring to Figure 3, this seems strange, since there is evidently no qualitative or step change in the nature of muonium centres on going from the octahedrally to the tetrahedrally co-ordinated lattices. Yet the hydrogen analogues have not been reported, either in the elemental (GroupIV) or compound (III-V) semiconductors; muonium appears to be a unique substitute for paramagnetic hydrogen in these cases.

Hydrogen is known to be present in diamagnetic states, however, especially in the compound semiconductors, and its existence in electronic grade material is now understood to be important. It is inevitably incorporated in material produced by the reaction of hydride gases and is in fact essential for the compensation or passivation of electrically active impurities - the non-observation of paramagnetic hydrogen is undoubtedly related to this point, which is examined in Section 14. The first exploratory μSR studies on an amorphous semiconductor (a-Si:H, which contains a lot of hydrogen) are summarised in Section 15.

6. MUONIUM DIFFUSION

A recent and particularly elegant contribution to this field is a study of the diffusion, as opposed to the spectroscopy, of muonium in solids. In these experiments the relaxation time T_1 of the muon Zeeman energy is measured as a function of temperature. The relaxation is induced by fluctuations of the superhyperfine interaction with neighbouring nuclei as the muonium moves from site to site. A minimum of T_1 (well known in conventional magnetic resonance) is expected when the hop rate matches a transition frequency of the muon spin. Two such minima are observed, moving apart in temperature as the working field is increased. This implies that the muonium mobility itself goes though a minimum at a particular temperature and increases above and below this temperature (Figure 4). Results are available for various alkali halides and for a compound semiconductor /9/; these constitute a striking demonstration of the diffusive behaviour of a "small polaron" defect and the measurements will undoubtedly be extended to many other systems.

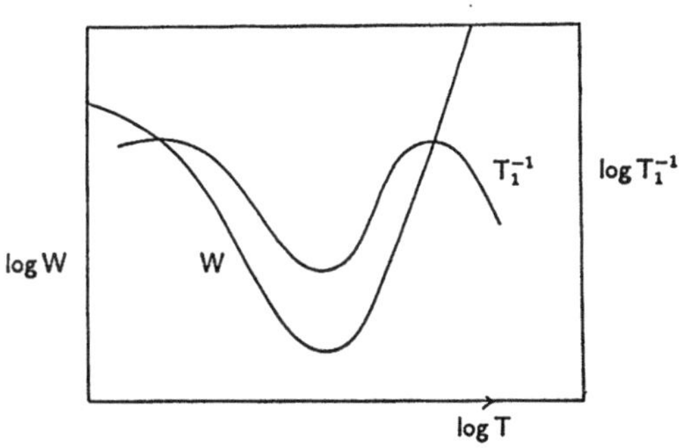

Figure 4. Behaviour of the muon spin-lattice relaxation rate T_1^{-1} for the case that the muonium hop rate W exhibits a minimum. (Sketched from /9/; the mobility minimum is at 50K in the "insulator" KCl and 90K in the "semiconductor" GaAs.

7. NATURE OF THE Mu' CENTRES IN SEMICONDUCTORS

The muonium defect centres represented in Figure 3 have muon-electron hyperfine couplings which are essentially isotropic. For a given material, their characteristic frequencies in the μSR spectra are the same for all orientations of a single-crystal sample and may be detected equally well in polycrystalline or powder samples. These centres were assumed by all authors in the early μSR literature to be located at the tetrahedral cage centre (Figure 5a), this being the only site with the necessary cubic symmetry. An objection was raised on chemical grounds by Cox and Symons (1986) /10/ who pointed out the tendency for muonium to react and bond with one of its nearest neighbours, weakening the in-line Si-Si bond and displacing the muon from the tetrahedral (T) site to an off-centre or antibonding (AB) site, as in Figure 5(b). The term "antibonding" emphasises how the unpaired electron delocalises via the vacant antibonding orbital on the Si-Si pair.

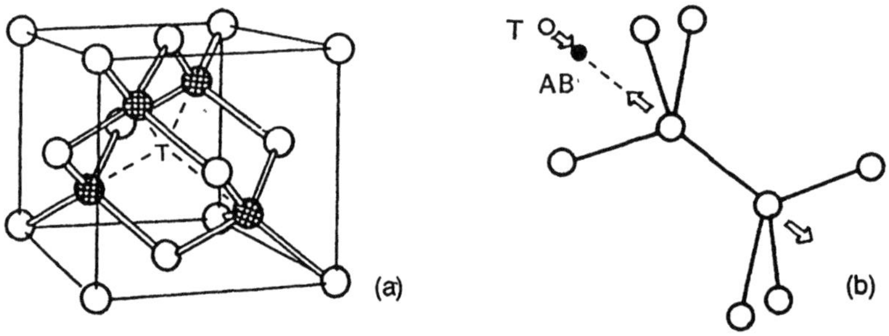

Figure 5. The diamond-type lattice showing a tetrahedral (body centre) interstitial site (a); the four nearest neighbours (nn) are shown shaded. Weakening of a Si-Si bond as the muonium (or hydrogen) is moved from the tetrahedral (T) to an antibonding (AB) site closer to one of these neighbours is detailed in (b).

There is at present no consensus on this point. In quantum chemical calculations performed on clusters of atoms, simulating the crystal lattice and defect centre, little relaxation of the tetrahedral cage is found when the muon or proton is placed and held at the cage centre - the T site /11/. Most authors have been content to leave it at that. Calculations of the potential surface on which the muon or proton move and seek their equilibrium position inevitably give a broad minimum at the T site, as in Figure 6(a), when relaxations of the neighbouring atoms are ignored or constrained to tetrahedral symmetry. Without this constraint, calculations become more expensive in terms of computer time, but tend to bear out the chemical expectation: some authors find local minima at the AB sites/12/; others, however, find only a saddle point /13/. The location and metastability (See Section 14) of the Mu' centres have therefore become open questions.

8. DYNAMICAL MODELS

An instantaneous location for the muon which is displaced from the T site would imply some sort of dynamical averaging to secure the effective isotropy of the hyperfine coupling. Various possibilities are illustrated in Figure 6. Either there is a broad minimum of potential energy at the T-site, or local minima at the AB sites which are not deep enough to localise the muon. It is interesting to conjecture whether they are deep enough to localise protons, however, for which

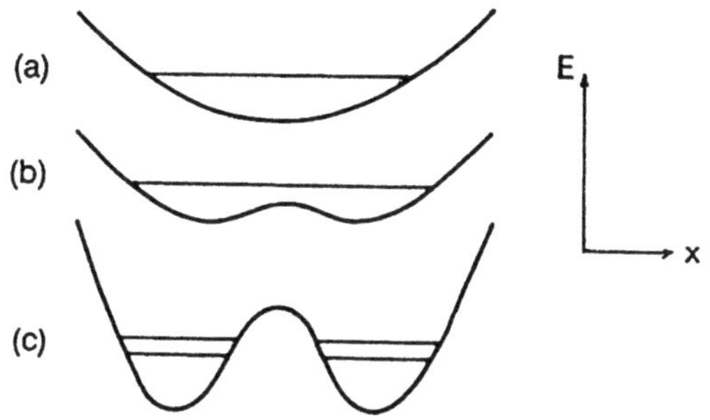

Figure 6. Potential energy surface and ground state energy for interstitial muonium. For simplicity, the surface is represented in one dimension with a broad minimum at the T site in (a), and with local minima of increasing depth off-set towards neighbouring atoms in (b) and (c). (There would be four such minima in the 3-D surface). In (a) and (b) the muon is delocalised over the cage by virtue of its zero point energy. In (c) the local minima are deep enough to contain the zero point energy but the degeneracy of levels in adjacent wells is lifted by tunneling.

the zero point of energy is lower by a factor $(m_p/m_\mu)^{1/2} \approx 3$; this would also explain why isotropic hydrogen centres have not been observed in semiconductors!

Hyperfine coupling, as well as equilibrium geometry and energy, can in principle be obtained in calculations which do not use spin-restricted wave functions. Of especial interest would be values which are correctly averaged over the spread in position of the muon or proton. Calculations in which the muon (or proton) is held static at the T-site consistently overestimate the contact interaction, often giving values larger than for the free atoms themselves. This is itself an indication of the necessity for dynamical models, since spin density on the muon (or proton) is expected to fall off as the particle moves towards a neighbouring atom. The vibrational averages should more realistically model the measured couplings, as well as the isotope effect between muonium and hydrogen.

9. THE Mu" CENTRES

The muonium centres detected in the cuprous halides appear from Figure 3 to have a fundamentally different character to those in other materials, especially when they are considered as I-VII compounds and contrasted with the alkali halides! The cuprous halides are unusual materials in several respects, including the fact that they have vacant cation s-states lying energetically close to the top of the valence band. A $CuMu^+$ radical-cation model has been proposed for these centres /10/. It also implies that the muon will be instantaneously displaced towards a $Cu+$ neighbour.

Cuprous chloride is interesting in that two distinct Mu" centres are observed, with slightly different hyperfine couplings - both isotropic /14/. Level crossing resonance experiments confirm that for *both* centres the muon has four equivalent copper neighbours /15/. This result is consistent with the situation represented in Figure 6(c), in which the two centres would correspond to the two distinct tunneling states.

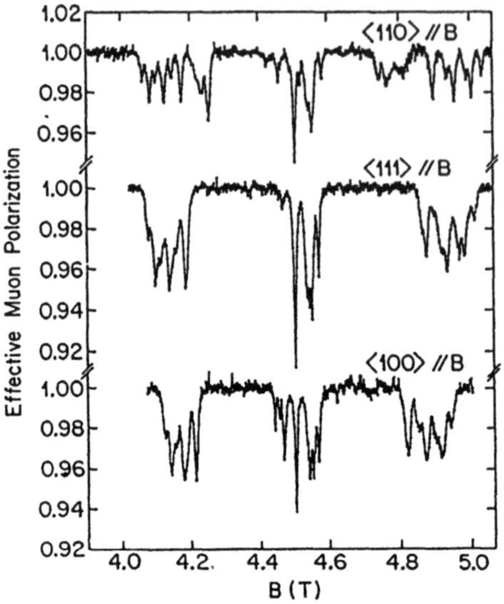

Figure 7. Level crossing resonance spectrum for one of the Mu" centres in CuCl reproduced from the thesis of Schneider (1990); see also Schneider et al. /15/. The spectrum represents resonant polarisation transfer from the muon to the neighbouring nuclei.

These spectra are surely the most beautiful in the μSR literature; an example is reproduced in Figure 7.

10. THE Mu* CENTRES IN SEMICONDUCTORS

As if the observation of hydrogen-like defect centres was not already enough, μSR experiments have also demonstrated that a second type of muonium defect centre is formed in some of the semiconductors /16/. This was quite unexpected; it constituted a real scoop for the μSR technique. For the second type of centre, known as Mu*, the hyperfine coupling is smaller again, and highly anistropic. A characteristic spectrum is shown in Figure 8.

Such centres are formed in the Group IV elements with the diamond-type lattice, i.e. Si and Ge as well as diamond itself. They are also formed in most of the compound semiconductors having tetrahedral co-ordination (the "zincblende" structure), namely the III-V's GaAs and GaP and the II-VI's ZnSe and (zincblende itself) ZnS /18/. Although dubbed "anomalous muonium" in the early literature, the Mu* centres are now thought to be to be more stable than the Mu'. That is, the isotropic centres are only metastable (both types are formed by muon implantation and they coexist at low temperature). Thermal conversion of Mu' to Mu* has been observed directly, both in diamond and in silicon /19/.

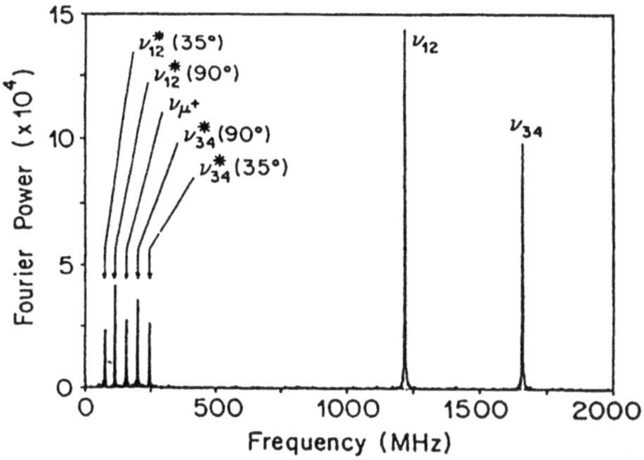

Figure 8. The μSR spectrum of GaAs (taken from /17/), showing frequencies corresponding to the Mu' and Mu* centres (a). The Mu' frequencies are independent of crystal orientation so that the hyperfine coupling is isotropic as implied in the Breit-Rabi diagram of Figure 2. The Mu* frequencies vary strongly with orientation, indicating an anisotropic coupling. /18/.

11. THE BOND-CENTRE MODEL FOR Mu*

Deduction of the spin Hamiltonian parameters for the various Mu* centres from their orientation-dependent μSR spectra is one of μSR's most elegant and important achievements /18/. It is instructive to go further, and to decompose the principal values of the hyperfine tensor into isotropic and dipolar terms:

$$A_{iso} = 1/3(A_{\parallel} + 2A_{\perp}) \text{ and } B = 1/3(A_{\parallel} - A_{\perp}).$$

The isotropic term is then found to be negligible compared with its value for the Mu' centres. There is therefore virtually no contact interaction - the electronic density $|\psi_e|^2$ is extremely small at the muon site. The dipolar term is surprisingly large, however, and indicates that the unpaired electron is nonetheless localised very close nearby - along a <111> direction from the muon.

These considerations led Cox and Symons /10/ to propose a bond-centre location for the muon, together with a molecular orbital model for the electronic structure which is illustrated in Figure 9(b). The singly occupied molecular orbital (SOMO) is non-bonding on the muon: it is antisymmetric about this location, which accounts for the negligible contact interaction. The unpaired electron is localised close by, as required, spending half its time on each nearest neighbour in this simplified 3-centre model. The SOMO may be thought of as deriving from the antibonding orbital of the original Si-Si pair (Figure 9a); the original bond must be stretched by about 40% to accommodate the muon so this orbital is lowered and detached from the conduction band. In other words, the neighbouring atoms relax from their normal positions and a localised electron state is created at the defect.

Although greeted intitially with considerable scepticism, the bond-centre model for Mu* is now thoroughly substantiated, both by experiment and theory! The experimental support is

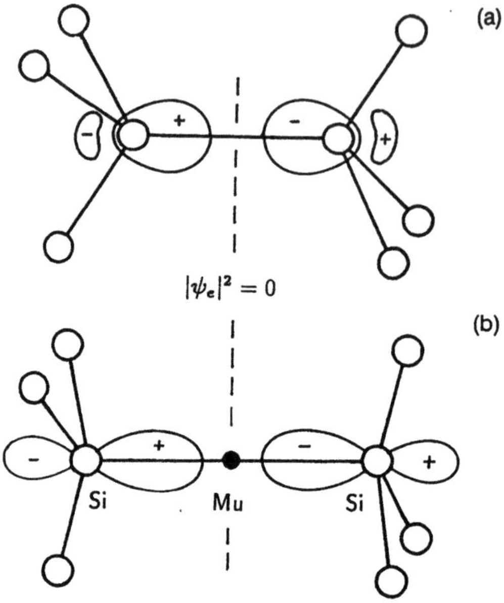

Figure 9. Bond-centre location for the muon in the Mu* centres depicting the undisturbed lattice before the arrival of the muon in (a), and the relaxation required to accommodate the muon in (b). The vacant antibonding orbital on the central Si-Si pair in (a) and the singly occupied molecular orbital for Mu* in (b) are sketched to show the nodal plane though this site.

summarised by Estle and Kiefl /20/, in a short compendium which nicely updates the comprehensive review of early work given by Patterson /18/.

12. LEVEL CROSSING RESONANCE

The most compelling evidence comes from Level Crossing Resonance (LCR) experiments in GaAs and GaP, which are able to map out the electronic spin density on the atoms neighbouring the defect centre /21/. The findings may be summarised as follows (orbital populations are estimated using $a^2_s = A_{iso}/A_0$ and $a^2_p = 2B/2B_0$, where A_0 and B_0 are atomic values):

Mu* in GaAs :

	nn Ga	nnAs
a^2_s	8%	2%
a^2_p	30%	43%
$a^2_s + a^2_p$	38%	45%
a^2_p/a^2_s	4	23

Mu* in Si :

	each nn Si
a^2_s	2%
a^2_p	19%
$a^2_s + a^2_p$	21%
a^2_p/a^2_s	10

The essential feature is that over 80% (38% + 45%) of the spin density is located on the two nearest neighbours (so that the 3-centre model is a remarkably good approximation). The p/s ratios should be compared with the value for sp^3 hybrid co-ordination; they give an indication of the local distortion of the lattice. The greater s-character on Ga and p-character on As is also understandable in this slightly ionic material.

In silicon, delocalisation onto the surrounding lattice is evidently greater: LCR accounts for only 50% of the spin density on the nearest and next nearest neighbours. These experiments are harder in this material since only 5% of the nuclei (^{29}Si) posess a nuclear spin; the low isotopic abundance may be turned to advantange in interpreting the intensities of satellite lines in the μSR precession spectra, however - a procedure which also confirms the bond-centre site by demonstrating that the muon has two equivalent close neighbours.

13. QUANTUM MECHANICAL CALCULATIONS

The experimental findings and the qualitative molecular orbital model have stimulated quantitative quantum mechanical calculations by a number of groups /22/. These include cluster calculations using ab initio (e.g. Hartree Fock) or local density functional methods as well as supercell calculations using local density functional methods. All of these methods give results for the equilibrium configuration and energy; the more detailed produce a three dimensional potential energy surface on which the muon (or proton) finds its equilibrium position. Those which are not spin restricted also give results for the hyperfine parameters. All authors agree that the bond-centre location is the global minimum for the neutral interstitial defect, muonium or hydrogen, both for the elemental and for the compound semicondutors.

The earlier calculations identified an energy barrier between the T site and the BC site, qualitatively consistent with the metastabilty of Mu' and the thermal conversion of Mu' to Mu*. In the light of the uncertainty now surrounding the precise location of Mu' (Section 7), details of the route to the BC site need to be re-evaluated. The basic picture is undoubtedly correct, however, namely that Mu' is a mobile species moving between the open spaces of the interstitial cages and carrying only a small lattice distortion, and that an energy barrier has to be surmounted before the more stable bond-centred location, with its huge associated relaxation, can be reached. This barrier is the energy required to stretch the intrinsic bond so that the muon can be accommodated.

14. HYDROGEN, METASTABILITY, AND ASSOCIATION WITH OTHER IMPURITIES

μSR experiments have focused on the paramagnetic (neutral) states of muonium since comparatively little information is available on the diamagnetic state or states. Theoretical treatments for hydrogen, on the other hand, have addressed all possible charge states. The findings of various authors /23/ are summarised in Figure 10. The bond-centre site remains stable in p-type silicon, but the defect becomes positively charged (H^+). The cage interstitial site is only stable in n-type material, and then only as the negative charge state (H^-) This explains how hydrogen is so efficient in the compensation of electrically active impurities in semiconductors, and gives a clue as to why the isotropic paramagnetic state corresponding to Mu' has not been detected: the microsecond timescale of the μSR techniques is required to display this metastable state. Even the neutral bond-centre state is electronically or structurally unstable in the presence of other impurities, which is probably why illumination of the sample is needed to display the hydrogen analogue of Mu* /24/.

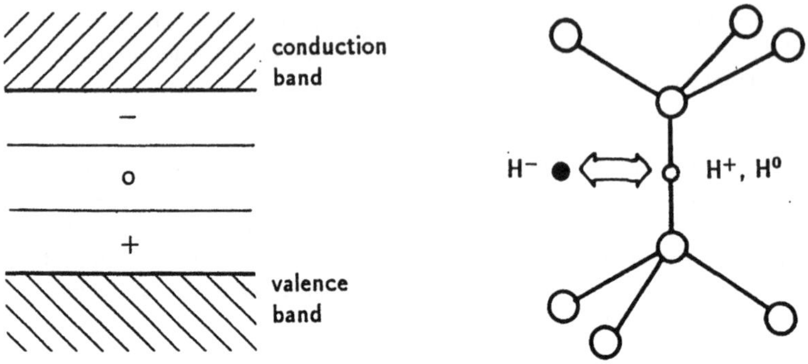

Figure 10. Variation of the location and charge state of interstitial hydrogen with Fermi energy /23/. The energy gap is divided into three zones. When the Fermi energy lies in the upper zone H^- is stable at a tetrahedral cage centre; when it lies in the lower zone, H^+ is stable at a bond-centre. H^0 is stable at a bond-centre in intrinsic material, i.e. when the Fermi energy is at mid-gap.

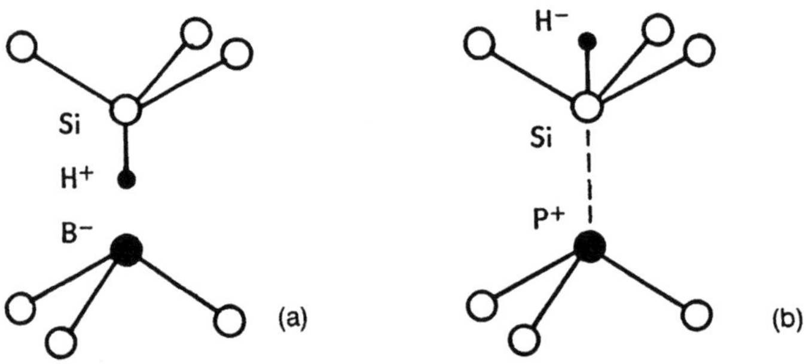

Figure 11. Hydrogen-inpurity complexes. An acceptor (eg B in Si) may be compensated by electron transfer from hydrogen at an adjacent bond-centre (a); a donor (eg P in Si) may be passivated by transfer of an electron to hydrogen at a nearby antibonding site (b) /23,26/.

Assuming these considerations carry over to pionium ($\pi^+ e^-$), ionisation provides an alternative explanation of the site change under illumination for pions in germanium, demonstrated by channeling experiments /25/. This result could simply represent the preferred locations of the various charge states.

These compensation processes can in principle occur even if the hydrogen or muonium is spatially remote from the other impurities - it is the "recombination" mechanisms and rates which determine the metastable lifetimes of the muonium states. Structural equilibrium in the presence of other impurities (as for annealed material) is illustrated in Figure 11. Here the hydrogen is physically adjacent to the other impurities. In the case of shallow donors, the impurities are passivated. That is, electrically active levels are removed entirely from the energy gap. The stability of these complexes is predicted by various authors and there is now direct evidence for their existence from IR spectroscopy /23,26/.

15. AMORPHOUS SILICON

What becomes of the Mu', Mu* and diamagnetic states on going from crystalline to amorphous material? The random network of amorphous silicon, for instance, might be expected to provide a variety of environments for Mu' (through the variety of cage sizes) and also for Mu* (through the existence of strained Si-Si bonds). The possible locations are illustrated in Figure 12. Preliminary experiments on amorphous hydrogenated silicon (a-Si:H) suggest that both the Mu* and diamagnetic fractions are enhanced at the expense of Mu' /27/. The increased Mu* fraction can be attributed to easier formation of the bond-centred defect in the presence of weakened Si-Si bonds. The increased diamagnetic fraction is unlikely to represent trapping at dangling bonds since very few Si atoms remain undercoordinated in hydrogenated material (this role of hydrogen in passivating electrically active centres is in fact essential in electronic grade amorphous silicon); probably the diamagnetic fraction in both the amorphous and crystalline materials is ionised Mu*. Evidence for a Mu' fraction in a-Si:H is to date inconclusive; it may be that the loss of periodicity greatly curtails the Mu 'mobility - to the point where it becomes stuck at a particular site and is no longer isotropic - or indeed that rapid conversion to Mu* is facilitated in the disordered structure.

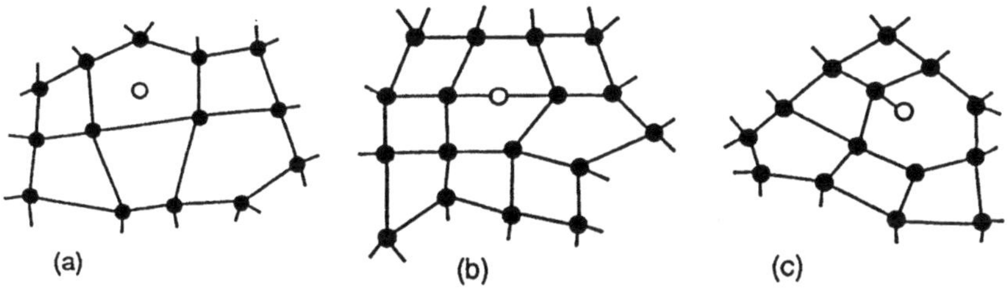

Figure 12. Schematic (2-dimensional) representation of amorphous silicon depicting proton or muon sites within cages (a), at bond-centres (b) or saturating dangling bonds (c)

REFERENCES

1. V W Hughes and T Kinoshita in "Muon Physics I: Electromagnetic Interaction" ed V W Hughes and C S Wu, Academic Press (New York, 1977) 11

2. S Chu, A P Mills, A G Yodh, K Nagamine, Y Miyke and T Kinga, Phys Rev Lett **60** (1988) 101

3. C J Oram, J M Bailey, P W Schmor, C A Fry, R F Kiefl, J B Warren, G M Marshall and A Chu, Phys Rev Lett **52** (1984) 910;
A Badertscher, S Dhawan, P O Egan, V W Hughes, D C Lu, M W Ritter, K A Woodle, M Gladisch, H Orth, G zu Putlitz, M Eckhouse, J Kane, F G Mariam and J Reidy, Phys Rev Lett **52** (1984) 914

4. D E Casperson, T W Crane, A B Denison, P O Egan, V W Hughes, F G Mariam, H Orth, H W Reist, P A Sonder, R D Stambaugh, P A Thompson and G zu Putlitz, Phys Rev Lett **38** (1977) 956

5. H Keller, this workshop

6. S F J Cox, J Phys C Solid State Physics 20 (1987) 3187

7. The criterion is given for instance by A M Stoneham "Theory of Defects in Solids" (Oxford, OUP: 1985)

8. H P Baumeler, K W Blazey, W J Choyke, T L Estle, H Keller, R F Kiefl, W Kündig, P F Meier, W Odermatt, B D Patterson, S L Rudaz, J W Schneider and C Schwab, SIN Newsletter **17** (1985) 69

9. R F Kiefl, R Kadono, J H Brewer, G M Luke, H K Yen, M Celio and E J Ansaldo, Phys Rev Lett **62** (1989) 792;
R Kadono, R F Kiefl, E J Ansaldo, J H Brewer, M Celio, S R Kreitzman and G M Luke, Phys Rev Lett **54** (1989) 1065;
R Kadono, R F Keifl, J H Brewer, G M Luke, T Pfiz, T M Riseman, B J Sternlieb, Proc μSR 90 (Hyperfine Interactions **64** (1990) 635)

10. S F J Cox and M C R Symons, Chem Phys Lett **126** (1986) 516

11. See for example S Estreicher, Phys Rev B **36** (1987) 9122;
N Sahoo, S K Mishra, K C Mishra, A Coker, T P Das, C K Mitra, L C Snyder and A Glodeanu, Phys Rev Lett **50** (1983) 913

12. Dj Maric, unpublished (1990);
P R Briddon and R Jones, Proc μSR 90 (Hyperfine Interactions **64** (1990) 593) ; L Paresi, P Giannozzi and F K Reinhart (submitted to Phys Rev B: Rapid Communications, 1990)

13. C G Van de Walle, P J H Denteneer, Y Bar-Yam and S T Pantelides, Phys Rev **B39** (1989) 10791

14. R F Kiefl, W Odermatt, H-P Baumeler, J Felber, H Keller, W Kündig, K W Blazey, T L Estle and C Schwab, Phys Rev **B34** (1986) 1474

15. J W Schneider, Thesis, University of Zurich (1989);
J W Schneider, M Celio, H Keller, W Kündig, W Odermatt, B D Patterson, B Pümpin, I M Savic, H Simmler, T L Estle, C Schwab, R F Kiefl and D Renker, Phys Rev B**41** (1990) 7254;
J W Schneider, H Keller, W Odermatt, B Pümpin, I M Savic, H Simmler, S A Dodds, T L Estle, R C DuVarney, K Chow, R Kadono, R F Kiefl, Q Li, T M Riseman, H Z Zhou, R L Lichti and C Schwab, Proc μSR 90 (Hyperfine Interactions **64** (1990) 543)

16. J H Brewer, K M Crowe, F N Gygax, R F Johnson, B D Patterson, D G Fleming and A Schenk, Phys Rev Lett **31** (1973) 143

17. R F Kiefl, J W Schneider, H Keller, W Kündig, W Odermatt, B D Patterson, K W Blazey, T L Estle and S L Rudaz, Phys Rev **B32** (1985) 530

18. B D Patterson, Rev Mod Phys **60** (1988) 69, plus references therein

19. W Odermatt, H-P Baumeler, H Keller, W Kündig, B D Patterson, J W Schneider, J P F Sellschop, M C Stemmet, S Connell and D P Spencer, Phys Rev **B38** (1988) 4388;
E Westhauser, E Albert, M Hamma, E Recknagel, A Weidinger and P Moser, Hyp Int **32** (1986) 589

20. R F Kiefl and T L Estle, in "Hydrogen in Semiconductors" (Semiconductors and Semimetals Vol 34) eds J I Pankove and N M Johnson, Academic Press (1991) ch 15

21. R F Kiefl, M Celio, T L Estle, G M Luke, S R Kreitzman, J H Brewer, D R Noakes, E J Ansaldo and K Nishiyama, Phys Rev Lett **58** (1987) 1780;
R F Kiefl, M Celio, T L Estle, S R Kreitzman, G M Luke, T M Riseman and E J Ansaldo, Phys Rev Lett **60** (1988) 224

22. T A Claxton, A Evans and M C R Symons, J Chem Soc Faraday Trans 2 **32** (1986) 2031;
T L Estle, S Estreicher and D S Maryrick, Phys Rev Lett **58** (1987) 1547;
S Vogel, M Celio, Dj M Maric and P F Meier, J Phys Condens matter **1** (1989) 4729;
Dj Maric, S Vogel, P F Meier and S K Estreicher, Phys Rev **B40** (1989-II) 8545;
C G Van de Walle, Phys Rev Lett **64** (1990) 669

23. R Jones, at the Chelsea Meeting on Amorphous and Liquid Semeiconductors (December 1989); M Stutzman, at the Chelsea Meeting on Amorphous and Liquid Semiconductors (December 1989); see also C G Van de Walle, P J H Deuteneer, Y Bar-Yam and S T Pontelides, Phys Rev B**34** (1986) 1474

24. Yu V Gorelkinskii and N N Nevinnyi, Pis'ma Zh Tekh Fiz **13** (1987) 105 [Sov Tech Phys Lett **13** 45]

25. G Flik, J N Bradbury, D W Cooke, R H Heffner, M Leon, M A Paciotti, M E Schillaci, K Maier, H Rempp, J J Reidy, C Boekema and H Daneil, Phys Rev Lett **57** (1986) 563; see also K Maier, this workshop

26. P J H Denteneer, C G Van de Walle and S J Pantelides, Phys Rev **B39** (1989) 10809; K J Chang and D J Chadi, Phys Rev **B40** (1989-I) 11644; G G De Leo, W B Fowler, T M Sudols and K J O'Brien, Phys Rev **B41** (1990-1) 7581; M Stavola, S J Pearton, J Laporte and W C Bautrement-Smith, App Phys Lett **50** (1987) 1086; K Bergman, M Stavola, S J Pearton and J Loparte, Phys Rev **B37** (1988) 2770

27. S F J Cox, E A Davis, M C R Symons, W Hayes, A Wright, A Singh, F L Pratt, T A Claxton and F Jansen, Proc μSR 90 (Hyperfine Interactions **64** (1990) 551)

Muonium Substituted Molecules

S.F.J. Cox

ISIS Pulsed Muon Facility, Rutherford Appleton Laboratory,
Chilton, Oxfordshire, OX11 0QX, UK

The manner in which Muon Spin Rotation and Level Crossing Resonance are used to characterise muonium substituted organic radicals is described, and illustrated with spectra for the ethyl radical and related species. Comparison with ESR data for the unsubstituted radicals reveals significant structural and hyperfine isotope effects which can be traced to the effects of zero point motion. The first comparable results for a diamagnetic species, exhibiting a quadrupole isotope effect by comparison with conventional NQR data, are presented and discussed.

1. ANALOGIES

Implanted in matter, whether in the solid, liquid or gas phase, positive muons cannot remain as free particles. They seek to lower their energy by the acquisition of electrons. In metals they attract a screening charge of conduction electrons. In gases they stick to atoms (even rare gas atoms) or molecules to form molecular cations or else they strip electrons from the host atoms to form the hydrogen-like atom, μ^+e^-, called muonium. In solids which are insulating or semiconducting, muonium can also be formed, though its properties are modified significantly from those of the free atom; in fact, the stable state may be thought of as the result of a chemical reaction with the host. Likewise in molecular materials, including organic liquids, muonium can be formed - and subsequently react chemically.

In this "chemical" behaviour, the positive muon resembles the proton. Whereas the negative muon plays the role of a heavy electron the positive muon does not play the role of a heavy positron! It adopts localised rather than propagating states in crystalline lattices; it adopts the position of an atomic nucleus rather than an "orbital" state in molecules. It is even worth mentioning that it is used experimentially to probe spin density (compare proton NMR) rather than charge density (contrast positron annihilation) and that its lifetime is unaffected by its chemical state or environment.

Likening the positive muon to a proton would horrify nuclear or high energy physicists! The point is that as far as solid state and chemical physics are concerned, where the phenomena of interest are characterised by energies $\lesssim$ 10 eV, the fact that the muon does not experience the Strong Interaction is usually unimportant. Chemical behaviour is determined by the electromagnetic interaction. The fact that the muon decays via the Weak Interaction is the basis of μSR spectroscopy!

Springer Proceedings in Physics, Vol. 59 Exotic Atoms in Condensed Matter
Editors: G. Benedek · H. Schneuwly © Springer-Verlag Berlin, Heidelberg 1992

2. NOMENCLATURE

Muonium is an unfortunate name, which should have been reserved for $\mu^+\mu^-$. However, we are stuck with it. The muonium - protium analogy can be further underlined by the contrast with positronium, Ps = e^+e^-, which has no central nucleus and only half the binding energy of the hydrogen isotopes. Since the electron in muonium has a reduced mass only 0.5% smaller than in protium, muonium in fact has a size and binding energy very similar to that of hydrogen. Whether it can formally be called a light isotope of hydrogen is perhaps a pedantic question; certainly this is how it behaves chemically /1/. The nomenclature Mu has been adopted both to denote the atomic state and as the chemical symbol. In the following, Mu^+ is used for any cationic species in which the positive muon is associated with a molecule or cluster; μ^+ is reserved for the energetic free particle.

Much of this paper is concerned with muonium which is covalently bonded in organic radicals. These paramagnetic molecules are particularly amenable to study. They are often formed when positive muons are implanted in (unsaturated) organic compounds. They display characteristic frequencies in the muon spin rotation spectra from which the muon-electron hyperfine coupling can be measured and the radical itself identified. That is, the position adopted by muonium in the molecule can be determined. These molecules are called "muonium substituted" in the sense that they each have counterparts in which hydrogen occupies this position, the great majority known previously to conventional radical chemistry and ESR spectroscopy. Thus the ethyl radical $\dot{C}H_2.CH_3$ has a muonium substituted version $\dot{C}H_2.CH_2Mu$. The muon hyperfine couplings are expected to be governed by the same factors as the proton hyperfine constants in "ordinary" radicals and so may be interpreted in the same manner in terms of the electronic and conformational structure of these molecules.

3. DYNAMICAL ISOTOPE EFFECTS

In this respect, muonium substitution may be compared with the more familiar process of deuteration. Of interest is the effect of isotopic substitution on the various 1-electron molecular properties. Deuteration rarely causes these properties to change by more than a few percent, however. The effects are greatly enhanced by muonium substitution, by virtue of the much higher isotopic mass ratio, $m_\mu/m_p \simeq 1/9$. Such a ratio has been described as "unprecedented in the chemical literature" /2/.

Especially it is the existence of a light isotope of hydrogen which gives this enhancement, and favours the observation of quantum dynamical effects. Hyperfine couplings are arguably the easiest 1-electron property to measure with precision and in μSR studies of molecular species they have to date received almost exclusive attention. Hyperfine isotopic effects are displayed by a comparison of μSR data on the muonium substituted radicals with ESR data on their unsubstituted counterparts. This is the subject of Sections 4-14.

The first measurement of a quadrupole isotope effect is reported in Sections15-16, where the prospects for future studies in this and other areas are also examined.

4. ESR AND μSR SPECTROSCOPY

The different manner in which hyperfine couplings are measured by ESR and μSR is illustrated in Figure 1. ESR makes use of transitions of the electron spin, which are shown as light arrows. These are detected at constant frequency and the proton hyperfine constant deduced from the separation of the resonant fields. μSR makes use of the transitions shown as bold arrows; these are analogous to ENDOR transitions in conventional magnetic resonance.

Figure 1 is the Breit-Rabi diagram. This was originally conceived for atomic hydrogen but serves equally well for muonium. It also serves for muonium-substituted radicals - provided the muon-electron hyperfine coupling is isotropic and the hyperfine couplings to other nuclei can be neglected. This is surprisingly often the case! Anisotropy of the couplings is not detected in the liquid (or gas) phase, since it is averaged out by the rapid tumbling motion of the molecules. Muon (or proton) hyperfine couplings in organic radicals rarely exceed 10% of their value in free muonium (or hydrogen) so the Paschen-Back regime is reached in correspondingly lower fields - typically a few kilogauss /3/. When all the nuclear spins present are decoupled, the μSR spectrum collapses to two lines, as illustrated in Figure 2. (The

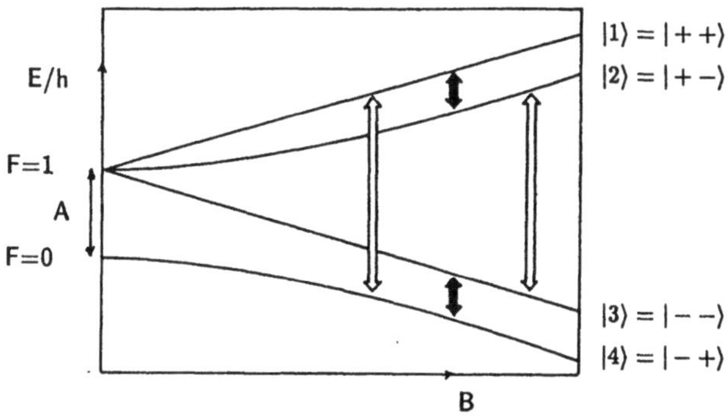

Figure 1. Breit-Rabi diagram for one electron spin (S = ½) and one nuclear spin (I = ½) with an isotropic hyperfine coupling **AI.S**. Light arrows indicate the ESR transitions and bold arrows the μSR or ENDOR transitions.

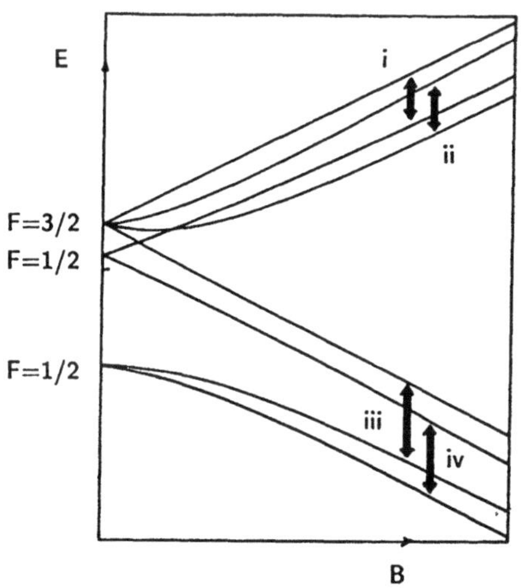

Figure 2. Zeeman diagram for a radical with isotropic couplings between the electron and two spin-½ nuclei. The transition frequencies (i) and (ii) become equal in high field, as do (iii) and (iv).

complexity of the ESR spectrum, as more nuclei are "added", can readily be imagined). The muon hyperfine coupling constant A_μ is given (exactly) by the sum of the frequencies of the two transitions - or their difference, according as $A_\mu/2$ is greater or less than the muon Larmor frequency in the applied magnetic field:

$$A_\mu = \nu_{12} \pm \nu_{34} \qquad\qquad (A_\mu/2 \gtrless \gamma_\mu B/2\pi).$$

5. ESR SPECTRUM OF THE ETHYL RADICAL

The ethyl radical and its muonium substituted counterpart have been exhaustively studied by ESR and μSR respectively and are almost understood! As prototype or parent of all the alkyl radicals they serve as a good example of how the various hyperfine couplings are measured and interpreted.

The unsubstituted radical $\dot{C}H_2CH_3$ and its ESR spectrum are illustrated in Figure 3. Major spin density is located on the trigonal carbon atom and is depicted in 3(a) as localised in an atomic $2p_z$ orbital. Actually, the singly occupied molecular orbital (the SOMO) spreads over the whole molecule. The methylene or α protons lie on a nodal plane of this orbital, so their couplings are due to spin polarisation and are expected to be negative. The corresponding splittings may be seen in 3(b)), although their sign cannot in fact be determined from the simple ESR spectrum.

Admixture of the methyl group C–H bonding orbitals in the SOMO amounts to a leakage of spin density onto the β protons. Their hyperfine coupling must therefore be positive; it corresponds to the larger splitting in 3(b). The delocalisation is known as hyperconjugation. The equivalence of the three β protons indicates that the methyl group is free to rotate about the C–C bond (at all temperatures).

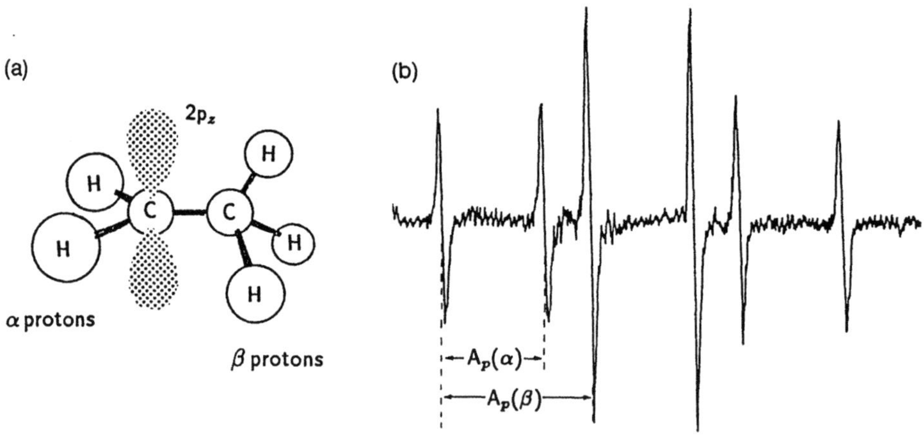

Figure 3. Structure (a) and ESR spectrum (b) (recorded in the liquid phase /4/) of the ethyl radical.

6. FORMATION OF THE RADICALS

The ethyl radical is formed by γ-irradiation of ethene under the correct conditions - a hydrogen atom is removed radiolytically from one molecule and reacts by addition with another, breaking the carbon-carbon double bond /4/ (here the dot is used to emphasise the presence and position of the unpaired electron spin):

$$CH_2{=}CH_2 + H\cdot \rightarrow \dot{C}H_2{-}CH_3$$

Its muonium substituted counterpart is formed by muon-irradiation of ethene.The possible processes involved following muon implantation (in organic liquids) are considered elsewhere /5/. Formally at least, the net result corresponds to the addition of muonium in an analogous fashion:

$$CH_2{=}CH_2 + Mu\cdot \rightarrow \dot{C}H_2CH_2Mu.$$

7. μSR SPECTRUM OF THE SUBSTITUTED RADICAL

A μSR spectrum for $\dot{C}H_2CH_2Mu$, showing the two characteristic "high field" lines (plus a third from muons reaching a diamagnetic state) is given in Figure 4. Only the muon hyperfine coupling can be deduced from such a spectrum. Fortunately another technique exists to determine the other couplings - including their sign (Section 8). This is just one example of the large number of muonium substituted radicals which have now been identified /3/ - over 300 to date /6/. The majority have been studied in the liquid phase.

The remarkable feature of this spectrum is that it was detected in the gas phase! To put this achievement in context, no ESR spectrum has been reported for the ordinary ethyl radical (or for any other of such high molecular weight) in the gas phase. Coupling to total angular momentum splits gas phase ESR spectra into a multitude of lines corresponding to different rotational states so that they become impossible to detect except for diatomic and a few triatomic species.

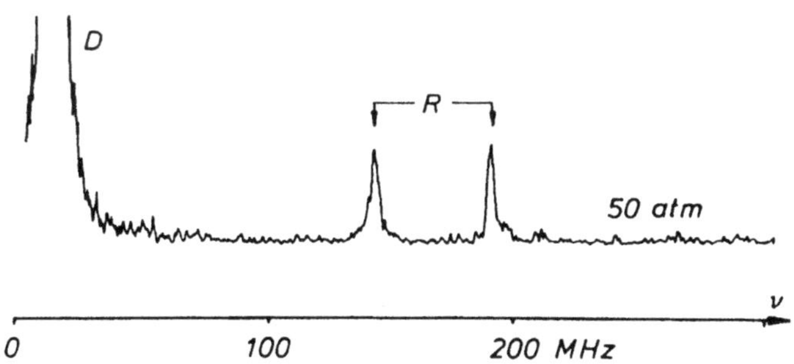

Figure 4. μSR spectrum of the muonium substituted ethyl radical, $\dot{C}H_2CH_2Mu$, in gaseous ethene at 50atm (from Roduner and Garner /7/.) Such spectra are still visible with good signal to noise ratio down to 1atm pressure /8/!

8. LEVEL CROSSING RESONANCE

Of the two frequencies displayed in a high field μSR spectrum, one corresponds to the hyperfine field at the muon adding to the applied magnetic field, the other to it subtracting. The same is true for ENDOR-type transitions of other nuclear spins. It is therefore always possible to tune the applied field so that the transition frequencies of the muon and another nucleus exactly match. This puts the muon and the other nucleus "on speaking terms" so that they may exchange polarisation by cross-relaxation.

This is the principle of the technique known as Level Crossing Resonance /9/ as applied to paramagnetic species /10/. Purists refer to it as Avoided Level Crossing /11/ since the cross relaxation can only occur, of course, if the degeneracy of the crossing is lifted - as illustrated in the insert to Figure 5.

A spectrum for the muonium substituted ethyl radical is given in Figure 6. Knowing the muon hyperfine coupling A_μ (from data such as that in Figure 4), the α and β proton couplings in $\dot{C}H_2CH_2Mu$ can be determined with great precision from the position of the two resonances. For each set of equivalent nuclei the condition is of the form /10,11/

$$B_{res} = \pi[(A_\mu - A_n)/(\gamma_\mu - \gamma_n) - (A_\mu + A_n)/\gamma_e]$$

$$\simeq \pi(A_\mu - A_n)/(\gamma_\mu - \gamma_n)$$

so that even the relative signs of the couplings are available. Again, most work to date has been on liquid samples, but Figure 6 is a gas-phase spectrum chosen as an example of recent developments, and to show that Level Crossing Resonance is equally powerful in the gas phase. (In fact, it is easier than ordinary μSR - the requirement on radical formation rate is less stringent).

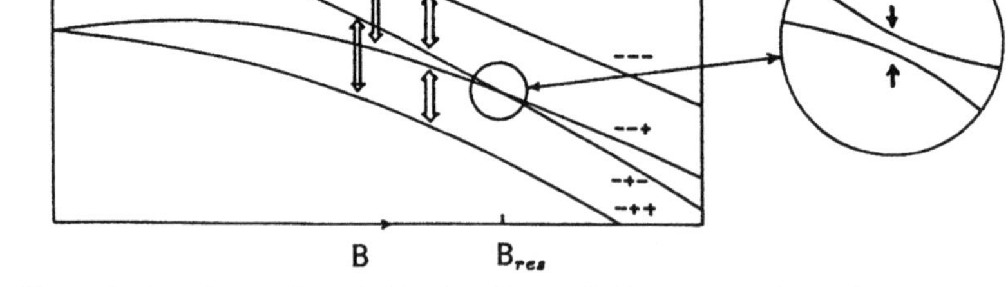

Figure 5. Level crossing and its "avoidance". The energy level diagram of Figure 2 is redrawn to show the resonant coincidence of muon and proton transition frequencies. The resonance condition can always be met - if the sign of the proton coupling is reversed the level crossing occurs in the upper group of levels instead! The splitting shown in the insert is a measure of the cross relaxation or polarisation transfer rate.

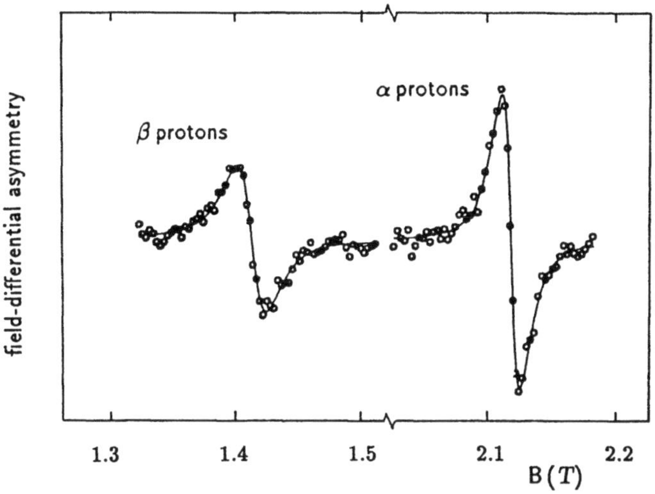

Figure 6. LCR spectrum of CH_2CH_2Mu in ethene gas at 6.5atm (from /8,13/).

The combination of (transverse field) μSR and (longitudinal field) LCR is therefore able to give a complete characterisation of the muonium substituted radicals. It could even be said to better ESR in giving the signs of the couplings and extending measurements to the gas phase!

9. THE HYPERFINE ISOTOPE EFFECT

In order to compare the muon and proton hyperfine couplings, determined from such spectra, it is customary to scale down the muon value by the ratio of the particles' magnetic moments, and use the "reduced" value:

$$A'_\mu = (\mu_p/\mu_\mu)A_\mu = 0.314A_\mu$$

It has been known for some time that the reduced muon coupling in the ethyl radical, as in the other alkyl radicals, is greater than the normal proton coupling /3/. This indicates that the substituted group CH_2Mu is not free to rotate like the ordinary methyl group but adopts instead a preferred orientation, maximising overlap or hyperconjugation of the C–Mu bonding orbital with the unpaired electron wavefunction (Figure 7a). The temperature dependence of A_μ allows the effective barrier to rotation to be determined/12/.

The LCR data (A_p in Figure 7b) confirm this behaviour. When the muon effectively eclipses the $2p_z$ orbital (giving a high value of A_μ) the two protons are held near the nodal plane (giving a low value of A_p). Torsional oscillation becomes important as the temperature is raised and, as free rotation is approached, spin density on the muon becomes more nearly equal to that on the CH_2Mu protons.

Recent data shows the curves of Figure 7(b) extending smoothly through the freezing and boiling points. There is no significant change to the internal (intramolecular) barrier to rotation in $\dot{C}H_2$–CH Mu, therefore, at the phase transitions /13/.

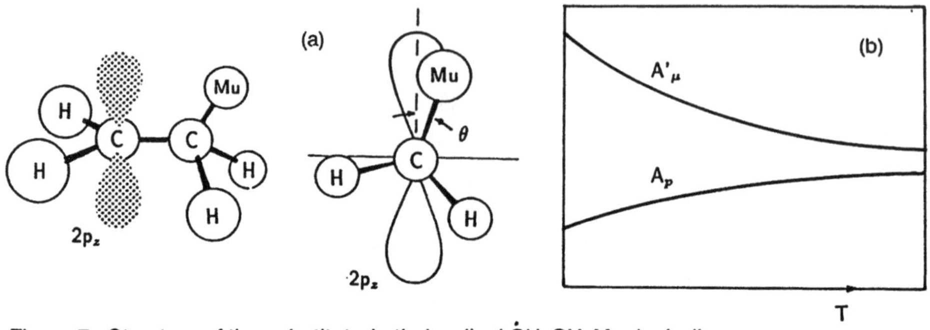

Figure 7. Structure of the substituted ethyl radical $\dot{C}H_2CH_2Mu$, including an end-on view illustrating the 'eclipse' of the $2p_z$ orbital by the C–Mu σ-bond (a). As temperature increases the amplitude of thermal libration about $\theta = 0$ increases, resulting in variation of the β couplings sketched qualitatively in (b).

10. THE TERTIARY-BUTYL RADICAL: LIQUID AND SOLID PHASE DATA

A nice illustration of the same effect is given by liquid phase data for the tertiary butyl radical. (This species can be considered derived from the ethyl radical by replacing the two α protons with methyl groups, to give a more symmetric molecule: Figure 8a). LCR spectra are shown in Figure 8(b). The stronger resonance is from protons in the CH_3 groups (which rotate freely at all temperatures); the weaker is from protons in CH_2Mu groups. The resonances move closer together with increasing temperature, as free rotation of the CH_2Mu group is approached. (The overall shift of both resonances reflects the variation of the muon coupling).

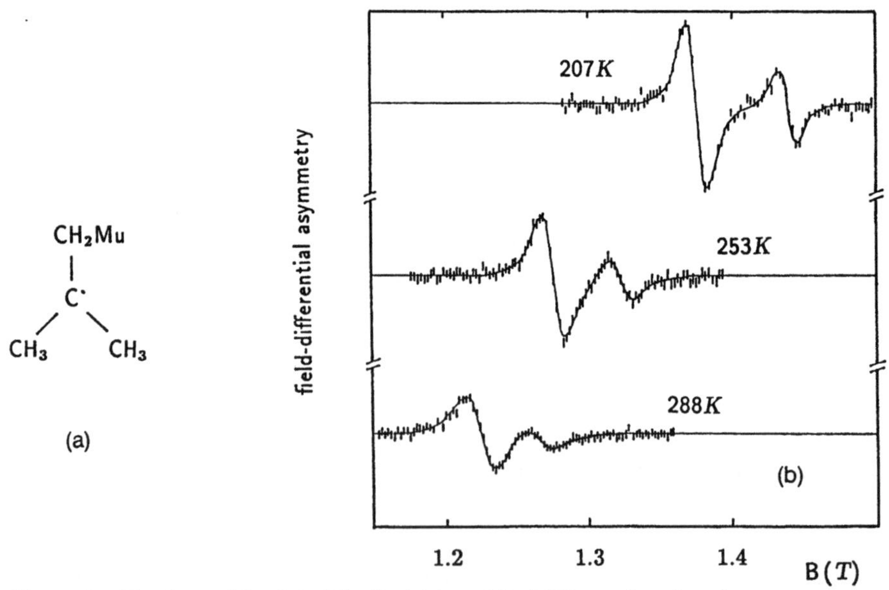

Figure 8. Structure of the t-butyl rdical (a) and its LCR spectra at various temperatures (b) /14/.

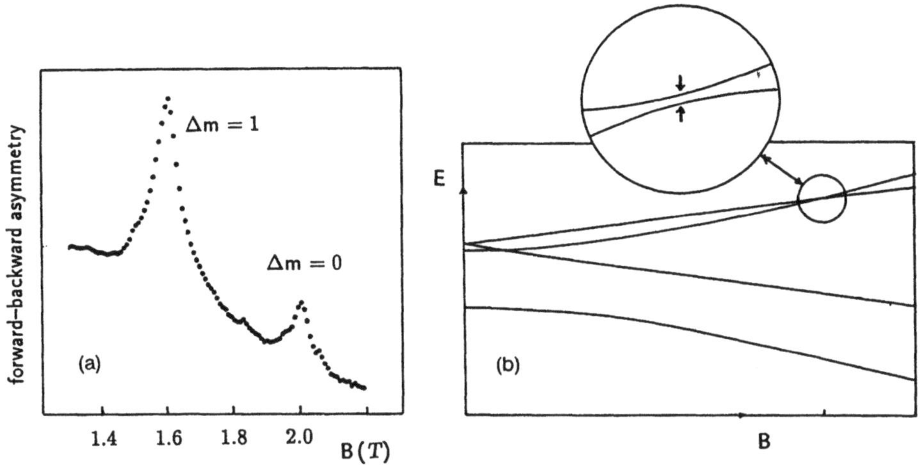

Figure 9. LCR spectra for $(CH_3)_2\dot{C}.CH_2Mu$ in solid isobutene (a), taken from/15/, and the simplified Breit-Rabi diagram drawn with a slightly anisotropic coupling to show the "pure muon" ($\Delta m=1$) level crossing resonance (b).

Information is also available on molecular motion in these studies - in this case intramolecular reorientation. Unlike the ethyl radical, t-butyl shows a step change in A_μ at the melting point of the host (isobutene). The bonding to the trigonal carbon is thought to be slightly pyramidal, and the coupling between inversion at the radical centre and rotation of the peripheral methyl groups to be altered between the solid and liquid phases /15/.

The LCR spectrum of the same radical in the solid state is shown in Figure 9(a). It shows an additional feature which corresponds to a transition of the muon spin alone ($\Delta m=1$). This resonance becomes allowed in the solid state because the intrinsic anisotropy of the couplings is preserved. The transverse component of the hyperfine field cannot be "tuned out" by the applied field at the level crossing depicted in Figure 9(b); the muon spin precesses about this transverse component and depolarises.

"Pure muon" resonances of this sort are potentially important in determining A_μ (not to mention its anisotropy) in the solid phase, where transverse field μSR spectra tend to be broadened (in the manner of powder spectra) beyond detection in non-crystalline media. That is, LCR alone can in principle provide full characterisation of radicals in the solid state.

11. ZERO POINT ENERGY

The above sections serve to illustrate the spectroscopic techniques and the type of data which is available. Some aspects of this data, notably the comparison of muon and proton hyperfine couplings, are interpreted below in terms of the different zero point energies of the two particles. This section serves to give a feeling for the energies involved.

A legitimate starting point for treating most problems in quantum chemistry and chemical physics is the adiabatic or Born-Oppenheimer approximation, namely the separation of

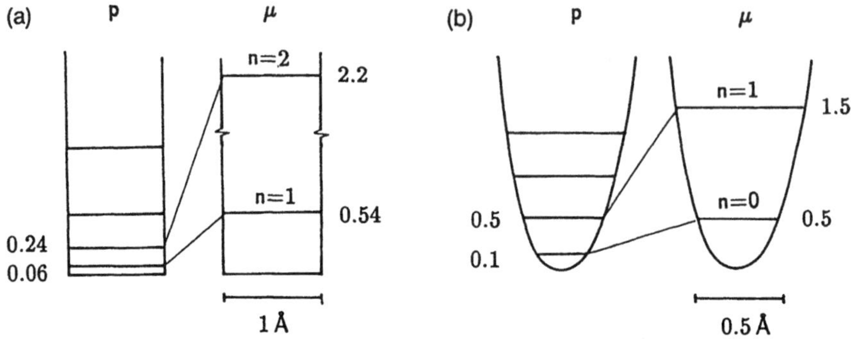

Figure 10. Energy levels for protons and muons confined within three dimensional square-well (a) and harmonic (b) potentials. Energy levels are calculated in eV for a cubic box of side 1Å in (a) and for a force constant of 25eV/Å ($400Nm^{-1}$ - this is about the value for a C-H bond) in (b). For conversion to other units, 1eV = 3.7x10 a.u = 96kj.mol = $1.2x10^4$K.

nuclear and electronic wave-functions. A drastic further assumption is commonly adopted in considering ground-state molecular properties: this is to treat the nuclei as fixed at their equilibrium positions! This neglect of the Uncertainty Principle where nuclei are concerned rarely has detectable consequences for molecules containing heavy atoms. For these, a probability distribution characterising the nuclear positions is only necessary to describe excited vibrational states. For molecules containing light atoms, however, the nuclear positions are certainly not fully determinate; that is, the effects of zero point energy are detectable.

To take the simplest example, the hydrogen molecule H_2 must dissociate slightly more readily than D_2, the difference in zero point energy between the protons and deuterons decreasing the "static" binding energy. Similar effects exist for all molecules containing H or D and are expected to be greatly enhanced when hydrogen is substituted by its light isotope Mu.

Figure 10 shows the "text-book" cases of particles confined within (a) a square well and (b) a harmonic potential, with energies for protons and muons calculated from the standard formulae. For the cubic box, the energy levels vary as the inverse of the particle mass (equation 1) so that the ground state for a muon lies higher than that for a proton by a factor $m_p/m_\mu \simeq 9$! These levels vary as the inverse square of the confinement length L; for the fictitious 1Å box depicted, the fractional population of the first excited states, in thermal equilibrium at room temperature, would be small (10^{-3}) for the proton, and quite negligible (10^{-28}) for the muon.

$$E_n = n^2h/8mL^2 \text{ per dimension } (n = 1,2\ 3...) \qquad (1)$$

A square well may be a reasonable approximation to some of the potentials encountered by muonium when trapped in interstitial cages in crystals. For the present case of muonium covalently bonded in molecules a harmonic oscillator potential, $V = ½kx^2$, is generally more representative. For this, the energy levels are equally spaced and their mass dependence is weaker (equation 2). The ground state energy, as well as the vibrational frequencies, vary as

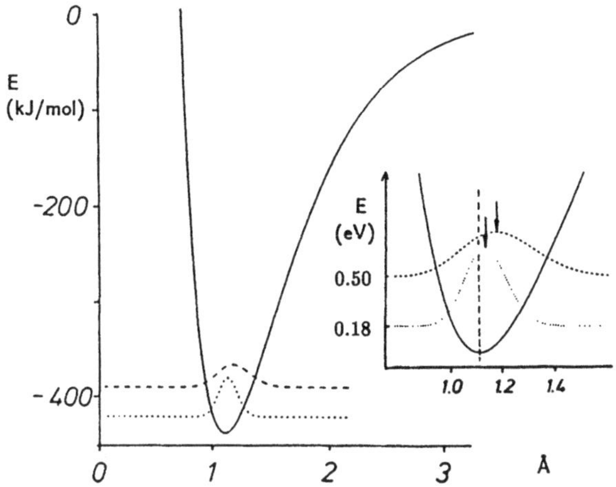

Figure 11. Morse curve for the C–X bond, zero point energies and bond length distributions (density distributions for the light nucleus if carbon is considered stationary) for X=H and X=Mu and, at right, the effective or dynamically averaged bond lengths (taken from /16/).

$m^{-1/2}$ so that each level for the muon lies about 3 times higher than the corresponding level for the proton.

$$E = (n + \tfrac{1}{2})\hbar\omega = (n + \tfrac{1}{2})\hbar(k/m)^{1/2} \text{ per dimension } (n = 0,1\ 2...) \quad (2)$$

Roughly speaking, the muon ground state is on a level with the first excited state for the proton.

For the force constant used in Figure 10(b), the ground state rms displacement (equation 3) is 0.4Å for a muon and 0.23Å for a proton. The equilibrium populations at room temperature of the excited levels are negligible for both particles.

$$\langle x^2 \rangle^{1/2} = (mk)^{-1/4}\ (\hbar/2)^{1/2} \quad (3)$$

Figure 11 shows the situation in a realistic potential or Morse curve for the C–X bond (X=H,Mu). Zero point energy makes the bond slightly weaker for X=Mu than for X=H; anharmonicty of the potential makes it on average slightly longer as well. Roduner and Reid /16/ estimate an extension of 0.06Å or 5% of the equilibrium separation.

12. THE EFFECT ON CONFORMATION

The origin of the barrier to internal rotation in $\dot{C}H_2CH_2Mu$ can also be traced to the muon zero point energy. Claxton and Graham /17/ calculate the total energy of the molecule in the two extreme conformations, i.e. when the C–Mu bond adopts the eclipsed position (dihedral angle constrained to $\theta = 0$) and the staggered position ($\theta = 90°$). They also calculate the change in energy for small displacements of all nuclei about their equilibrium positions, i.e. the force constants for all bonds, and find that the molecule is slightly "stiffer" in the staggered conformation. Addition of the nuclear zero point energies then tips the energy balance in favour of the eclipsed conformation. The methylene protons even bend slightly out of the plane, towards the muon, as in Figure 12(a).

The case of internal rotational freedom admittedly presents a rather delicate energy balance, but such a pronounced isotope effect on molecular conformation is nonetheless remarkable.

13. THE RESIDUAL ISOTOPE EFFECT

Even when the orientational preference is allowed for, the spin density on the muon is nonetheless greater than on a proton in the same position, by about 20%. Referring to Figure 7(b), A'_{μ} does not become exactly equal to A_p even in the free rotation limit. This is best demonstrated by comparing the average coupling for the CH_2Mu group,

$$\bar{A} = 1/3(A'_{\mu} + 2A_p) = 79\text{MHz},$$

with the value 75 MHz for the CH_3 protons in the normal radical /8/.

14. DYNAMICAL AVERAGING OF THE HYPERFINE COUPLINGS

The origin of this difference is illustrated in Figure 12. As the C–X bond is stretched, spin-density on X increases rapidly as a non-linear function of C–X separation. In a qualitative (valence bond) description, this corresponds to an increasing mixture of (CH_2CH_2 + Mu(↑)) as opposed to C(↑)H_2CH_2Mu. In the limit, a muonium atom is withdrawn from the molecule – the reverse of its formation! The quantitative dependence has been calculated recently using ab initio (Unrestricted Hartree-Fock) methods by Claxton et al /18/.

The dynamically averaged spin density on X is therefore greater for X = Mu than for X = H, by virtue both of the (anharmonic) bond extension and the greater rms (harmonic) displacement of the lighter atom. The perpendicular normal modes also contribute to the net effect.

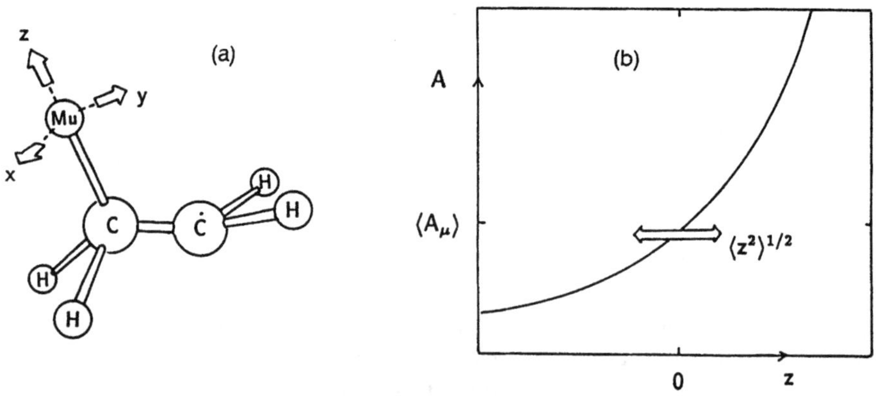

Figure 12. The eclipsed conformer of the ethyl radical (a) indicating the three normal modes of the muon and the variation of hyperfine coupling (b) with the stretch mode (sketched from /18/).

At risk of perpetuating a description which is no longer strictly necessary, this result may be taken as confirming the notion of enhanced hyperconjugation for the muon /19/. It would seem reasonable to retain qualitative "chemical" explanations and models as long as they provide reliable insight and correct predictions!

15. DIAMAGNETIC MOLECULES

In most substances, some fraction of the muons implanted reach diamagnetic states. There is no signature of their identity in the μSR spectra, however, comparable with the hyperfine frequencies for paramagnetic species. These muons precess at close to their Larmor frequency. Chemical shifts are in general too small to resolve or measure. The diamagnetic fraction must represent muonium which is substituted in ionic or molecular species (or, where appropriate, in hydrogen bonds). LCR can in principle be used to identify the species, and the position adopted by the muon, in favourable cases.

The first such identification in a molecular species is illustrated in Figure 13. In this experiment, muons are implanted in ice, and the muon Zeeman energy is tuned to a quadrupole splitting on ^{17}O. The sample is enriched to about 50% in $H_2{}^{17}O$ for this purpose. In a diamagnetic species, the polarisation transfer can only be mediated by the muon-nuclear dipole-dipole interaction, which is effective only at short range. Detection of the resonance implies that the muon is chemically bonded to oxygen /20/: the species identified is the muonium-substituted or "ultralight" water molecule, HMuO. This is believed to be formed in the two stage process /21/:

$$H_2O + \mu^+ \rightarrow H_2MuO^+$$

$$H_2MuO^+ + H_2O \leftrightarrows HMuO + H_3O^+.$$

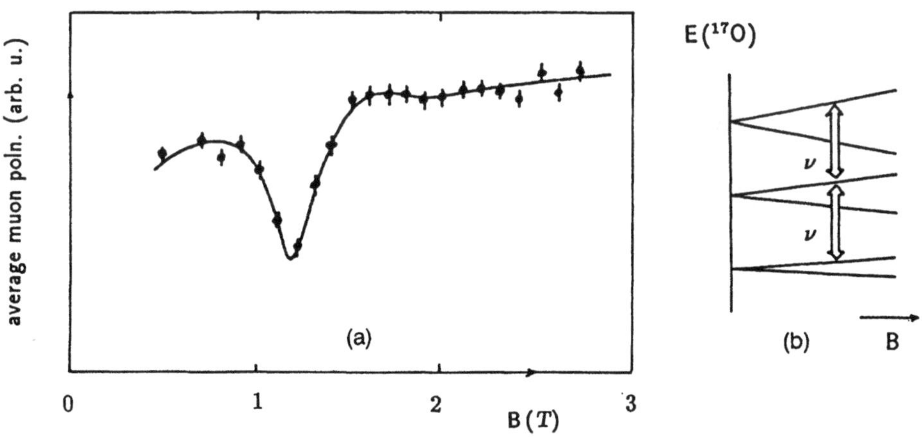

Figure 13. LCR spectrum for $HMu^{17}O$ in ice (a) and the (approximately equal) quadrupole splittings involved (b) (taken from /22/).

16. THE QUADRUPOLE ISOTOPE EFFECT

The quadrupole splittings ν on ^{17}O in ordinary ice are known from the conventional NQR literature. These can be used to predict the position of the resonance:

$$B_{res} = 2\pi\,\nu/(\gamma_\mu - \gamma_O).$$

The resonance is found at a field which is 5–10% lower than this prediction, implying a correspondingly smaller value for the quadrupole coupling in the muonium substituted molecule. This is the first such measurement of a quadrupole isotope effect by μSR.

It is interesting that the electric field gradient at the oxygen nuclei already decreases by some 30% between vapour and ice - an unusually large change which is attributed to the effects of hydrogen bonding in the condensed phase. The further reduction on going from H_2O to HMuO in ice may be understood in terms of an enhancement of these effects in the vicinity of a muonium bond, namely a greater polarisation of the substituted (and neighbouring) molecules, more nearly tetrahedral bonding and therefore greater symmetry of the charge distribution surrounding the oxygen nuclei involved. This is explained in terms of the anharmonicity of the proton (muon) potential in a hydrogen (muonium) bond, together with the larger zero point energy of the lighter particle. The mechanism is illustrated in Figure 14.

This intuitive picture remains to be confirmed by quantum-chemical calculations. Since the available (e.g. ab initio) methods are able to compute the distribution of charge density more reliably than they are that of spin density, measurements of quadrupole isotope effects may prove to be a valuable source of data with which to check these methods, and perhaps capable of a more certain interpretation than the hyperfine isotope effect in radicals of the muon and proton dynamics.

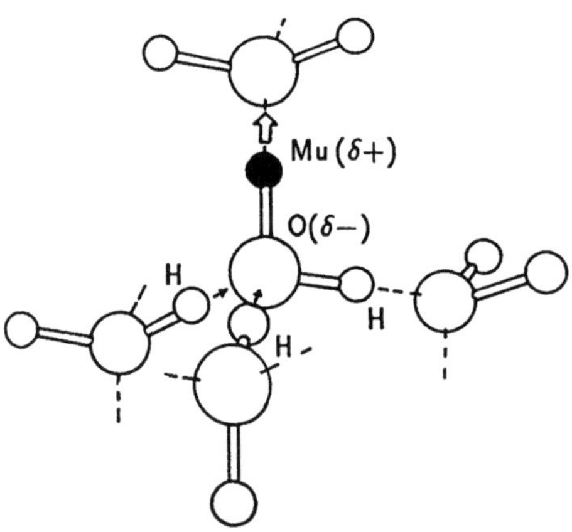

Figure 14. Possible effects of muonium substitution on bond lengths and polarisation in ice /22/.

17. FUTURE POSSIBILITIES

The possibility of measuring muon-proton dynamic isotope effects in other molecular properties should not be overlooked. Dipole moment is an obvious candidate.

The various contributions to these effects would be easier to unravel if the normal modes associated with the muon motion could be isolated with laser spectroscopy. This may prove possible: the characterisic frequency of the stretching mode of C–X or O–X bonds should be greater for X = Mu than for X = H by a factor of approximately $(m_p/m_\mu)^{1/2} \simeq 3$, which is sufficient to detach these modes from the rest of the vibrational spectrum. A form of trigger detection could be envisaged, in which a particular μSR or LCR transition is monitored whilst the laser frequency is swept: a resonant loss of signal would indicate a vibrational excitation and allow its characteristic frequency to be determined.

REFERENCES

1. D C Walker, "Muon and muonium chemistry" Cambridge (1983)

2. D G Fleming, D M Garner, L C Vaz, D C Walker, J H Brewer and K M Crowe, "Positron and Muonium Chemistry" Advances in Chemistry Series, ed H J Ache (Am Chem Soc, Washington DC) **175** (1979) 279

3. E Roduner, W Strub, P Burkhard, J Hochmann, P W Percival, H Fischer, M Ramos and B C Webster, "Chem Phys **67** (1982) 275 (for more recent references, see also E Roduner: "The positive muon as a probe in free radical chemistry", Springer Verlag Lecture Notes in Chemistry **49** (1988)

4. M C R Symons, "Chemical and biochemical aspects of ESR spectroscopy" Van Nostrand Reinhold (1978)

5. S F J Cox and M C R Symons, Radiat Phys Chem **27** (1986) 53

6. Estimated by E Roduner: private communication (1990)

7. E Roduner and D M Garner, Hyp Int **32** (1986) 733

8. P W Percival, J-C Brodovitch, S-K Leung, D Yu, R F Kiefl, D M Garner, D J Arseneau, D G Fleming, A Gonzalez, J R Kempton, M Senba, K Venkataswaran and S F J Cox, Chem Phys Lett **163** (1989) 241

9. A Abragam, CR Acad Sci (Paris) **299** (1984) 95

10. R F Kiefl, S R Kreitzman, M Celio, R Keitel, G M Luke, J H Brewer, D R Noakes, P W Percival, T Matsuzaki and K Nishiyama, Phys Rev **A34** (1986) 681

11. M Heming, E Roduner, B D Patterson, W Odermatt, J Schneider, H Baumeler, H Keller and I M Savic, Chem Phys Lett **128** (1986) 100

12. M J Ramos, D McKenna, B C Webster and E Roduner, J Chem Soc Faraday Trans I **80** (1984) 255 and 267

13. P W Percival, J-C Brodovitch, S-K Leung, D Yu, R F Kiefl, D M Garner, G M Luke, K Venkataswaran and S F J Cox, "Intramolecular motion in organic radicals", IOP Short Meetings Series (IOP Bristol) **22** (1989) 99

14. P W Percival, R F Kiefl, S R Kreitzman, D M Garner, S F J Cox, G M Luke, J H Brewer, K Nishiyama and K Venkataswaran, Chem Phys Lett **135** (1987) 465

15. P W Percival, J-C Brodovitch, S-K Leung, D Yu, R F Kiefl, G M Luke, K Venkataswaran and S F J Cox, Chem Phys **127** (1988) 137

16. E Roduner and I D Reid, Israel J Chem **29** (1989) 3

17. T A Claxton and A M Graham, J Chem Soc Faraday Trans I **83** (1988) 2307

18. T A Claxton, S F J Cox, A M Graham, Dj Maric, P F Meier and S Vogel, Proc μSR 90 (Hyperfine Interactions **65** (1990) 913)

19. M C R Symons, Hyp Int **17–19** (1984) 771, and S F J Cox, T A Claxton and M C R Symons. Radiat Phys Chem **28** (1986) 107

20. S F J Cox, G H Eaton, J E Magraw and C A Scott, Chem Phys Lett **160** (1989) 85

21. P W Percival, Radiochimica Acta **26** (1979) 1

22. S F J Cox, J A S Smith and M C R Symons, Proc μSR 90 (Hyperfine Interactions **65** (1990) 993)

Magnetic Order and Fluctuations as Detected by Positive Muons

C. Bucci

Dipartimento di Fisica, Università di Parma, I-43100 Parma, Italy

Abstract. This paper describes some experiments in which local magnetic fields are investigated by polarized positive muons in a variety of magnetic materials. Transition metals, rare earth intermetallic compounds, oxides, halides, spin-glass systems and high T_c superconductors will be presented by considering a few examples only : the phenomenology is quite broad on the one hand and, on the other, the present lecture is not conceived as a thorough review. In μSR research a preliminary objective is the understanding of the state of the positive muon after implantation, its specific localization, if any, and the understanding of the link between the local magnetic field at the muon site and the magnetic moments of the system under investigation. Examples of this studies are given for MnF_2 and CoF_2 single crystals. Results concerning the static properties in the magnetically ordered state as well as the fluctuations in the disordered state are confronted with the available results from Nuclear Magnetic Resonance and neutron diffraction and scattering.

1. General Remarks on μSR

General and specialized papers are available on the various aspects of the μSR techniques, typically in the proceedings of the μSR Conferences [1] and in the book by A.Schenck [2]. Here we assume, in addition, that details on the technique are presented elsewhere in the workshop and we limit ourself has to recalling the major features which are relevant for studies in magnetism.

From the instrumental point of view, μSR includes a variety of arrangements which provide rather complete information regarding the local fields sensed by the muon. Dealing with spin-polarized muons, obvious differences exist between experiments performed in the absence and in the presence of an external magnetic field and, in the latter case, between a transverse and a longitudinal field with respect to the initial muon polarization. Transverse field (TF-μSR), zero field (ZF-μSR) and longitudinal field (LF-μSR) are the most commonly used versions of μSR for investigating magnetic systems. Other recent developments

 © Springer-Verlag Berlin, Heidelberg 1992

such as radio- frequency resonance (RF-μSR) and avoided level crossing (ALC-μSR) have been also applied in some cases to study magnetic materials.

In TF-μSR the data display directly the Larmor precession which is characterized by a frequency and a decay. The precession frequency is $\omega_\mu = \gamma_\mu |\mathbf{B}_\mu|$ where $\mathbf{B}_\mu$ differs, in general, from the externally applied field. Such a difference represents the response of the material to the external field provided that one can account correctly for the perturbation introduced by the muon itself. Paramagnetic and diamagnetic shifts have been studied widely both in magnetic and in superconducting materials. In magnetically ordered systems $\mathrm{B}_\mu = |\mathbf{B}_\mu|$ may have a non-vanishing average value even in the absence of external fields, in which case ω_μ can be related directly to the order parameter in the system. The decay in the precession amplitude reflects the existence of local random fields, both static and fluctuating and it can be seen as the μSR equivalent of the pulsed NMR free-induction decay following a 90 degrees pulse. The actual shape of the TF-μSR decay depends on the features of the local fluctuations: one can find particularly simple extreme cases of gaussian shape (for static distribution of local random fields) or lorentzian shape (for local fields rapidly fluctuating in time) although intermediate or more complex situations are not rare.

In ZF and LF-μSR with positron detectors in the forward-backward geometry referred to the initial muon polarization, one takes advantage of the possibility of measuring directly the muon polarization itself and its time evolution by combining the raw data according to:

$$P(t) = \frac{N_{for}(t) - N_{back}(t))}{N_{for}(t) + N_{back}(t)} \tag{1}$$

where $N_{for}(t)$ and $N_{back}(t)$ are the actual muon decay histograms and $P(t)$ represents the effective muon polarization which arises from the product of two factors, one instrumental (the effective positron decay asymmetry) and one, the most interesting, which is the longitudinal muon spin correlation function $G^{\mu}_{zz}(t) = \langle \mathrm{S}^z_\mu(t)\mathrm{S}^z_\mu(0)\rangle / \mathrm{S}^2_\mu$ and is related to the system's spin correlation functions [3]. Here again it must be recalled that a non-vanishing average local field in an ordered system may be detected, under suitable circumstances, as a precession frequency in $P(t)$. In order to establish, however, the general link between $P(t)$ and the static and dynamic correlations among the magnetic moments of the system one must determine the nature of the local field at the muon site(s) (dipolar, transferred hyperfine etc.).

2. Static Magnetic Order

We shall examine here examples taken from magnetic metals and insulators. For the localization and diffusion properties of muons in metals we refer the reader to a specialized review paper [4]. For the case of insulators we shall describe the localization studies on magnetic fluoride single crystals, together with their static magnetic properties as seen by the muons.

2.1 Magnetic Metals

The local field at the muon site can be described in terms of hyperfine and dipolar contributions; in the ordered state we consider its non-vanishing static average values, $\langle B_\mu(T)\rangle$, while its fluctuating components, $\Delta B_\mu(t,T) = B_\mu(t,T) - \langle B_\mu(T)\rangle$, will be examined in the next section. Therefore:

$$\langle B_\mu(T)\rangle = \langle B_{hf}(T)\rangle + \langle B_{dip}(T)\rangle \tag{2}$$

where the hyperfine contribution includes a contact field from the spherically symmetric part of the polarized electron cloud around the muon and a pseudodipolar field from its non-spherical part; the dipolar term is usually split into three parts: the dipole sum within the Lorentz sphere, the Lorentz field and the shape-dependent demagnetization field. The dipolar contributions may have different relative weights under particular circumstances: typically in a multidomain structure the demagnetization field vanishes and so does the dipole sum term when the local magnetic moments are arranged with cubic symmetry around the muon site.

Once the localization of the muon is determined and the dipolar contributions evaluated, one can single out from the experimental data in ZF-μSR the hyperfine contribution which is related to the electronic polarization at the muon site. The latter is an important piece of information regarding the magnetization of the system at an interstitial site, while other hyperfine studies by NMR, Mössbauer, PAC and neutron diffraction determine primarily the magnetization at regular lattice sites. In addition to this "complementarity", comparison between μSR and the other techniques must focus on the temperature dependence of the measured spontaneous magnetization.

A remarkable example among transition metals is the case of cobalt, which exhibits both a structural and an anisotropy transition in its ordered phase. μSR results follow in detail all these evolutions of the magnetic system, as shown in Fig. 1 [5].

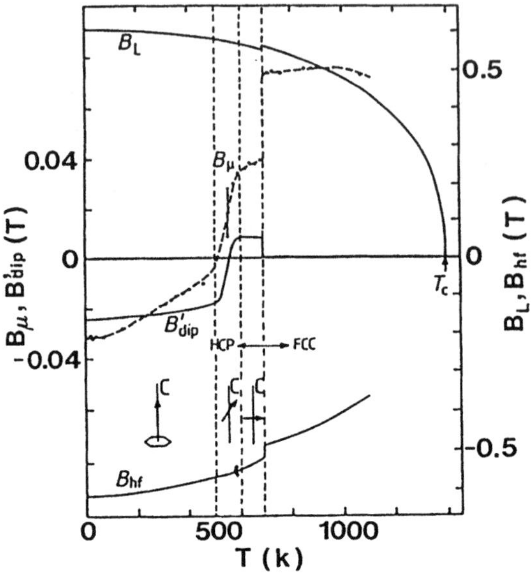

Figure 1 Temperature dependence of $\langle B_\mu(T)\rangle$ in Co. The contributions to the dipolar field are calculated for octahedral site occupancy and one can extract $\langle B_{hf}(T)\rangle$ from the μSR data (dots).

Similar experiments performed on Fe and Ni, together with Co, indicate that the the values of the hyperfine fields at T=0 are affected by the electronic and lattice distortions introduced by the presence of the positive muon [6]. Theoretical work by several authors provides a good basis for understanding the whole picture, which has, as main ingredients, the *s* and *d*-electrons participating in the screening cloud around the muon, a local spin density determined self-consistently [7], appropriate spin-dependent screened potential for the muon, lattice dynamics ,thermal expansion and muon zero-point energy [8].

In view of the existing distortions introduced by the muon, it is not surprising that the temperature dependence of the hyperfine fields determined by μSR differs, at low temperatures, from the spontaneous magnetization determined by different methods. An important experimental test for the theoretical models adopted to interpret the muon's distortions is the measure of the shift of the hyperfine field at the muon under hydrostatic pressure [9], which simulates the effects of lattice thermal expansion. The results indicate that lattice contributions alone can account for a large fraction of the discrepancy between the actual hyperfine field at the muon site and that calculated from the spontaneous magnetization.

Due to the strong interest in technological applications of hard magnetic materials for permanent magnets and soft materials for magnetic recording, many investigations are underway in which different techniques, including μSR, are applied to a variety of rare earth intermetallic compounds, cubic Laves phases, and amorphous alloys [10] [11] [12]. The role of the magnetic anisotropy in stabilizing complex magnetic structures and the involvement of the rare earth atoms in this process [13] represent the central physical aspect to be understood and μSR has a promising future in this field.

2.2 Magnetic Insulators

Due to the absence of conduction electrons and to the sharper localization of the magnetic moments, the information yielded by μSR on magnetic insulators is more direct than in metals.

Oxides represent a large family of systems investigated by μSR and an interesting common feature is that the localization of muons is related to the oxygen. As a result of studies on transition metal oxides and rare earth orthoferrites, it has been recognized that the covalent oxygen-muon bond provides the bridge for super-transferred hyperfine fields from the magnetic metal ion to the muon [14]. It is worth mentioning that similar conclusions seem to apply to the localization of muons in the high T_c superconducting oxides. The local field at the muon is therefore the result of hyperfine and dipolar fields; the latter is particularly sensitive to the orientation of the spontaneous magnetization with respect to the crystal axes and, as was seen in the case of Co, it can monitor very sharply any possible spin orientation transition. Here again we present only one example, such as the case of α-Fe_2O_3, which has an orientation transition at 263 K (the Morin transition) from a structure with Fe spins parallel to the c axis to a structure with Fe spins perpendicular to it [15]. As is shown in Fig. 2, the change in magnetic structure causes a sudden ZF-μSR frequency change which is accounted for by the change in dipolar fields.

Another interesting example concerns simple antiferromagnets in the family of the transition metal fluorides. In particular for CoF_2 and MnF_2 the static and dynamic magnetic properties are relatively well known from NMR and neutron scattering; they represent therefore an ideal test for μSR, particularly if experiments are performed on good single crystals. Here again the first goal is to determine the muon site and the nature of the field generated by the magnetic moments localized on the transition metal atoms. Both in the paramagnetic and in the antiferromagnetic phase TF-μSR as a function of the external field orientation relative to the crystal axes, as is shown in Fig. 3, reveals the symmetry of the muon site and a definite assignment is possible

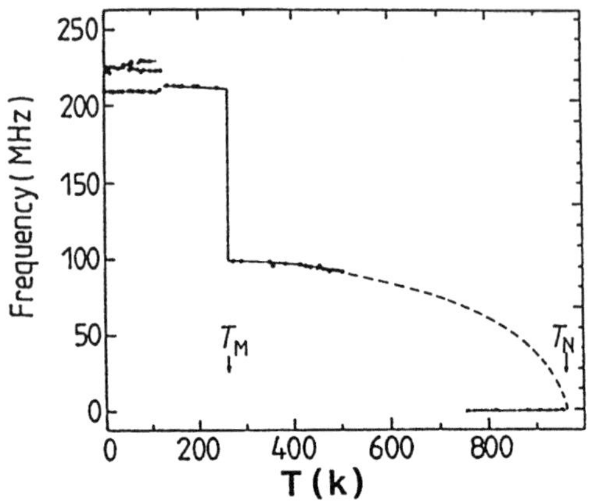

Figure 2 ZF-μSR determination of the muon precession frequencies in the ordered phase of α-Fe_2O_3. The collapse of the three frequencies into a single one at low temperature reflects a change in the muon stopping sites. The frequency drop at the Morin transition indicates the corresponding variation of the dipolar fields at the muon site.

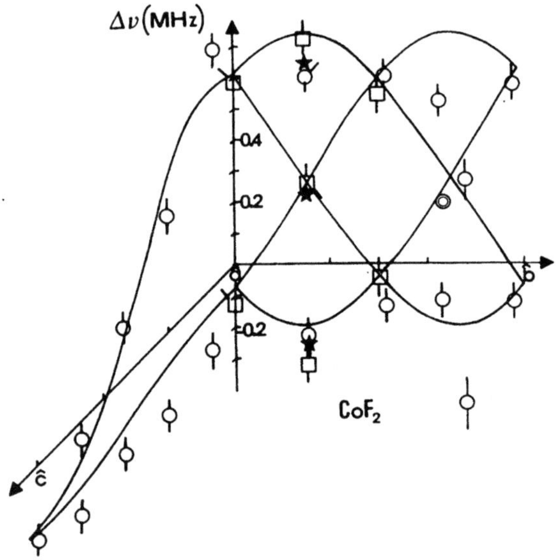

Figure 3. μSR frequency shifts and splitting, as a function of the relative angle between the external field ($B_{ext} = 0.3T$) and the crystal axis, *a*, *b*, and *c* for CoF_2 The continuous line is the calculated angular dependence for the octahedral site with dipolar fields only.

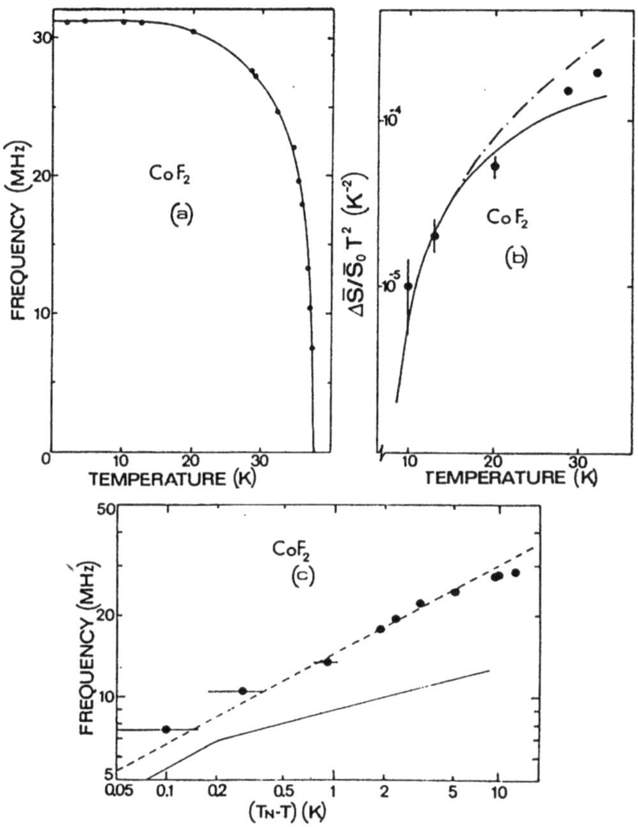

Figure 4. Temperature dependence of the ZF-μSR in CoF_2. a) The continuous line represents the spontaneous magnetization from neutron diffraction, normalized to the μSR data at T=0. b) Low temperature spin deviation compared to NMR data - broken curve - and to spin wave calculations - full line. c) Power law temperature behaviour of ω_μ upon approaching the transition temperature, compared to the neutron data - full line.

by properly taking into account the combined tensorial character of the magnetic sublattices susceptibility and dipolar fields [16]. The resulting site which is located at (1/2,0,0) and has octahedral symmetry, also fully justifies, in terms of dipolar fields alone, the ZF-μSR frequency measured at low temperatures, in the ordered phase of CoF_2. It is also interesting to note, as a side product of this study, that TF-μSR is a direct means to measure the sublattice susceptibility tensor in anisotropic systems, provided single crystals are available.

The temperature dependence of $\langle B_\mu \rangle$ over the entire ordered phase of CoF_2 scales well with the sublattice magnetization measured by neutron diffraction, as is shown in Fig. 4a [17]; close to the phase transition, however, μSR data follow better the power law with critical exponent $\beta = 0.3$ predicted for the 3-D Ising model, as is shown in in Fig. 4c . At the other extreme of the magnetization curve, i.e. at the lowest temperatures, the normalized frequency change ($\Delta\omega_\mu/\omega_\mu(T{=}0)$) corresponds to the normalized spin deviation and, as seen in Fig. 4b, it agrees well with the calculations performed within spin-wave theory [18].

2.3 Antiferromagnetic Perovskites

Most of the high T_c superconducting materials become antiferromagnetically ordered upon appropriate depletion of the carriers (either holes or electrons). The properties of these magnetic "counterparts" and the interplay of magnetic correlations with superconductivity has attracted much attention recently. We mention this topic here very briefly considering the dedicated lectures presented in the workshop. With reference to the three systems whose antiferromagnetism is believed to be unperturbed, La_2CuO_4, $YBa_2Cu_3O_6$ and $Bi_2Sr_2YCu_2O_{8+\delta}$, ZF-$\mu$SR at temperatures below T_N show that muons localize in sites where the average local field is non-zero. In analogy to the magnetic oxides discussed above, it is believed that muon-oxygen bonds are formed and indirect evidence supports this ; a direct experimental study of the localization is not available mainly due to the lack of good single crystals. The link between the muon sites and their local fields seems reasonably understood in terms of dipole fields generated by the copper spins, although the contribution of transferred hyperfine fields and its magnitude still needs to be clarified. All this considered, the temperature dependence of the spontaneous magnetization seen by the muon, the determination of T_N, its dependence on the carrier concentration and the resulting magnetic phase diagram are all vital pieces of information for the understanding of the frustration of the magnetic order by the extra holes or electrons . From the neutron scattering studies so far available one learns that these systems are 2D Heisenberg antiferromagnets with an inter-plane exchange interaction J_2 nearly five orders of magnitude weaker that the intra-plane Cu-Cu exchange J_1 and indeed the observed Neél temperatures agree with the calculated values for such a model:

$$T_N = \frac{J_1}{k_B \ln(J_2/J_1)} \qquad (3)$$

The temperature dependence of the spontaneous magnetization in many compounds of this family presents an interesting problem: different experimental methods (namely NMR, neutron scattering and μSR) show different behaviours and, apparently, none of them follows precisely

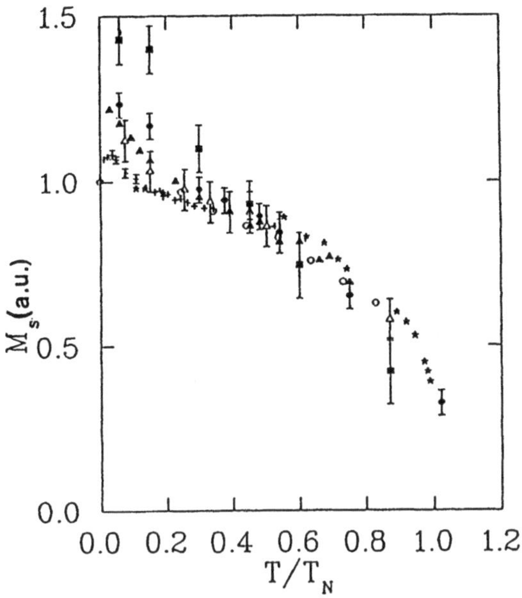

Figure 5. Normalized magnetization curves as obtained from neutrons (stars), NMR (open circles) and μSR (crosses) on $YBa_2Cu_3O_6$, together with μSR data from equivalent end-members of other high T_c families at zero or low carrier doping.

the theoretical predictions. An attempt at summarizing a variety of such data is shown in Fig. 5 which contains one curve for $YBa_2Cu_3O_6$ from neutron diffraction, one for the same material from NMR and several ZF-μSR curves for a variety of compounds with different carriers concentrations all scaled to the same T_N. The purpose of this figure is to point out that the interaction between extra holes (or electrons) and copper moments may be seen with different sensitivity by methods such as NMR, which probes the local magnetization at the Cu (or O) nuclei from typical non-local methods such as neutron scattering and from μSR, which is somewhat in between.

3. Fluctuations and Disordered Systems

Here we examine examples of μSR studies in which the loss of the initial muon polarization can be related to the static and dynamic fluctuations of the magnetic moments. In particular we refer to experiments near the phase transitions and studies of correlations in alloys. In precession experiments, typically TF-μSR, it is important to define the link between the precession damping (or the linewidth in frequency space) and the correlation function for the spin system. In strict analogy with NMR, we can refer to the Moriya expression for the linewidth [19]:

$$\Delta\omega_\mu = \frac{\gamma_\mu}{2}\int_0^\infty dt[\langle \Delta B_z(t)\Delta B_z(0)\rangle + \frac{1}{2}\langle \Delta B_x(t)\Delta B_x(0) + \Delta B_y(t)\Delta B_y(0)\rangle \cos\omega_\mu t] \tag{4}$$

where the first term on the right hand side represents the static correlations of the local fields longitudinal with respect to the external field and the second term represents the components of the transverse correlations at the muon Larmor frequency. When, at one extreme, the field at the muon site is the hyperfine (or transferred hyperfine) field due to a single magnetic moment, it is possible to express the linewidth in terms of the scattering function of the magnetic system $S(q,\omega)$:

$$\Delta\omega_\mu = \frac{1}{2N}(\gamma_\mu B_{hf})^2 \sum_q \{S^{zz}(q,0) + \frac{1}{2}[S^{xx}(q,\omega_\mu) + S^{yy}(q,\omega_\mu)]\} \tag{5}$$

When, at the other extreme, the muon senses the dipolar field due to all the surrounding magnetic moments, it can be easily seen [17] that the correlation functions of the local fields still transform into scattering function $S^{\alpha\beta}(q,\omega)$ multiplied by the fourier transform of the dipolar tensor $F_{\alpha\beta}(q)$. In general, therefore, if the nature of the local field at the muon site is known, the μSR linewidth gives the scattering function integrated over all q values with a weight function whose selectivity in q-space depends on the type of local fields probed by the muon. In a high temperature paramagnetic regime one expects isotropic magnetic fluctuations and one can still find anisotropic linewidths as a mere consequence of the tensorial character of the dipolar fields.

Close to phase transitions the fluctuations themselves may become anisotropic, even for isotropic magnetic Hamiltonians, as a result of any breaking of their continuous symmetry by a - regardless how small - anisotropic interaction. One then observes a linewidth whose anisotropy is related to that of the fluctuations. A typical case of dominant uniaxial anisotropy is that of CoF_2 (although the very small anisotropy of MnF_2 produces similar effects) where sufficiently close to T_N the scattering function along the c axis is much larger than its other components and one obtains:

$$\Delta\omega_\mu^c = \frac{1}{4N}\sum_q F_{cc}(q)\, S^{cc}(q,0) \tag{6a}$$

$$\Delta\omega_\mu^{a,b} \simeq \frac{1}{4N}\sum_q F_{cc}(q)\,\frac{1}{2}S^{cc}(q,0) = \frac{\Delta\omega_\mu^c}{2} \tag{6b}$$

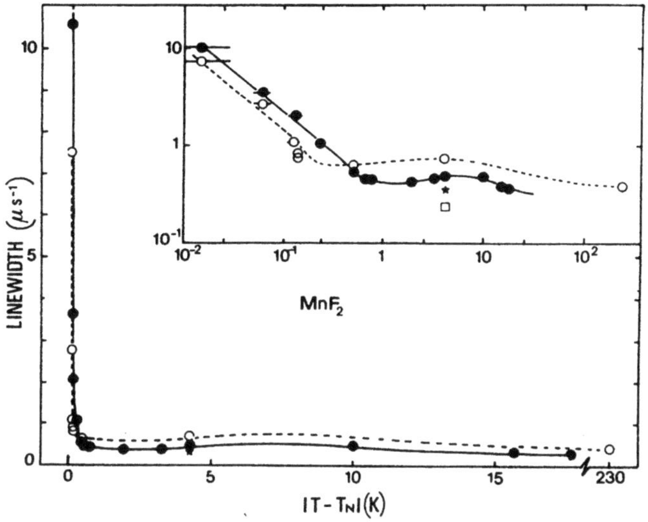

Figure 6. Temperature dependence of the linewidth for MnF_2 in an external field of 0.3 T parallel to the c axis (close circles) and to the a axis (open circles). The anisotropy of the linewidth has an inversion at $T = T_N + 0.4$ K where a crossover takes place between isotropic and anisotropic fluctuation regime.

In TF-μSR experiments one can actually observe the crossover from isotropic to anisotropic fluctuation regimes, as is shown in Fig. 6 for a single crystal of MnF_2.

An interesting manifestation of the magnetic fluctuations, often studied also by μSR, is the critical divergence of the linewidth upon approaching the transition temperature (an example is shown in the insert of Fig. 6). It must be pointed out that in the critical region one can predict the critical exponents by analyzing the essential features of the scattering function. Under very general assumptions and by using the fluctuation-dissipation theorem in its semiclassical form, we can factorize $S(q,\omega)$ in the real part of the wave-vector dependent susceptibility, $\chi(q)$, and the inverse life-time of the correlations, $\Gamma(q)$:

$$S(q,\omega) \simeq \frac{\chi(q)}{\Gamma(q)} \tag{7}$$

By using the appropriate expressions for $\chi(q)$ and $\Gamma(q)$ it is possible to predict the critical exponents for the μSR linewidth. Critical exponents, for short range interaction systems, depend only on the dimensionality and the symmetry of the problem; they are reported and discussed, for instance, in ref [20]. With reference to the μSR experiments on MnF_2

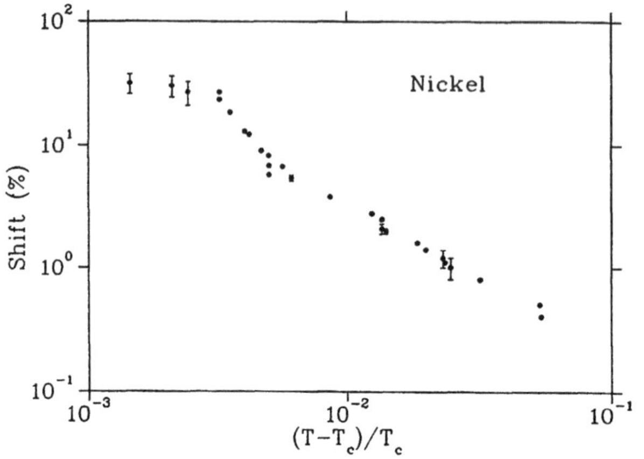

Figure 7. Temperature dependence of the ferromagnetic local field and paramagnetic shifts at the phase transition of Nickel.

and CoF_2, one finds that the critical exponents agree with the 3D Ising model, thus confirming the crucial role of anisotropy, also shown by NMR and neutron scattering experiments.

Another example of μSR investigations of fluctuations in the paramagnetic phase and in the neighbourhood of the transition temperature is shown in Fig. 7 [21] for Nickel. Here again it is possible to extract from the frequency shifts a critical exponent of 1.27, which agrees satisfactorily with the susceptibility data, while interesting questions remain open on the interpretation of the linewidth.

The paramagnetism of rare earth intermetallic compounds is widely investigated, the main goal being the understading of the coupling mechanisms between the magnetic momemts and the persistence of strong correlations above the ordering temperature, T_c. Typical systems are the cubic Laves phase (R.E.Al_2, R.E.Ni_2) or hexagonal intermetallics (R.E.Ni_5). Already in the early experiments, contrary to the expectations of independent rare earth spin fluctuations and of a consequent temperature independent μSR linewidth, significant correlations among the rare earth spins [22] were discovered. A large variety of rare earth compounds of this family is under thorough investigation [23] in order to single out and better define the role of crystal field effects and related magnetocrystalline anisotropy.

Regarding rare earths metals, Erbium provides a clear cut example of totally anisotropic fluctuations detected by ZF-μSR on a single crystal oriented with the c axis either parallel or perpendicular to the muon spin polarization [24]. The strong transverse relaxation in the paramagnetic region exists well above T_N in contrast to the weak, temperature in-

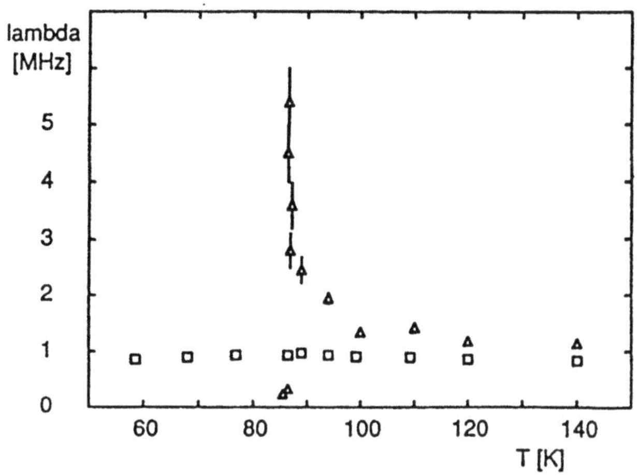

Figure 8. ZF- μSR relaxation rate in Er single crystal as a function of temperature above and below the transition temperature, $T_N = 85K$: squares are for muon polarization parallel to the c axis, triangles for polarization perpendicular to the c axis.

dependent relaxation in the longitudinal geometry, which indicates that the perpendicular components of the Er spins are uncorrelated. These recent results are shown in Figure 8.

Acknowledgments

The author gratefully acknowledges the suggestions and the critical discussion of the manuscript by Prof. Germano Guidi and Dr. Roberto De Renzi.

References

[1] Proceedings of the Topical Meeting on Muon Spin Rotation
  a) Eds. F.N.Gygax, W.Kuendig and P.F.Meier, Hyperfine Interactions, 6 (1979)
  b) Eds. J.H.Brewer and P.W.Percival, Hyperfine Interactions,8 (1981)
  c) Eds. T.Yamazaki and K.Nagamine, Hyperfine Interactions, 17-19 (1984)
  d) Eds. O.Hartmann and R. Wäppling, Hyperfine Interactions, (1988)
  e) Eds. S.F.J.Cox and G.Eaton, Hyperfine Interactions, (1990)

[2] A.Schenck "Muon Spin Rotation Spectroscopy" , Adam Hilger Ltd, Bristol and Boston, (1985)
[3] R.S.Hayano, Y.J.Uemura, J.Yamazato, N.Nishida, T.Yamazaki and R.Kubo, Phys. Rev. B 20 (1979) 19
[4] E.Karlsson, Physics Reports 82 **5** (1982)
[5] H.Graf, W.Kuendig, B.D.Patterson, W.Reichart, P.Roggwiller, M. Camani, F. N. Gygax, W. Ruegg, A. Schenck, H. Schilling and P. F. Meier, Phys. Rev. Letters 37 (1976) 1644
[6] H.Katayama, K.Terakura and J.Kanamori Solid State Comm. 25 (1979) 431
[7] B.Lindgren and D.E.Ellis Phys.Rev.B 26 (1982) 636
[8] S.Estreicher and P.F.Meier Phys.Rev.B 25 (1982) 297
[9] T.Butz, J.Chappert, J.F.Dufresne, O.Hartmann, B.Lindgren, L. O. Norlin, P.Podini, and A.Yaouanc Phys. Lett. 75A (1980) 321
[10] A.Yaouanc et al, J.Magn.Magn.Mat. 67 (1987) L286
[11] S.Barth et al., Phys. Rev. B33 (1986) 430
[12] K.Aggarwal et al. in ref.[1] e)
[13] G.Asti, F.Bolzoni and L.Pareti, IEEE Trans. on Magnetics MAG-23 ,5 (1987)2521
[14] C.Boekema, Hyperfine Interactions 17 (1984)
[15] K.Ruegg, C.Boekema, W.Kuendig, P.F.Meier and B.D.Patterson Hyperfine interactions 8 (1981) 547
[16] R.De Renzi, G.Guidi, P.Podini, R.Tedeschi, C.Bucci, S. F. J. Cox Phys. Rev. B 30,1 (1984) 186
[17] R.De Renzi, G.Guidi, P.Podini, R.Tedeschi, C.Bucci, S. F. J. Cox Phys. Rev. B 30,1,(1984) 197
[18] M.E.Lines, Phys. Rev 137 (1965) A982
[19] T.Moriya, Progr. Theor. Phys. 28 (1962) 371
[20] C.Bucci and G.Guidi, Phys. Rev. B9,7 (1974) 3053
[21] K.Nishiyama, K.Nagamine, T.Natsui, S.Nakajima, K.Ishida, Y. Kuno, J. Yamazato, H. Nakayama, T. Yamazaki and E. Yagi, Jour. Magn. Magn. Materials 31 (1983) 695
[22] L.Asch, J.Chappert, O.Hartmann, G.M.Kalvius, E.Karlsson, L. O. Norlin, R.Wappling and A.Yaouanc, Jour. Magn. Magn. Materials 31 (1983) 697
[23] P.Dalmas de Reotier et al. in [1] e)
[24] O.Hartmann et al. in [1] e)

High-T_c Superconductors: A Background for μSR Measurements

C.E. Gough

Superconductivity Research Group, University of Birmingham, Birmingham B15 2TT, UK

Abstract. A brief review of HTC superconductors and their properties in the normal and superconducting states is presented as a background to μSR measurements in these compounds. Emphasis is placed on those properties that provide fundamental information about the electronic states of the layered cuprate structures. The importance of μSR measurements in elucidating mixed state properties is stressed and possible explanations for the apparent universal relationships between relaxation rate and transition temperature are discussed. Phase homogenity is seen as an underlying problem when interpreting the doping dependence of any HTC property.

1. Introduction

Bednorz and Muller's Nobel Prize-winning discovery of superconductivity in $La_{2-x}Ba_xCuO_4$ [1] not only dramatically increased superconducting transition temperatures to ~35K, but, more importantly, identified a new class of superconducting compounds – the quasi two-dimensional layered cuprate structures, now known as HTC superconductors.

Following the initial discovery, Chu and coworkers raised the transition temperature of the $La_{2-x}Sr_xCuO_4$ variant of the original compound to above 50K by the application of pressure [2]. The same groups then discovered superconductivity above liquid nitrogen temperatures in a multiphase YBaCuO sample, the superconducting phase subsequently being identified as $YBa_2Cu_3O_7$ – the 123 phase of YBCO [3]. Almost a year later, Maeda et al. [4] discovered superconductivity in a multiphase BiSrCaCuO compound containing $Bi_2Sr_2Ca_2Cu_3O_{10}$ with T_c~108K. The highest superconducting transition temperature yet achieved is 125K for a thalium version of the bismuth structure, $Tl_2Ba_2Ca_2Cu_3O_{10}$, discovered by Sheng and Hermann [5]. Many superconducting variants of these structures have been synthesised and are described by Raveau [6].

The scientific challenge and potential technological importance of these materials has resulted in a plethora of over 10,000 research publications and numerous review articles, workshops and

© Springer-Verlag Berlin, Heidelberg 1992

conferences. The object of this brief and inevitably selective and arbitrary overview is to describe some of the more important normal and superconducting properties and to provide the relevant background to μSR and other HTC measurements reported elsewhere in this volume [7–10].

A useful overview of the magnetic and electrical properties of the layered HTC cuprates in their normal and superconducting states has been given by Batlogg [11]. Other specialised reviews are to be found in recent workshop and conference proceedings [12–15], which are extensively referred to in this paper as convenient references to a much wider literature than can be cited here.

μSR has turned out to be an extremely powerful technique for investigating the magnetic properties of the layered cuprate structures in both their insulating–antiferromagnetic [16–18] and metallic–superconducting [18,19] states. The merit of μSR is that it enables the magnitude and variation of magnetic field to be determined within the bulk of a superconductor on both microscopic and macroscopic scales. Useful measurements of intrinsic properties can therefore even be made on powdered or sintered material. For HTC superconductors this is a particular advantage, as it is much easier to achieve uniform chemical or oxygen stoichiometry in small powder grains than in even modestly sized single crystals.

Excellent reviews of μSR measurements on HTC superconductors have been given by Uemura [17] and Keller [18]. In the antiferromagnetic insulating state, μSR measurements complement neutron diffraction [20,21] and NMR/NQR [22,23] measurements. In the superconducting mixed state, the μSR relaxation rate provides a measure of the modulation of the spatially varying fields associated with the flux lines and hence a measure of the magnetic penetration depth $\lambda(T)$. Another recent application has been to place an upper limit to any microscopic internal field in the superconducting state arising from possible d–state pairing of electrons [24], as predicted from the anyonic model for HTC superconductivity [25].

In this review, we first consider the chemical structures of the layered cuprate superconductors and describe the dependence of their normal and superconducting properties on doping. The extreme anisotropy reflects the essentially 2–dimensional character of electron conduction in the CuO_2 planes common to all HTC superconductors. The relationship between the observed experimental properties and theoretical models for HTC superconductivity is briefly considered in parallel with experiment. The superconducting mixed state properties are described in terms of the Ginsburg–Landau model involving the temperature dependent superconducting coherence length, $\xi(T)$, and magnetic penetration length, $\lambda(T)$. Finally, we consider the apparently universal relationships between μSR relaxation rate and T_c for HTC superconductors. Two rather different explanations are considered. The first implies a specific model for HTC superconductivity while the second attributes the observed behaviour to sample inhomogeneities.

2. The structure of the layered cuprate superconductors

2.1 General features of the layered cuprates

Figs.1a–1d illustrate idealised structures of several of the more important HTC superconductors discovered to date. In practice the structures are often complicated by orthorhombic distortions of the unit cell, extended superstructures and inhomogeneities in cation and oxygen stoichimetry. The ubiquitous presence of defects and superstructures in many of these compounds often presents problems when attempting to determine the crystalline structure with the accuracy required for detailed electronic band–structure calculations.

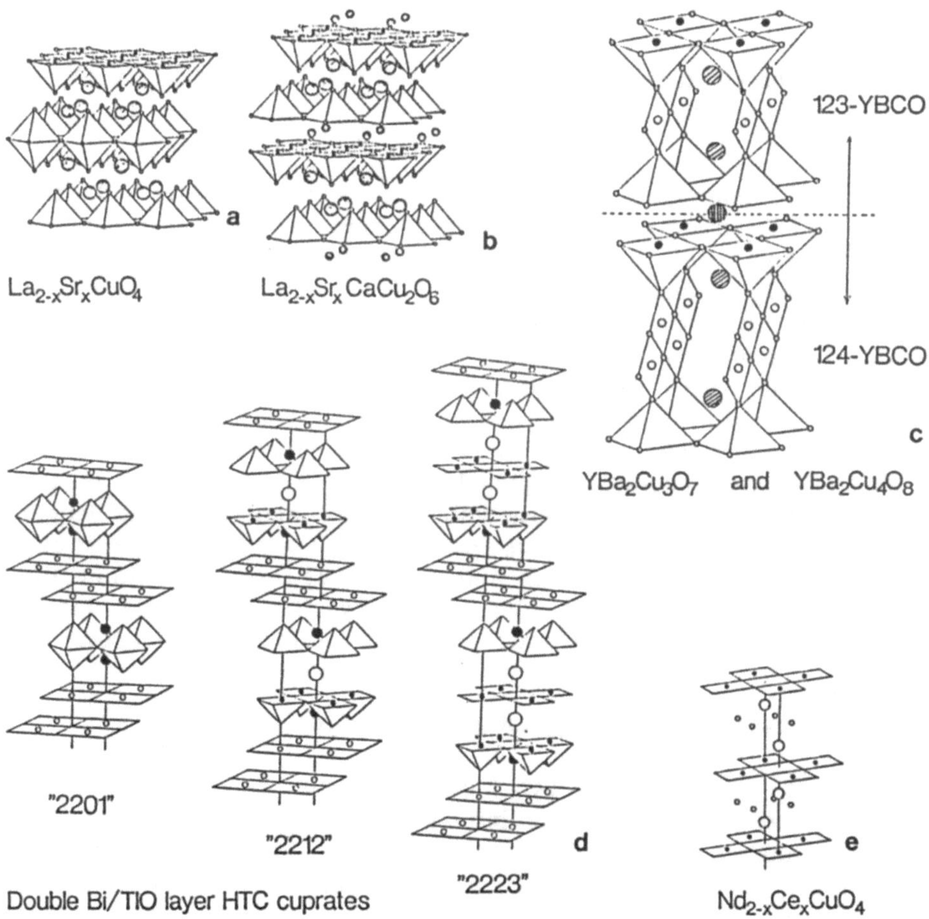

Fig. 1

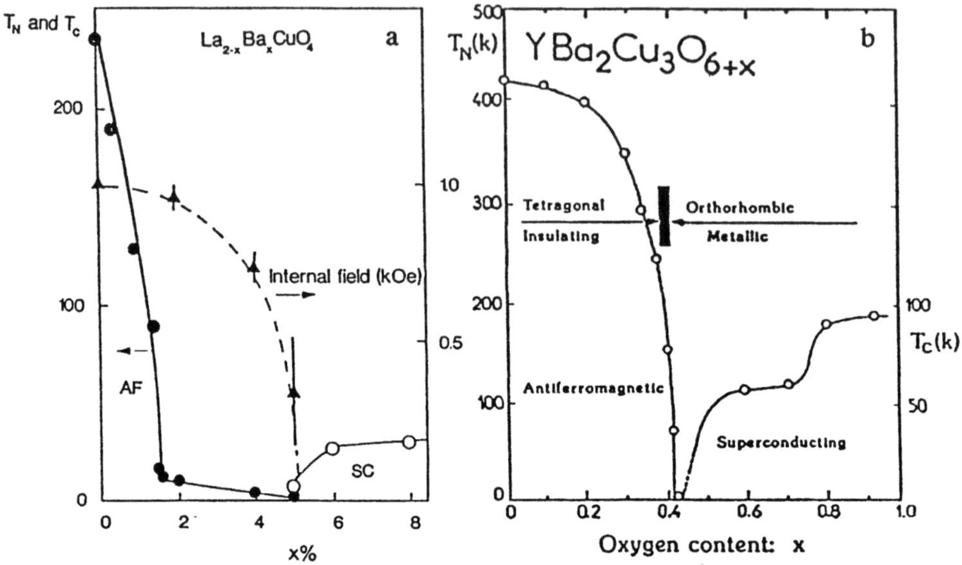

Fig. 2 Phase diagram of $La_{2-x}Ba_xCuO_4$ and $YBa_2Cu_3O_{6+x}$ (from refs. 22 and 47).

All HTC compounds are related to perovskite structures. In the simplest HTC structure, each copper atom is surrounded by six oxygen atoms in an octahedral arrangement, as in the $La_{2-x}Ca_xCuO_4$ compound shown in Fig. 1a. On substitutional doping, this compound undergoes an insulating–antiferromagnetic to metallic–superconducting transition, Fig.2a, which appears to be a feature of all HTC superconductors. The associated metallic to semiconducting transition was investigated by Russian [26] and French [27] groups well before the discovery of HTC superconductivity.

In the layered cuprate compounds the Cu–O bond lengths in the CuO_2 planes (~1.9 Å) are considerably shorter than the bond length to apycal oxygen atoms above or below the Cu atoms. The strong overlap of the copper $3d_{x^2-y^2}$ and oxygen $2p_\sigma$ orbitals within the CuO_2 planes accounts for the essentially 2–dimensional, highly anisotropic electrical and magnetic properties of these compounds.

2.2 Single layer $La_{2-x}(Ba,Sr)_xCuO_4$ compounds

The parent undoped La_2CuO_4 compound is believed to be a Mott–Hubbard antiferromagnetic insulator with a Néel temperature of ~240K, the conduction band being split by strong electronic correlations [28]. The magnetic properties are those of a 2–dimensional, spin–1/2, Heisenberg antiferromagnet, which had not been investigated experimentally prior to the discovery of the layered cuprate compounds. Zero-point energy fluctuations of the spin-1/2 moments strongly influence the magnetic properties [29].

Electron transfer from the CuO_2 planes, on substitutional replacement of trivalent La atoms by divalent Sr,Ba or Ca atoms, results in hole-conduction in the planes, as confirmed by the sign of the Hall coefficient [11, 30]. The holes are believed to be largely concentrated on the oxygen atom sites [31] but with some hybridisation to the copper atoms [32].

Introducing additional spins onto the oxygen sites via doping results in magnetic frustration [33] and a rapid decrease in Néel temperature [34], as illustrated in Fig.2a. Although long range AF order is completely suppressed for $x \sim 0.015$, neutron diffraction [34], NMR [35] and μSR [36] measurements all indicate the continued existence of a local moment on the Cu atoms at least until $x \sim 0.05$. In this range the localised moments are believed to form a magnetic spin glass [33]. On further doping the compound undergoes an insulating to metallic transition with a superconducting transition rising from zero to a maximum of ~35K for $x \sim 0.15$. For higher doping levels, the transition temperature drops. It remains unclear whether this reflects an intrinsic decrease of T_c on doping or whether increased doping results in a structural instability and the appearance of second phases leading to a reduction in T_c.

For this compound, the highest transition temperature occurs close to the orthorhombic to tetragonal phase transition. Incipient phase transitions are known to lead to the highest critical temperatures for conventional superconductors, such as in Nb_3Ge. A similar instability, perhaps associated with a localised defect, might yet prove to be the origin of superconductivity in the HTC compounds [37].

Pairing of electrons via interactions with the spin moments on the copper sites remains a strong contender as a possible alternative mechanism for superconductivity. It is therefore important to consider the evidence for a fluctuating spin moment on the Cu atoms within the superconducting state. Unfortunately, published μSR measurements are contradictory. Uemura [38] finds no evidence for a non-zero Cu moment persisting into the superconducting state, whereas Weideinger et al. [39] and others claim to observe a residual moment throughout the superconducting region. The existence of regions of second phase material or of different stoichiometries might possibly explain the latter observations. The problem of ensuring phase homogeniety is endemic to all investigations of substitutional or oxygen vacancy doping in HTC compounds.

2.3 Double layer YBCO compounds

The $YBa_2Cu_3O_{7-\delta}$ or 123–YBCO structure involves two closely spaced planes of CuO_2 on either side of the yttrium atoms, the planes being formed from two facing pyramids formed from the original ocatahedra with one apycal oxygen removed. The double CuO_2 layers are connected by CuO_3 chains along the b–direction, as illustrated in the upper half of Fig.1c.

Surprisingly, substitution of yttrium by most other rare earth elements has little effect on T_C [40], even those with a large magnetic moment. The only exceptions are Pr, Ce and Tb, which tend to form tetravalent rather than trivalent states. Substituted magnetic rare earth atoms undergo an AF transition at low temperatures [41]. The lack of interaction between the magnetic rare earths and the superconducting electrons implies the rare earth atoms are close to a node of the electronic wave functions of charge carriers in the CuO_2 planes.

The magnetic and superconducting properties of $YBa_2Cu_3O_{7-\delta}$ are highly dependent on oxygen stoichiometry, as illustrated schematically by the phase diagram shown in Fig. 2b [47], which should be compared with that for the $La_{2-x}Ba_xCuO_4$ system in Fig.2a.

Both systems involve a transition on doping from an insulating–antiferromagnetic state to a metallic superconducting state. This is often cited in support of the view that HTC superconductivity is in some way connected with AF spin–correlations, presumably persisting into the superconducting state. However, it might equally well be argued that superconductivity is being destroyed by such interactions on approaching the antiferromagnetic phase transition.

On initially decreasing the oxygen stoichiometry from $YBa_2Cu_3O_7$, oxygen vacancies are introduced into the basal plane chain sites, while some oxygen atoms transfer to the previously empty sites in the basal plane [42,43]. Initially this produces little change in T_C but, for $\delta \sim 0.2$, T_C starts to fall rapidly. For oxygen stoichiometries in the range $YBa_2Cu_3O_{6.6-6.7}$ there is a plateau in T_C at around 60K. It is believed to arise from an ordering of oxygen vacancies in the basal plane chain sites, with every other oxygen atom vacant [44] or with alternating full and empty chains [6,45]. This is often referred to as the 60K YBCO phase. However, detailed TEM studies suggest the plateau region may result from the thermodynamic equilibrium of several different states of oxygen ordering [46].

It has even been suggested that the fully oxygenated $YBa_2Cu_3O_7$ is intrinsically unstable [37], the structure possibly being stabilised by an ordered oxygen vacancy state, such as with every fourth oxygen atom site along a chain empty [44]. This could explain why it is difficult to prepare fully oxygenated samples of the 123–YBCO compound.

For oxygen stoichiometry ~ $YBa_2Cu_3O_{6.4}$, the compound undergoes a transition to the semiconducting– antiferromagnetic state, though not all groups agree on the exact composition at which such a transition takes place, almost certainly because of problems of phase homogeneity[6]. $YBa_2Cu_3O_6$ is an insulating, spin–1/2, Heisenberg antiferromagnetic with very similar magnetic properties [47] to those of La_2CuO_4. However, unlike $La_{2-x}Sr_xCuO_4$, the 123–YBCO system does not appear to pass through an intermediate magnetic spin glass state between the AF and superconducting phases.

Like all HTC superconductors, 123–YBCO is synthesised via a high temperature solid–state reaction, which tends to expel oxygen from the structure. To optimise T_c, the oxygen has to be reintroduced by annealing in air or oxygen. Oxygen is taken up preferentially onto the chain sites resulting in a tetragonal to orthorhombic phase transition on cooling [46]. To accommodate the associated changes in lattice dimensions, the material undergoes a twinning transformation. The presence of such twinning, which is easily seen under a polarising optical microscope, can be used as a qualitative measure of the extent to which oxygen has been taken up by the structure.

A large number of other double CuO_2 layer HTC structures have been discovered. Of particular interest is $YBa_2Cu_4O_8$, the 124–YBCO structure, originally discovered by epitaxial thin film growth [48] but which can now be synthesised in the bulk [49]. This structure is shown in the lower half of Fig. 1c and involves a double CuO chain structure parallel to the b–axis. T_c of the 124 compound can be raised from ~ 60K to ~ 105K by doping with calcium [50] or by pressure [50b]. Unlike the 123–structure, $YBa_2Cu_4O_8$ is tetragonal and is also thermodynamically stable in air up to 800°C. This has a number of potential advantages for technologial applications. Cava et al. [51] discovered $Pb_2Sr_2Y_{0.5}Cu_{0.5}Cu_3O_8$ with a T_c~60K having a large cage–like PbO structure separating the double CuO_2 planes. Very recently, the same group [52] have succeeded in demonstrating superconductivity at ~60K in $La_{2-x}Sr_xCaCu_2O_{6-\delta}$, with the structure shown in Fig.1b. This is the double CuO_2 layer equivalent of the original $La_{2-x}Ba_xCuO_4$ compound discovered by Bednorz and Muller. Double layered CuO structures are also important members of the Bi–and Tl–based multi–layer structures described in the next section.

2.4 Bi and Tl multilayer cuprate superconductors

Several BiSrCaCuO and TlBaSrCuO HTC structures are illustrated in Fig. 1d. The generic chemical composition of these compounds can be written as $(BiO)_mSr_2O_2Ca_{n-1}(CuO_2)_n$ and $(TlO)_m$ Ba_2O_2 $Ca_{n-1}(CuO_2)_n$, where n is the number of CuO_2 layers separated by Ca atoms and m is the number of intercalating rock-salt layers of BiO or TlO; the additional Sr,Ba and O atoms are required for chemical balance and structural stability. Such compounds are often conveniently referred by their cation stoichiometry (e.g. the m = 2, n = 2, $Bi_2Sr_2CaCu_2O_8$ structure is the 2212-bismuth phase).

It is extremely difficult to prepare ideal single phase samples of these compounds that are free from intergrowths of other phases having different n or m values on an atomic scale. Their crystallographic structures are also complicated by extended superstructures [6,53].

In these compounds T_c initially increases with the number of CuO_2 layers. For the thalium compounds T_c is around 20K for the single CuO plane 2201 structure, it rises to ~ 80–100K for the two layer 2212 structure and is ~ 125K for the 3 layer 2223 structure. The end point of this series with $n \to \infty$ corresponds to $CaCuO_2$, which has to be stabilised by the addition of a small amount of Sr [54]. Unfortunately, instead of having a very high critical temperature, the compound is an insulator. T_c probably reaches a maximum at around 3–4 layers [55].

2.5 Electron doped HTC superconductors

The normal state and superconducting properties of all the above compounds are determined by the density of holes in the CuO_2 planes, which can be varied by cation subsitution or control of the oxygen stoichiometry. These superconductors are therefore known as hole-doped HTCs.

Of fundamental importance was the discovery of a new class of electron-doped HTC layered cuprate superconductors by Tokura et al [56] based on the T' phase of the parent Nd_2CuO_4 structure illustrated in Fig.1e. In this structure both pyramidal apical oxygen atoms are absent leaving simple CuO_2 planes. Electron doping of the planes is achieved either by replacing the trivalent neodymium with tetravalent cerium or by replacing oxygen with florine as in $Nd_{1.85}Ce_{.15}CuO_4$ and $Nd_2CuO_{3.5}F_{.5}$, both of which have a T_c ~ 25K [57].

The electronic properties of the electron-doped materials are not well understood, as electron energy loss (EELS) measurements of the oxygen 1s absorption edge implies the coexistence of simultaneous hole doping [58]. This may again be associated with chemical inhomogeneities involving coexisting regions of electron- and hole-doped material.

2.6 Problems of phase homogenity

The properties of all the above compounds are very sensitive to the exact oxygen stoichiometry. Doping by cation substitution often results in compensating changes in the oxygen stoichiometry. It is therefore always important to monitor the oxygen stoichiometry carefully when attempting to investigate the influence of cation doping on the normal state and superconducting properties.

Because of the difficulty of ensuring phase homogeneity in many of the HTC structures, much research directed towards elucidating the underlying physics has tended to concentrate on compounds exhibiting a maximum in the superconducting temperature (e.g. $YBa_2Cu_3O_7$, $La_{1.85}Sr_{.15}CuO_4$ and $Bi_2Sr_2CaCu_2O_8$, preferably in single crystal form), where small inhomogeneities might not be expected to change the superconducting properties appreciably. Single crystals of the Bi-based HTC superconductors have proved particularly useful for surface studies, such as electron tunnelling and photo-electron emission, as crystals can be cleaved at liquid helium or nitrogen temperatures to expose what are hopefully atomically clean surfaces.

3. Normal and superconducting state electronic properties

3.1 Normal state properties

An understanding of the normal state electronic properties of the layered cuprate compounds is clearly a prerequisite to the development of any successful microscopic theory of the superconducting state. However, to date, no universally accepted theory for the electronic normal state properties exists.

The room temperature normal state electrical conductivity of single crystals of HTC superconductors is highly anisotropic, reflecting the essentially 2-dimensional conduction in the CuO_2 planes. For YBCO the conductivity parallel to the CuO_2 planes is almost 100 times larger than in the perpendicular direction [11,59]. For the multi-layered bismuth based cuprates the ratio may be as high as 1000 or even larger [11].

The electrical resistance parallel to the CuO_2 planes is a remarkably linear function of temperature, with a near zero intercept even for HTC superconductors with T_c's as low as 20K. Measurements on high quality single crystals suggest an intrinsic temperature dependence of the form $\rho \sim \alpha T$, where α has approximately the same value, ~1 $\mu\Omega$ cm K^{-1}, for all HTC superconductors. This should be contrasted with conventional metals, where the resistance at low temperatures flattens off. Opinion is divided on whether the linear temperature dependence results from a strong electron coupling to a very low energy lattice vibration [37] or whether it reflects an intrinsic property of the electron charge carriers in the CuO_2 planes.

The latter view inspired Anderson's early and very influential resonating valence bond (RVB) model [60, 61]. The original RVB model also predicted a conductivity perpendicular to the planes proportional to 1/T, qualitatively consistent with experiment [11]. However, reliable measurement of the conductivity perpendicular to the planes in these highly anisotropic materials remains difficult and is very sample dependent [11]. Varma [62] has developed a phenomenological "marginal fermi-liquid" model for charge carriers in the layered cuprate compounds, which accounts for the observed temperature dependence of the conductivity and many other normal state properties. Very recently, Millis et al. [63] have proposed an antiferromagnetic Fermi liquid model, consistent with NMR measurements on YBCO in the normal state [64], which accounts for the observed linear temperature of the conductivity in terms of electron scattering from anti-ferromagnetic spin fluctuations.

Recent room temperature scanning tunnelling microscope measurements by Hasewaga and Kitizawa [65] appear to confirm that electronic conduction is via charge carriers in the CuO_2 planes. In these measurements, an atomic-scale fine tip is moved across the surface and its height adjusted to maintain a constant vacuum tunnelling current for a given applied voltage. This enables the surface structure to be determined on an atomic scale. On a freshly cleaved $Bi_2Sr_2CaCu_2O_8$ crystal exposing a surface perpendicular to the CuO_2 planes, the individual atoms in the CuO_2 and BiO planes could easily be resolved. By temporarily removing the feedback mechanism controlling the height of the tip above the surface, the voltage current-tunnelling characteristics of electrons from the BiO and CuO_2 planes could be determined. Measurements on the bismuth plane sites implied a semi-conducting gap in the electronic density of states, while those on the copper sites implied metallic properties. The vacuum tunneling characteristics of the CuO_2 plane sites were highly non-linear. They could be interpreted as providing evidence for a density of electronic states increasing linearly with distance from the fermi-surface, in contrast to the nearly constant density of states expected for a normal metal on the energy scales involved. This is consistent with Varma's marginal Fermi-liquid model.

Very similar conclusions have been drawn from conventional tunnelling measurements across surface barriers between HTC and normal metals. Initially, the results of such measurements were very inconclusive with little agreement from sample to sample or between groups working with different types of junctions. However, over the last year or so, a number of groups [66–68] have obtained reproducible and virtually identical characterisitics for different junction geometries. The resulting characteristics are therefore assumed to reflect the intrinsic properties of HTC superconductors. The non-ohmic normal state tunnelling characteristics imply a density of states increasing linearly as $|E-E_F|$, quite unlike that of a normal metal.

Evidence is therefore growing that the electronic charge carriers in the layered cuprate compounds have quite different properties form those of normal metals. It is still not established whether electronic quasi-particles in these materials can be treated as extended waves with a well-defined $\epsilon(k)$ relationship, implying a fermi-surface, or whether the electronic excitations are effectively localised by the repulsive coulombic interactions between d-electrons of opposite spin occupying the same transition metal site. This is the starting point for many theories of the cuprate superconductors based on variants of the Hubbard model. This involves an on-site repulsive coulomb energy U and an energy t associated with electrons hopping from site to site (see ref. 69 for a useful compilation of relevant, theoretical papers). The relative size of U and t determines whether the electrons are essentially free or localised. In practice, an intermediate situation can be anticipated.

The most convincing experimental evidence for the existence of a fermi surface in HTC superconductors comes from measurements of the angular-resolved photo-emission of electrons from freshly cleaved surfaces of 2212 BiSCCO [71,73–76]. At certain energies and angles relative to the ab-planes strong features in the spectrum of excited electron occur consistent with a particular branch of the band structure crossing the fermi surface, as illustrated in Fig.3a. However, the band structure deduced from the angular and energy dependence of the photo-emission [70,71], imply much narrower electronic bands than predicted by band-structure calculations [72]. Almost certainly, this can be attributed to strong electron correlations, which are neglected in conventional band-structure calculations.

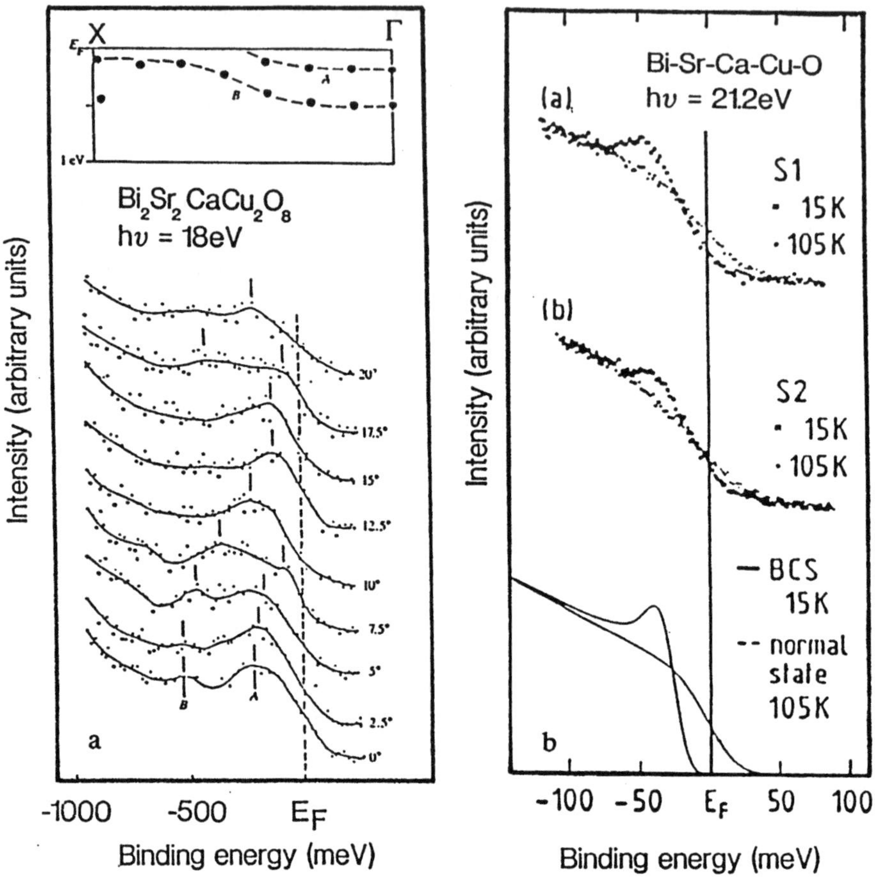

Fig. 3 Angular resolved photon emission a) in normal state showing a feature corresponding to a band crossing the fermi surface [70] and b) in the normal and superconducting states [74].

3.2 Superconducting state properties

Angular resolved photo-emission measurements [71,73–76] have also provided the most convincing evidence to date for the existence of an energy gap at the fermi surface on entering the superconducting state, as indicated in Fig. 3b. Differences between the spectra above and below the superconducting transition temperature demonstrate the opening up of a gap of ~20meV with the expected increase in density of states just below the gap. Fitting to a BCS model, these measurements give $2\Delta(0)/kT_c \sim 7.5 \pm 1$, well above the predicted weak-coupling value of 3.5.

The interpretation of electron tunneling measurements in the superconducting state remains somewhat problematical, though several groups [66–68] are now obtaining reproducible results with estimated values of $2\Delta(0)/kT_c \sim 5 \pm 1$, somewhat lower than deduced from the above electron-emission measurements but still consistent with strong-coupling within the context of a BCS description.

If superconductivity involves an attractive BCS-like attraction via the vibrational states of the oxygen atoms, a large isotope effect, $T_c \sim M^{-1/2}$, should be expected on replacing O^{16} by O^{18}. A small shift in T_c is indeed observed for the single-layer La-based cuprates [77] but a much smaller shift in 123-YBCO [78]. This was initially taken to imply that an electron-lattice phonon mechanism for superconductivity was unlikely. However, the full value of the predicted isotope effect is rarely observed even for conventional superconductors [79], so that not too much weight should given to this observation.

It is interesting to note that the transition to superconductivity does involve a change in lattice vibrational frequencies. In YBCO the 340 cm^{-1} Raman line shifts by about 3% [80] and smaller but significant shifts in other vibrational modes can be detected by Raman [81] and infra-red spectroscopy [82]. The Raman-active phonon peaks are observed on an anomalous broad background, which appears to be characteristic of electrons in the layered cuprate superconductors [80]. There is clearly an interaction between electrons in the superconducting state and lattice vibrations but that does not necessarily imply that electron-phonon coupling is the origin of superconductivity, though it cannot be ignored in any complete theory.

An important prediction of the BCS model for superconductivity was the initial increase in nuclear spin relaxation rate, T_1^{-1}, on entering the superconducting state [83]. Despite some early reports, this does not occur in the layered cuprate superconductors [64,84]. Furthermore the relaxation rate at low temperatures exhibits a power law temperature rather than the expected $e^{-\Delta/kt}$ dependence. This may be associated with defects and strains, microcracking and chemical inhomegeneities on the surface of the powder grains used in these measurements.

At T_c a discontinuity in heat capacity is observed associated with the second-order phase transition to the superconducting state (see Phillips [85] for a review of heat capacity measurements). The rapid decrease in size of the discontinuity on departing from full oxygen stoichiometry [87] should be noted in relation to later discussion of μSR relaxation rates. In sufficiently pure single crystals an additional λ-point anomaly associated with thermodynamic fluctuations has also been observed [87, 88]. Thermodynamic fluctuations are very important in HTC superconductors because of their very short coherence lengths and high critical temperatures. Fluctuations lead to a rounding of the resistive transition extending to temperatures well above T_c [89].

4. The superconducting mixed state

Irrespective of the nature of the microscopic model required to account for HTC superconductivity, it is reasonable to expect that the magnetic properties on a macroscopic scale can be described by the Ginsburg–Landau theory [90]. The superconducting properties are assumed to be described by a macroscopic wave function $\Psi e^{-i\Theta(r)}$, where $|\Psi|^2 = n_s(T)$ is the temperature dependent density of superconducting electrons, and $\varphi(r)$ is a position dependent phase factor describing current flow. GL theory assumes that the free–energy difference between the superconducting and normal states can be written as

$$F_s - F_n = \alpha(T-T_c)|\Psi|^2 + \beta|\Psi|^4 + 1/2m^* \; |-i\hbar\nabla\Psi - e^*\mathbf{A}\Psi|^2 +$$

$$+ \text{ magnetic energy terms.}$$

For BCS superconductors close to T_c, Gorkov [91] derived the above expansion from microscopic theory proving that $m^*=2m$ and $e^*=2e$, reflecting the pairing of electrons.

In practice, the GL expansion is a rather good model for conventional superconductors for all temperatures and fields. Despite the current impossibility of providing any kind of microscopic justification for HTC superconductors, the GL expansion appears to give a fairly good account of the observed thermodynamic properties provided thermal fluctuation effects are taken into account. However, if superconductivity involves p– or d–state electron pairing, the GL expansion and associated order parameter will need to be generalised to reflect the underlying symmetry of the wave function.

The Ginsburg Landau model involves two characteristic lengths; a temperature dependent coherence length $\xi(T)$, which is the length scale over which the superconducting order parameter, $|\Psi|^2 = n_s$, can vary, and the London penetration depth, $\lambda(T)$, which is the length over which diamagnetic screening currents flow and magnetic fields decay within the bulk. Much earlier, the London brothers [92] had shown that, in order to account for the exclusion of flux within a superconductor, the electrodynamic equations had to be modified in such a way to give a characteristic magnetic decay or penetration length, $\lambda(T) = (n_s e^2/\mu_0 m^*)^{-1/2}$, where n_s is the density of superconducting electrons and m^* their effective mass. Close to T_c, the Landau Ginsburg theory predicts both ξ and λ to vary as $(1-T/T_c)^{1/2}$. However, over an extended temperature range, $\lambda(T)$ is often observed to vary as $\lambda(0)/\surd(1-(T/T_c)^4)$, which is referred to as the empirical 2–fluid or Gorter–Casimir temperature dependence [93]. This differs from BCS theory in the weak coupling limit [94] but approximates to the predicted dependence in the strong coupling limit.

The magnetic behaviour of a superconductor depends on the relative magnitudes of ξ and λ. HTC superconductors are extreme type-2 superconductors with $\lambda >> \xi$. On increasing the magnetic field, flux is initially excluded from the bulk – the Meissner effect – until the lower critical field, H_{c1}, is reached. At H_{c1}, Abrikosov [95] showed that magnetic flux enters the superconductor in the form of individual current vortices or flux lines with magnetic flux h/2e, the flux quantum. In an ideal superconductor the flux lines would be arranged on a triangular lattice with a spacing proportional to $(\frac{\Phi_0}{B})^{-1/2}$, where B is the average magnetic induction within the superconductor. This is known as the mixed state or Abrikosov/Shubnikov phase. When the flux line spacing becomes comparable with ξ, superconductivity is suppressed with a transition to the normal state at the upper critical field $H_{c2} = \Phi_0/2\pi\xi^2$.

Measurements of $H_{c2}(T)$ for fields directed perpendicular and parallel to the CuO_2 planes can therefore be used to determine the in-plane and perpendicular to plane coherence lengths ξ_{ab} and ξ_c (see Forgan [96] for a discussion of the influence of HTC anisotropy on ξ and λ). Typical coherence lengths for YBCO-123 extrapolated to zero temperature are ξ_{ab} ~15 Å and ξ_c~3 Å [97]. Problems associated with thermally acivated flux creep and variations in sample quality probably account for the rather wide discrepancies in published values for the coherence length. The very short coherence lengths of these materials implies that defects on a microscopic scale can have a significant effect on local superconducting properties.

The London penetration depth $\lambda(T)$ can be determined either by direct measurement of the penetration of magnetic field at the surface of a superconductor or from the amplitude of field modulation within the superconducting mixed state. Direct measurement of field penetration requires very clean surfaces, the intrinsic temperature dependence of $\lambda(T)$ apparently only being observed in carefully selected single crystals [98]. Measurement of the spatial field modulation within the Abrikosov or mixed state by μSR are likely to give more reliable estimates of $\lambda(T)$ because they are bulk measurements and should not be affected by surface imperfections.

For fields significantly greater than H_{c1}, flux lines are sufficiently closely spaced that the spatial field modulation, δH, between lines is relatively small. It can then be shown [99] that $\delta H \sim (\frac{d}{\lambda})^2 H \sim \Phi_0/\lambda(T)^2$, where $d \sim (\Phi_0/\mu_0 H)^{1/2}$ is the flux line spacing. Measurement of the amplitude of field modulation therefore gives $\lambda(T)$, which in turn provides information on the ratio n_s/m^*.

Magnetic decoration measurements, in which a fog of very fine ferromagnetic powder is allowed to deposit on the surface of a superconductor cooled in a field [100], have provided very

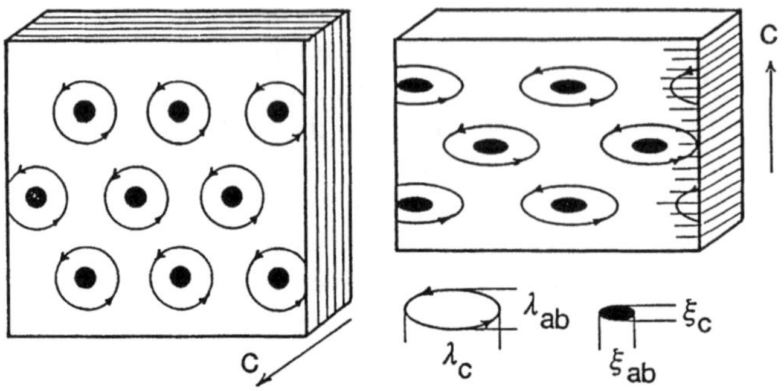

Fig. 4 Anisotropy and nomenclature of coherence lengths and penetration depths and associated influence on flux–line lattice. Note the penetration length along the c–direction is associated with current flow in the ab plane – hence λ_{ab}.

convincing evidence for the existence of a flux–line lattice in HTC superconductors [101]. Such measurements also illustrate the strong anisotropy of the flux line lattice with respect to field direction relative to the CuO_2 planes.

For fields directed perpendicular to the CuO_2 planes, the circulating vortex currents flow parallel to the planes – the easy current flow direction. When the field is directed parallel to the planes, there has to be a component of current flow perpendicular to the planes. Because of the weaker superconducting properties perpendicular to the planes, supercurrents flowing in this direction are less effective at screening field variations, so that the associated penetration depth is larger. Fig.4 illustrates the way that the anisotropy affects the shape of flux–lines for the two orientations of field [96]. For arbitrary angles of field direction the extreme anisotropy can lead to significant deviations from the intuitively expected flux line structure [102].

Decoration measurements provide information on spatial field variations on the surface rather than within the bulk of the superconductor. They may therefore be affected by surface morphology and defect structures. Furthermore, the limited spatial resolution of the technique limits measurements to rather small fields of ~ 20 gauss.

The flux line lattice can, in principle, be studied by low angle neutron scattering. Only recently have crystals of sufficient size and quality become available to demonstrate diffraction from the flux line lattice [103]. Forgan [8] describes these measurement elsewhere in this volume, where comparisons are also made with μSR measurements on the same crystal.

5. μSR measurements in the superconducting mixed state.

Polarised muons implanted into a superconductor are ideal probes of the spatially varying magnetic fields within the mixed state. Unlike measurements of magnetic properties in the insulating anti-ferromagnetic state, the interpretation of μSR data in the mixed state does not rely on knowing the trapping site of the muon within the unit cell as the mixed state field modulations are on a very much larger scale.

In the mixed state, muons precess at different rates depending on their position relative to the centre of the flux line cores. This leads to a dephasing of the resultant time-averaged precession signal, which is assumed to decay as $\exp(-\sigma^2 t^2)$, where σ is the relaxation rate (see Keller [7, 19] for a discussion of the relation between spatial field modulation and observed damping rate). The relaxation rate σ is inversely proportional to the amplitude of spatial field modulation, which varies as $\varphi_0/\lambda(T)^2$. Values for $n_s(T)/m^*$ can therefore be determined from such measurements.

Measurement of the μSR relaxation rate as a function of field direction relative to the CuO_2 planes enables the anisotropy of m^* to be determined. Such measurements were originally confined to magnetically oriented powder samples though more recently such measurements have been extended to single crystal samples [7, 104]. For YBCO, the measured anisotropy of anisotropy of σ gives $\lambda_c/\lambda_{ab} \sim 5$ corresponding to an effective mass ratio $m_c/m_{ab} \sim 25$. The large anisotropy again reflects the quasi 2-dimensional character of electronic conduction in the CuO_2 planes.

Because the anistropy is so large, averaging relaxation rates over all possible grain orientations in a powder sample gives an effective relaxation rate not too different from the value for fields directed along the c-axis [105]. Measurements on powders can therefore be used to provide valuable information on the dependence on doping and temperature of $\lambda_{ab}(T)$ and hence of $n_s(T)$.

Representative measurements of the temperature dependence of $\sigma(T)$ are given by Keller [7] in this volume. The temperature dependence can be described by the empirical 2-fluid model [93] again implying strong coupling within the framework of BCS theory. The absence of any marked temperature dependence of $\sigma(T)$ at low temperatures provides the most convincing evidence yet for s-state rather than p- or d-state pairing of the superconducting electrons. s-state pairing implies a constant energy gap for all directions in k-space, whereas p- or d-state pairing would lead to nodes in the energy gap along lines or at points on the fermi surface. This would result in a strong power law dependence on temperature at low temperatures, which is not observed in HTC superconductors.

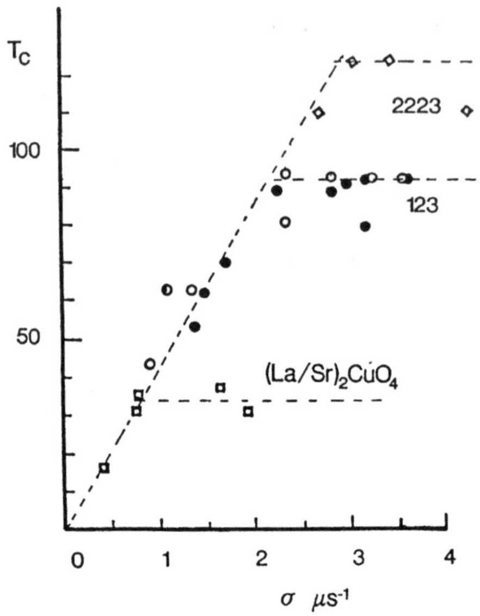

Fig. 5 Uemura plot of μSR relaxation rate, σ, as a function of T_c measured resistivity. For simplicity we have only plotted values for $(La,Sr)_2CuO_4$, $YBa_2Cu_3O_{7-\delta}$ and the Tl "2223" compound (data from Uemura [18, 106] and Keller [7].

However, a slight note of caution may be necessary, as ac measurements of the diamagnetic exclusion of flux from finely divided powdered samples always show a marked temperature dependence of the penetration depth below $T_c/2$, which is inconsistent with a BCS or 2-fluid model dependence of $\lambda(T)$. We have also noted that the NMR relaxation rate appears not to decrease as rapidly as the predicted $e^{-\Delta/kT}$ temperature dependence and that, in direct measurements of $\lambda(T)$, the supposedly intrinsic behaviour is only observed for carefully selected crystals. These differences in behaviour may simply reflect surface problems. It would be interesting to compare the temperature dependence of $\lambda(T)$ with measurements made on the same powder sample by a variety of techniques including μSR.

Finally, we consider the relationship between critical temperature and μSR relaxation rate first pointed out by Uemura et al. [18, 107] and illustrated in Fig.5. We have also included recent 123–YBCO measurements reported by Keller [7] in this volume. For a given class of HTC compound, the measurements all appear to lie on a set of universal curves with an initial linear relationship followed by a plateau region. For example, for the $La_{2-x}Sr_xCuO_4$ compounds the increase in T_c on doping is initially proportional to the observed relaxation rate until a plateau region at ~35K is reached, where the relaxation rate increases but the critical temperature remains approximately constant. Similarly, if we start from the right hand side of the plot, T_c for 123–YBCO initially remains remarkably constant on introducing oxygen vacancies even though the relaxation rate and hence derived "$n_s(T)$" are significantly decreased. The μSR relaxation provides a sensitive measure of the oxygen stoichiometry. When the oxygen vacancy concentration is sufficient to decrease T_c, the measurements appear to fall on the same linear relationship observed for the single–layered lanthanum compounds. Similar behaviour is observed for the 3–layer compounds. The drop in T_c for samples to the extreme right of the plateau regions may be intrinsic but it is equally likely to arise from phase instabilities on attempting to exceed optimium doping levels.

The initial linear increase of relaxation rate with critical temperature has been taken to imply an intrinsic mechanism for superconductivity with T_c proportional to n_s, the density of superconducing electrons in the CuO_2 planes. Emery [107] has pointed out that this is just the dependence expected for an extended BCS model in which the fermi energy of the electrons is smaller than the energy of the excitations responsible for the attractive pairing interaction. For such a model, $kT_c \sim \epsilon_F \exp(-1/\lambda)$, where λ is a measure of the coupling energy. Provided λ does not depend on n_s, this gives the required linear dependence of T_c on n_s, since in 2–dimensions $\epsilon_F \propto n$. The plateau region is assumed to arise from a breakdown of the model at large doping concentrations.

However, the existence of the plateau region and the eventual drop in relaxation rates could be explained by a completely different mechanism. Let us suppose that the intrinsic superconducting properties are represented by the points to the extreme right of the plateau regions. For the 123–YBCO structure this would correspond to the nearly fully oxygenated state. On reducing the oxygen stoichiometry, the oxygen vacancies could well form local regions of ordered, oxygen deficient phase, having a lower T_c, as indicated in fig. 6. A macroscopic supercurrent could still pass between these domains so that T_c would be unchanged. However, if the off–stoichiometry regions are distributed on a scale $< \lambda(T)$, the quantised circulating currents associated with flux lines will have to pass through both optimised and off–stoichiometry material. This will lead to a decrease in the effective value of n_s and hence the observed decrease in μSR relaxation rate. T_c will eventually drop when there is no longer a percolative path through regions of optimised stoichiometry. The μSR relaxation rate must clearly go to zero with T_c, though for this model one would have to assume that the observed similarity in initial slopes for all HTC superconductors was fortuitous.

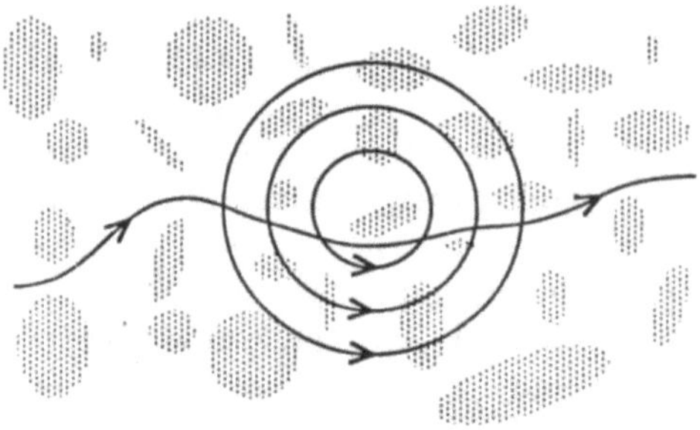

Fig. 6 Schematic representation of localised oxygen vacancy ordering to form a second phase. T_c measured by bulk supercurrent is initially unchanged but effective n_s is decreased.

It is important to consider how such a model might also affect other properties. For example, we have already remarked that, for the 123–YBCO compound, the discontinuity in heat capacity at T_c is reduced on moving away from near–full oxygenation while T_c remains unchanged [87, 108] – see fig.5. This is consistent with a decrease in the number of superconducting electrons in the planes, as expected from the above model. A comparison of μSR measurements and ΔC_{T_c} as a function of doping would clearly be of interest. Another measurement, that can in principle provide useful information about local variations in charge carriers within the CuO planes, is the measurement of NMR relaxation rate of Cu atoms in the planes. On moving away from full stoichiometry, the NMR signal is appreciably broadened with components from copper in–plane atoms corresponding to at least three different environments [109], presumably with a decreased superconductivity transition temperature.

We are therefore placed with two rather different explanations for the form of the Uemura plots. Further experimental studies including μSR measurements on samples that have been investigated using other relevant techniques will hopefully resolve which explanation is correct.

Although μSR measurements alone are unlikely to resolve the problem of establishing a satisfactory model for HTC superconductors, such measurements have already provided much valuable information that must ultimately be explained by any theory claiming to solve the HTC problem.

6. Acknowledgments

I should like to thank Giorgio Benedek and Hubert Schneuwly for inviting me to the Workshop on Exotic Atoms in Condensed Matter, which provided the motivation for writing this brief review. I am also grateful to Ted Forgan, Steve Cox, A. Schenk and Y. Uemura for helping me to understand the role of μSR in elucidating the properties of HTC superconductors.

References

1. J.G. Bednorz and K.A. Muller, Z.Phys. B64, 189 (1986)
2. P.H. Hor, L. Gao, R.L. Meng, Z.J. Huang, Y.Q. Wang, K. Foster, J. Vassilious, C.W. Chu, M.K. Wu. J.R. Ashburn and C.J. Torng, Phys.Rev.Lett. 58, 911 (1987)
3. M.K. Wu, J.R. Ashburn, C.J. Torng, P.H. Hor, R.L. Meng, L. Gao, Z.J. Huang, Q.Y. Wang and C.W. Chu, Phys.Rev.Lett. 58, 908 (1987)
4. M. Maeda, Y. Tanaka, M. Fukutomi and T. Asano, Jpn.J.Appl.Phys. 27, 209 (1988)
5. Z.Z. Sheng and A.M. Hermann, Nature 332, 138 (1988)
6. B. Raveau, C. Michel, M. Hervieu, J. Provost and F. Studer in *Earlier and Recent Aspects of Superconductivity*, ed. Bednorz and Muller (Springer-Verlag, Berlin, Heidleberg, New York, 1990) pp 66–95
7. H. Keller, *Exotic Atoms in Condensed Matter Physics -Erice, May 1990* (Springer-Verlag, Berlin, Heidleberg, 1990/1)
8. E.M. Forgan, ibid.
9. K. Nagamine, ibid.
10. A. Bianconi, ibid.
11. B. Batlogg in *High Temperature Superconductivity: The Los Alamos Symposium-1989. Ed. K. Bedell, D. Coffey, D. Meltzer, D. Pines and J.R. Schrieffer* (Addison-Wesley Publishing, 1990) 37–82
12. IBM J.Res.Develop. 33(3), May 1989
13. *Earlier and Recent Aspects of Superconductivity*, ed. J.G. Bednorz and K.A. Muller (Springer-Verlag Berlin, Heidelberg 1990)
14. Proc. Int Conference M^2S-HTC, Stanford, 1989. Ed. R.N. Shelton, W.A. Harrison and N.E. Phillips (Elsevier, North-Holland 1989) published as Physica C162–164
15. *High Temperature Superconductivity: The Los Alamos Symposium-1989. Ed. K. Bedell, D. Coffey, D. Meltzer, D. Pines and J.R. Schrieffer* (Addison-Wesley Publishing, 1990)
16. N. Nishida et al., Jpn.J.Appl.Phys. 26, L1856 (1987)

17. D.R. Harshmann, G. Aeppli, G.P. Espinosa, A.S. Cooper, J.P. Remeika, E.J. Ansaldo, T.M. Riseman, D.L. Williams, D.R. Noakes, B. Ellman, T.F. Rosenbaum, Phys.Rev. B38, 852 (1988)
18. Y.J. Uemura et al, Journal de Physique, Suppl.C8, 49, 2087 (1988): Y.J. Uemura, L.P. Le, G.M. Luke, B.J. Sternleib, J.H. Brewer, R. Kadano, R.F. Kiefl, S.R. Kreitzman and T.M. Riseman, Physica C162–164, 857 (1989)
19. H. Keller in *Earlier and Recent Aspects of Superconductivity*, ed. J.G. Bednorz and K.A. Muller (Springer–Verlag Berlin, Heidelberg 1990) pp 66–95.
20. R.J. Birgeneau et al, IBM J.Res. 33(3), 270 (1989)
21. J. Rossat–Mignod in Int.Seminar on High Temperature Superconductivity, ed V.L. Akesnov, N.N. Bogolubov and N.M. Plakida (World Scientific, Singapore 1990) pp 74–85
22. Y. Kitaoka, K. Ishida, K. Fujiwara, K. Asayama, H. Katayama–Yoshida, Y. Okabe and T. Takahishi, IBM J.Res. 33(3), 277 (1989)
23. A. Rigamonti, F. Borsa, M. Corti, T. Rega, J. Ziolo and F. Waldner, *Earlier and Recent Aspects of Superconductivity*, ed. J.G. Bednorz and K.A. Muller (Springer–Verlag Berlin, Heidelberg 1990) pp 441–466
24. R.F. Kiefl et al., Phys.Rev.Lett. 64, 2082 (1990).
25. B.I. Halperin, J. March–Russell and F. Wilczek (to be published).
26. I.S. Shaplygin, B.G. Kakhan and V.B. Lazarev, Russian Journal of Inorganic Chemistry 24, 820 (1979)
27. N. Nguyen, L. Er–Rakho, C. Michel, J. Chosnet and B. Raveau, Mat.Res.Bull. 15, 891 (1980)
28. N.P. Ong, Z.Z. Wang, J. Clayhold, J.M. Tarascon, L.H. Greene, W.R. McKinnon, Phys.Rev., 35, 8807 (1987).
29. J.M. Tranquada in *Earlier and Recent Aspects of Superconductivity*, ed. J.G. Bednorz and K.A. Muller (Springer–Verlag Berlin, Heidelberg 1990) pp 422–440
30. A. Kitazawa in *Earlier and Recent Aspects of Superconductivity*, ed. J.G. Bednorz and K.A. Muller (Springer–Verlag Berlin, Heidelberg 1990) pp 45–65
31. T. Takahashi, H. Matsuyama, H. Katayama–Yoshida, Y. Okabe, S. Hosoya, K. Seki, H. Fujimoto, M. Sato and H. Inokuchi, Nature 334, 691 (1988)
32. A.J. Arko. Am.Phys.Soc. HTSC Symposium, St.Louis (March 1987)
33. A. Aharony, R.J. Birgeneau, A. Coniglio, M.A. Kastner and H.E. Stanley, Phys.Rev.Lett. 60, 1330 (1988)
34. G. Shirane, Jnl. de Physique, Suppl C8, 49, 2113 (1988)

35. I. Watanabe, K. Kamagai, Y. Nakamura, T. Kimura, Y. Nakamichi and H. Najima, P.Phys.Soc.Jpn. 56 3028 (1987)
36. J.I. Budnick, B. Chamberland, D.P. Yang, C. Niedermeyer, A. Golnik, E. Recknagel, M. Rossmanith, A. Weidinger, Europhysics Lett. 5, 651 (1988)
37. J.C. Phillips. *Physics of High-T_c Superconductors* (Academic Press, Singapore, 1990)
38. Y.J. Uemura, L.P. Le, G.M. Luke, B.J. Sternleib, J.H. Brewer, R. Kadano, R.F. Kiefl, S.R. Kreitzman and T.M. Riseman, Physica C162-164, 857 (1990)
39. A. Weidinger, C. Niedermeyer, A. Golnik, R. Simon, E. Recknagel, Phys.Rev.Lett. 62, 102 (1989)
40. D.W. Murphy, S. Sunshine, R.B. van Dover, R.J. Cava, B. Batlogg, S.M. Zahurak and L.F. Schiemeyer, Phys.Rev.Lett. 58, 1888
41. B.D. Dunläp, M. Slaski, Z. Sungaila, D.J. Hinks, K. Zhang, C. Segre, S.L. Malik and E.E. Alp, Phys.Rev. B37, 592 (1988).
42. J.D. Jorgensen, M.A. Beno, D.G. Hinks, L. Soderholm, K.J. Volin, R.L. Hitterman, J.D. Grace, I.V. Schuller, C.U. Segre, K. Zhang and M. Kleefisch, Phys.Rev.B36, 3608 (1987)
43. C.U. Segre, B. Dabrowski, D.G. Hinks, K. Zhang, J.D. Jorgensen, M.A. Beno and I.K. Schuller, Nature 329, 227 (1987).
44. D.J. Werder, C.H. Chen, R.J. Cava and B. Battlog, Phys.Rev. B38, 5130 (1988)
45. J.M. Tranquada in *Earlier and Recent Aspects of Superconductivity*, ed. J.G. Bednorz and K.A. Muller (Springer-Verlag Berlin, Heidelberg 1990) pp 422-440
46. R. Beyers, B.T. Ahn, G. Grman, V.Y. Lee, S.S.P. Parkin, M.L. Ramirez, K.P. Roche, J.E. Vazquez, T.M. Gur and R.A. Huggins, Physica C162-164 548 (1989)
47. J. Rossat-Mignod, J.X. Boucherle, P. Burlett, J.Y. Henry. J. M.Jurgens, G. Lapertot, L.P. Regnault, J. Schweizer, F. Tasset and C. Vettier in *Int. Seminar on High Temperature Superconductivity, Dubna, June, 1989,* (World Scientific, Singapore 1989) pp 74-85
48. A.Kapitulnik and K.Char, IBM J.Res.Develop. 33, 252 (1989)
49. J.Karpinski, E.Kaldis, E.Jilek, S.Rusiecki and B.Bucher, Nature 336, 660 (1988)
50. T. Miyatake, M. Kosuge, N. Koshiuka, H. Takahashi, N. Mori, S. Tanaka, Physica, C167, 297 (1990).
51. R.J.Cava, B.Batlogg, J.J.Krajewski, L.W.Rupp, L.F.Schneemeyer, T.Siegrist, R.B.Van Dover, P.Marsh,

W.F.Peck, P.K.Gallagher, S.H.Glarum, J.H.Marshall, R.C.Farrow, J.V.Waszczak, R.Hull and P.Trevor, Nature 336, 211 (1988)

52. P.J.Cava, B.Batlogg, R.B.vanDover, J.J.Krajewski, J.V.Waszczak, R.M.Fleming, W.F.Peck, L.W.Rupp, P.Marsh, A.C.W.P.James and L.F.Schneemeyer, Nature 345, 602 (1990)

53. A.W.Hewat, E.A.Hewat, P.Bordet, J.J.Capponi, C.Chillout, J.Chevenas, J.L.Hodeau, M.Marezio, P.Strobel, M.Francois, K.Yvon. P.Fischer, J.L.Tholence, IBM J.Res. Develop 33, 220 (1989)

54. L.F. Schneemeyer, R.B. Van Dover, S.H. Glarum, S.A. Sunshine, R.M. Fleming, B. Baylogg, T. Siegrist, J.H. Marshall, J.V. Waszczk, P.L.W. Rup, Nature, 332, 422 (1988).

55. H. Ihara, N. Terada, M. Jo, S. Watanabe, R. Sugise and T. Shimomura, MRS Spring Meeting 1989.

56. Y.Tokura, H,Takagi and S.Uchida, Nature 337, 345 (1989)

57. A.C.W.P.James, S.M.Zahurak and D.W.Murphy, Nature 338, 240 (1989)

58. J.Fink, J.Pfluger, T.Muller–Heinzerling, N.Nucker, B.Scheerer, H.Romberg, M.Alexander, R.Manzke, T.Buslaps, R.Claessen and M.Skibowski, in *Earlier and Recent Aspects of Superconductivity*, ed. J.G.Bednorz and K.A.Muller (Springer–Verlag Berlin, Heidelberg 1990) pp 377–406

59. S.W.Tozer,A.W.Kleinsasser, T.Penney, D.Kaiser and F.Holtzberg, Phys.Rev.Lett. 59, 1768 (1987)

60. P.W.Anderson, Science 253, 1196 (1987)

61. T.M. Rice, Jnl.Mag.Mat., 75, 542 (1988).

62. C.M.Varma, P.B.Littlewood, S.Schmitt–Rik, E.Abrahams and A.Ruckenstein, Phys.Rev.Lett.63, 1996 (1989)

63. AF Fermi liquid model Millis, Monien and Pines, PR B, July 1st 1990

64. M.Takigawa, P.C.Hammel, R.H.Heffdner, Z.Fisk, K.C.Ott and J.D.Thompson, Physica C162–164, 853 (1989)

65. T.Hasegawa and K.Kitazawa – Jap.Jnl.Appl.Phys 29, L434 (1990)

66. J.J.Geerk, X.X.Xi and G.Linker, Zeit, fur Physik B73, 329 (1989)

67. A.Fournel, B.Oujia, J.B.Sorbier, J.Noel, J.C.Levet, M.Potel and P. Gougeon, Eorophys. Lett. 6, 653 (1989)

68. M.Gurvitch, J.M.Valles, A.M.Cuculo, R.C.Dynes, J.P.Garno, L.F.Schneemeyer and J.V.Waszczak, Phys.Rev.Lett. 63, 1008 (1989)

69. *Strongly Correlated Electron Systems, Trieste 1989.* Int.Jnl.Mod.Phys., vol.3, December, 1989

70. T.Takahashi, H.Matsuyama, H.Katayama–Yoshida, Y.Okabe, S.Hosoya, K.Seki, H.Fujimoto, M.Sato and H.Inokuchi, Nature 334, 691 (1988): Phys.Rev. B39, 6636 (1989)
71. C.G.Olson, R.Liu, A.B.Yang, D.W.Lynch, A.J.Arko, R.S.List, B.W.Veal, Y.C.Chang, P.Z.Jiang, A.P.Paulikas, Science 245, 731 (1989)
72. S.Massida, J.Yu and A.F.Fremman, Physica C152, 251 (1988)
73. H.Krakauer and W.Pickett, Phys.Rev.Lett. 60, 1665 (1988)
74. J.M.Imer, F.Pathey, B.Dardel, W.D.Schneider, Y.Baer, Y.Petroff and A.Zettl, Phys.Rev.Lett. 62, 336 (1989)
75. Y.Chang, Ming Tang, R.Zanoni, M.Onellion, R.Joynt, D.L.Huber, G.Margaritondo, P.A.Morris, W.A.Bonner, J.M.Tarascon and N.G.Stoffel, Phys.Rev. B39, 4740 (1989)
76. J.Fink, J.Pfluger, T.Muller–Heinzerling, N.Nuker, B.Scheerer, H.Romberg, M.Alexander, R.Manzke, T.Buslaps, R.Claessen and M.Skibowski,in *Earlier and Recent Aspects of Superconductivity*, ed. J.G.Bednorz and K.A.Muller (Springer–Verlag Berlin, Heidelberg 1990) pp 377–406
77. B.Batlogg, G.Kourouklis, W.Weber, R.J.Cava, A.Jayaraman, A.E.White, K.T.Short, W.L.Rupp and E.A.Reitman, Phys.Rev.Lett. 59, 912, 1987; T.A.Faltens, W.K.Ham, S.W.Keller, K.J.Leary, N.J.Michaels, A.M.Stacy, H–C. zur Loye, D.E.Morris, W.T.Babtee, L.C.Bourne, M.L.Cohen, S.Hoen and A.Zettl, Phys.Rev.Lett. 59, 915 (1987)
78. K.J.Leary, H–C. zur Loye, S.W.Keller, T.A.Faltens, W.K.Hamm, J.N.Michaels and A.M.Stacy, Phys.Rev.Lett. 59, 1236 (1987)J.P.Franck, J.Jung, G.L.Salomons, W.A.Miner and M.A.K.Mohamed, Physica 162–164, 753 (1989) .
79. P.B.Allen, Nature 335, 396 1988
80. C.Thomsen, M.Cardona, R.Liu, T.Ruf, B.Gegenheimer and E.T.Heyen, Physica C 162–164, 1079 (1989)
81. S.L.Cooper, M.V.Klein, R.G.Pazol, J.P.Rice and D.M.Ginsberg, Phys.Rev.B37, 5920 (1988)
82. A.Mascarenhas, H.Katayama–Yoshida, J.Pankove ans S.K.Deb, Phys.Rev. B39, 4699 (1989)
83. L.C.Hebel and C.P.Slichter, Phys.Rev. 113, 1504 (1957)
84. Y. Kitaoka, K. Ishida, K. Fujiwara, K. Asayama, H. Katayama–Yashida, Y. Okabe and T. Takahishi, IBM Res. Dev. 33, 277 (1989).
85. J.E. Gordon, M.L. Tan, R.A. Fisher, N.E. Phillips, Solid State Comm., 69, 625 (1989).
86. S.E.Inderhees, M.B.Salamon, N.Goldenfeld, J.P.Rice, B.G.Pazol, D.M.Ginsberg, J.Z.Liu and G.Crabtree, Phys.Rev.Lett.60, 1178 (1988)

87. A.Junod, D.Eckert, T.Graf, G.Triscone, and J.Muller, Physica C162–164, 482 (1989)
88. K.Fossheim, O.M.Nes, and T.Laegreid, Int.J.Mod.Phys. B1, 635 (1988)
89. B.Oh, K.Char, A.D.Dent, M.Naito, M.R.Beasley, T.H.Geballe, R.H.Hammond, A.Kapitulnik and J.M.Graybeal, Phys.Rev., B37, 7861 (1988).
90. V.L.Ginsburg and L.D.Landau, Zh.Eksp.Teor.Fiz., 20, 1064 (1950)
91. Gorkov, Soviet Physics – JETP 9, 1364 (1959).
92. F. and H.London, Proc.Roy.Soc.(London) A149, 71 (1935)
93. C.J.Gorter and H.Casimir, Physica 1, 306 (1934)
94. B.Muhlschlegel, Z.Phys.155, 313 (1959)
95. A.A.Abrikosov, JETP 5, 1174 (1957)
96. E.M.Forgan, Nature 329, 483 (1987)
97. U.Welp, W.K.Kwok, G.W.Crabtree, K.G.Vandervoort and J.Z.Liu, Physica C162–164, 735 (1989)
98. *L.* Krusin–Elbaum, R.L. Greene, F. Holtzberg, A.P. Malozemoff and Y. Yeshurun, Phys.Rev. 62, 217 (1989).
99. E.H.Brandt, Phys.Rev. B37, 2349 (1988)
100. U. Essmann and H. Trauble, Phys.Letts A24, 526 (1967)
101. G.J. Dolan, G.V. Chandrashekhar, T.R. Dinger, C. Field and F. Holtzberg, Phys.Rev.Lett 62, 827 (1989):
G.J. Dolan, F. Holzberg, C. Feild, and T.R. Dinger, Phys.Rev.Lett.62, 2184 (1989).
102. A.M. Grishin, A.Yu. Matynovich and S.V. Yampolskii (to be published): A.I. Buzdin and D. Feinberg (to be published)
103. E.M. Forgan, D.McK. Paul, H.A. Mook, P.A. Timmins, H. Keller, S. Sutton and J.S. Abell, Nature 343, 735 (1990)
104. E.M. Forgan, S.L. Lee, S. Sutton, F. Wellhofer, J.S. Abell, C.E. Gough, S.F.J. Cox and C.A. Scott, Supercon. Sci and Tech. 3, 217 (1990)
105. W. Barford and J.M.F. Gunn, Physica C156, 515 (1988)
106. Y.J. Uemura et al., Phys.Rev.Lett. 62,2317 (1989)
107. V. Emery, IBM J.Res. Devel. 33, 246 (1989): Phys.Rev.Lett. 58, 2794 (1987)
108. J.W. Loram and K.A. Mirza, in Springer Ser. Solid–State Sci., Vol.99 (Springer, Berlin, Heidelberg 1991).
109. W.W. Warren, R.E. Walsteat, G.F. Brennert, R.J. Cava, B. Batlogg and L.W. Rupp, Phys. Rev.B 39, 831 (1989).

Neutron Versus Muon Measurements in the Mixed State of Superconductors

E.M. Forgan

School of Physics and Space Research, University of Birmingham, Edgbaston, Birmingham, B15 2TT, UK

Abstract. We describe briefly the mixed state of type II superconductors in a magnetic field and the relationships between the structure of this state and important properties of the superconductor. We show how the "field contrast" in the mixed state may be measured by neutron diffraction and describe the first observation by this technique of flux lines in the high-T_c superconductor YBCO. We compare the information obtainable from neutron and muon spin rotation measurements of mixed state properties and show that these two techniques are complementary.

1. Introduction

In the mixed state, adopted by type-II superconductors in a field larger than H_{c1}, magnetic flux enters in the form of singly-quantised flux lines, each carrying a flux equal to $\Phi_0 = h/2e$, with supercurrents flowing like vortices around the core of each flux line. The structure of isolated flux lines is shown schematically in figure 1(a). At large distances from the core, the field dies away approximately exponentially, over distances described by the "London penetration depth" λ_L. The velocity of the supercurrents causing the field variation is controlled by the quantisation condition

$$\oint p\,d\ell = h \tag{1}$$

for any path enclosing the flux line core.

At short distances r from the core, the electromagnetic contribution to the Cooper pair momentum is negligible, and equation (1) applied to a circular path leads to

$$2mv_s = h/2\pi r \tag{2}$$

where 2m is the Cooper pair mass and v_s is the superfluid velocity. If the supercurrent density J_s also varied as $1/r$ near $r = 0$ this would cause the field B to rise logarithmically to infinity at the vortex core. However, the superelectron wavefunction must be zero

Springer Proceedings in Physics, Vol. 59 Exotic Atoms in Condensed Matter
Editors: G. Benedek · H. Schneuwly © Springer-Verlag Berlin, Heidelberg 1992

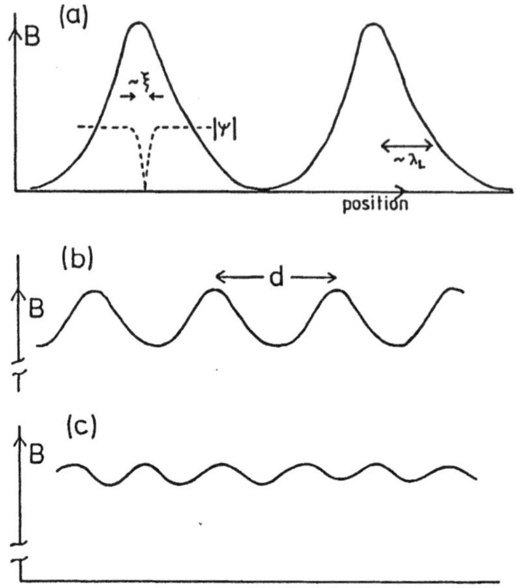

Figure 1 The structure of flux lines in the mixed state

(a) Isolated flux lines; the meanings of λ_L and ξ.

(b), (c) The change in the spatial variation of field as the average induction is increased.

on the axis of the flux line: this causes the pair density n_s to drop to zero over a distance ξ, called the coherence length, and cuts off the divergence in J_s and B. The coherence length may be thought of as the size of a Cooper pair, and is also the distance at which the supercurrent kinetic energy is comparable with the pair binding energy. This is why depairing occurs at the flux line core.

If the flux density in the superconductor is increased, the fields due to individual flux lines begin to overlap as shown in figure 1(b). Their mutual repulsion causes them to form a lattice, usually a triangular one. Finally, at sufficiently high fields, the flux line cores overlap, the field becomes uniform and superconductivity is destroyed at the field B_{c2}. At all fields in the mixed state, the flux line spacing d is given by the condition that there is one flux quantum per line: thus d is related to the average flux density B by

$$\Phi_o \approx Bd^2 \quad \rightarrow \quad d \propto B^{-\frac{1}{2}} \; . \qquad (3)$$

This equation is approximate for a triangular lattice and is exact for a square one.

High-T_c superconductors are characterised by $\lambda_L \gtrsim 1000$ Å and $\xi \lesssim 10$ Å i.e. $\lambda_L/\xi > 100$. Thus $B_{c2} >> B_{c1}$ and as long as $B << B_{c2}$, the magnetic field variation over most of space is controlled by λ_L; however, ξ controls the field value at the cores. Thus the investigation of the mixed state can give us two most important superconducting quantities: the size of the cooper pairs (ξ) and an indication of their density ($\lambda_L^2 \propto m/n_s$).

2. Methods of observation of the mixed state

At the surface of a superconductor, we may in principle observe either B(r) or n_s(r): the spatial variation of B has been made visible by "decoration" with small magnetic particles [1], but this technique has only been effective at very small fields. The drop in n_s at flux line cores has also been detected by scanning tunneling microscopy, but so far only in a low-T_c superconductor [2]. Within the bulk, which is preferable, we are limited to detecting B(r), so we need local field probes, typically neutrons or muons.

For a neutron of given spin orientation, a spatially varying B will give a spatially varying Zeeman energy; thus the magnetic moment of neutrons allows them to be diffracted by the "field contrast" between flux line cores and the spaces between them. If we take the symbol ΔB as a measure of this field contrast, then we may expect the intensity of the diffracted beam to be proportional to ΔB^2. In TF μSR, the spins of implanted positive muons precess in the local field B(r) after implantation, so their range of precession frequencies is a measure of ΔB. More precisely, random implantation (on a scale of λ_L) will sample the field distribution P(B') given by

$$P(B') = \int d^3r \ \delta \ (B(\underline{r}) - B') \ / \int d^3r \ . \tag{4}$$

3. A heuristic discussion of the factors controlling ΔB

Consider the application of a magnetic field parallel to the surface of a thin superconducting film of width d. If d is much smaller than λ_L, the Meissner effect is by no means complete: only a fraction ΔB of the field is excluded, with

$$\frac{\Delta B}{B} \gtrsim \left(\frac{d}{\lambda_L}\right)^2 \ . \tag{5}$$

The field variation within the film is qualitatively similar to that between two vortex centres, when the flux lines are sufficiently close for their fields to overlap. We may therefore expect equation (5) to give the field contrast in the mixed state, with d representing the intervortex spacing. This equation shows that the fractional field

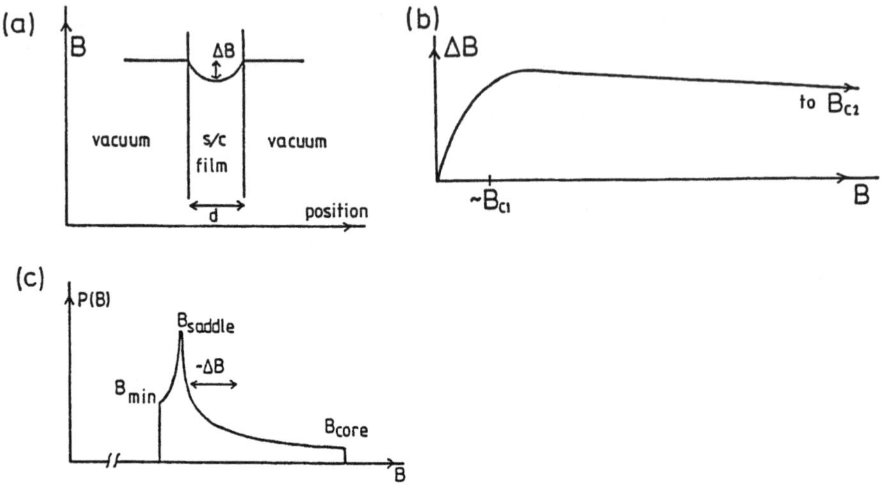

Figure 2 The field contrast in the mixed state
(a) Simple model to obtain an expression for ΔB.
(b) Variation of ΔB with average field.
(c) Probability distribution of B.

contrast becomes smaller as the flux lines are moved closer together. However, if we use equation (3) relating B and d, we find

$$\Delta B = \text{const}.\ \Phi_o/\lambda_L^2 \tag{6}$$

so that the absolute magnitude of the field contrast is independent of field. This surprising result is confirmed by explicit calculations [3]; which show that it applies so long as the field is not too close to B_{c2}, where the field contrast must vanish, nor too close to B_{c1}. The dependence of ΔB on B is sketched in figure 2(b). Of course, the field contrast (or its RMS value) is only one number describing the distribution (eqn.(4)) of values of field, which will have the general shape shown in fig. 2(c), [4].

4. Diffraction by a flux line lattice

The flux line spacing given by equation (3) is much larger than atomic spacings, e.g. ~ 1000 Å at B = 0.2T. Even with cold neutrons of wavelength λ_n ~ 10Å this gives a Bragg diffraction angle $\sim\frac{1}{2}^o$. Two possible diffraction geometries are shown in figure 3. In 3(a) the field is perpendicular to the neutron beam: the disadvantage of this geometry is that no diffraction occurs unless the flux lattice planes are in the correct orientation to the incoming beam. In the alternative arrangement of Fig. 3(b), the flux lattice

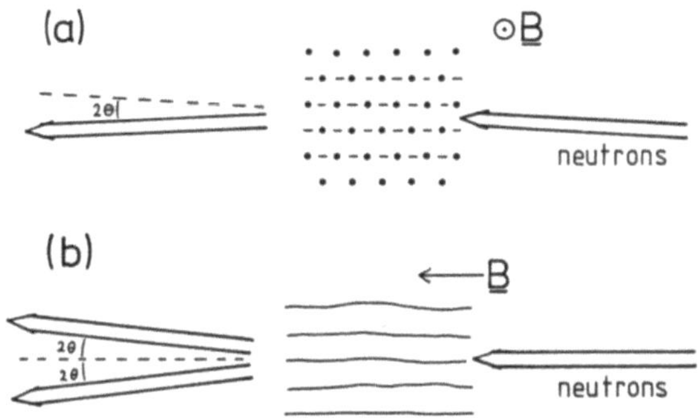

Figure 3 Geometry for observation of flux lines by diffraction
(a) B perpendicular to neutron beam.
(b) B almost parallel to beam.

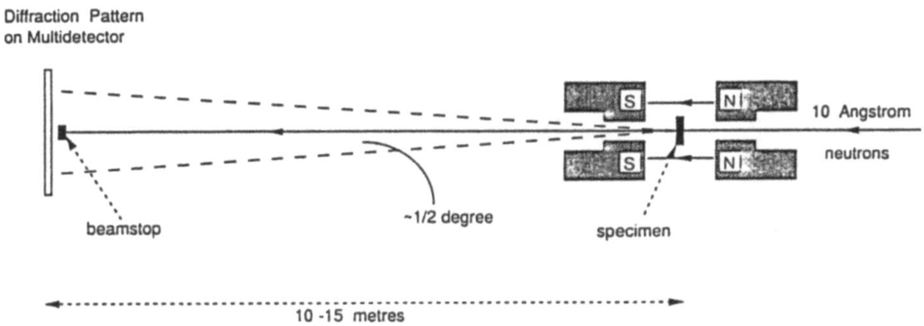

Figure 4 Experimental arrangement for small–angle neutron diffraction

planes are almost parallel to the neutron beam. Slight misalignments of the lattice planes ("mosaic spread") or divergence of the incoming beam then allow the Bragg condition to be satisfied, and diffraction to occur, whatever the orientation of the planes about the beam axis. Indeed, even a liquid–crystal–like arrangement of flux lines, having no orientational order, would give rise to a liquid–like diffraction ring. An experimental arrangement which we adopted to achieve the geometry of fig. 3(b) is shown schematically in figure 4. The field is supplied by an electromagnet with holes through its pole pieces to allow the passage of neutrons. The specimen is contained within a cryostat, and the diffracted neutrons fall on a (64 x 64) cm^2 multidetector.

This is a difficult experiment to perform in high–T_c superconductors because the diffracted intensity from the flux lattice is proportional to ΔB^2, which depends on $1/\lambda_L^4$ (equation (6)), which is a factor ~ $(\frac{1}{4})^4$ smaller than in conventional superconductors. In addition, high–T_c materials are available as much smaller crystals which are often full of defects, such as the twins in YBCO, which

Figure 5 Diffraction pattern from YBCO in the mixed state

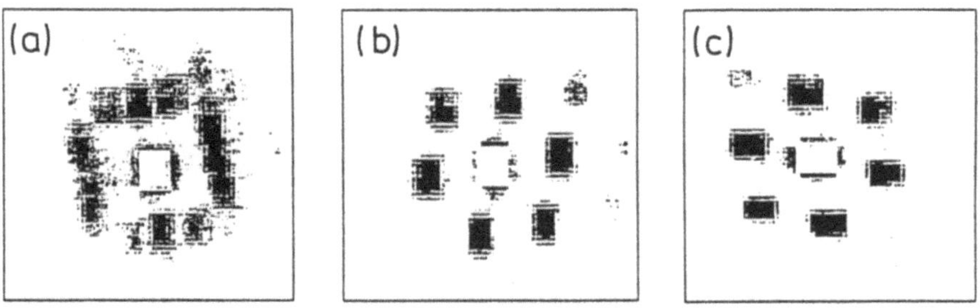

Figure 6 Diffraction patterns from Nb in the mixed state
(a) With B//(100): this is a sum of (b) and (c).
(b) One triangular flux lattice orientation: selected by tilting the field relative to the crystal axis.
(c) Pattern (b) rotated by 90°: the other possible flux lattice orientation.

cause extra Small Angle Neutron Scattering. In an experiment recently performed [5] on a 150mg crystal, the signal was much smaller than the background and could only be clearly detected after subtracting the scattering which was present when the sample was in zero field.

A representation of the difference results obtained with a field of 0.2T parallel to the crystal c-axis is shown in figure 5. It is clear that the extra scattering due to the flux lines lies in a ring; the diameter of the ring is that expected for diffraction by singly-quantised flux lines at a spacing given by the trapped field of 0.2T. It is notable however that the intensity is not uniform around the ring: the regions of high intensity show that there are certain preferred orientations of flux lattice planes, which is a proof that there is some lattice structure. However, the pattern does not show the sixfold symmetry expected if there is a triangular flux lattice extending throughout the sample.

If more than one flux lattice orientation is present within a sample, then a more complicated diffraction pattern can be created. This is illustrated by results from the low-T_c superconductor Nb: in figure 6(a) is shown a diffraction pattern obtained with the applied field parallel to a fourfold crystal axis. This arises because two equivalent triangular flux lattices exist in different parts of the sample. It is likely that similar effects lead to the pattern shown in figure 5. Further details of this experiment and consideration of the possibility that the flux lattice may "melt" at high temperatures are contained in [5].

5. Comparison of neutron diffraction and μSR in the mixed state

In low-T_c superconductors, both neutron diffraction and μSR give easily detectable signals capable of revealing considerable detail about the mixed state. In figure 7 are reproduced typical data obtained by each technique. Fourier synthesis of a neutron diffraction pattern gives the field as a function of position as shown in figure 7(a). However, diffracted intensities give only the magnitude of the fourier components and not their phase (which can be only 0° or 180° in our case): the two lines in figure 7(a) result from the two possible signs of a higher-order contribution. Of course, these correspond to

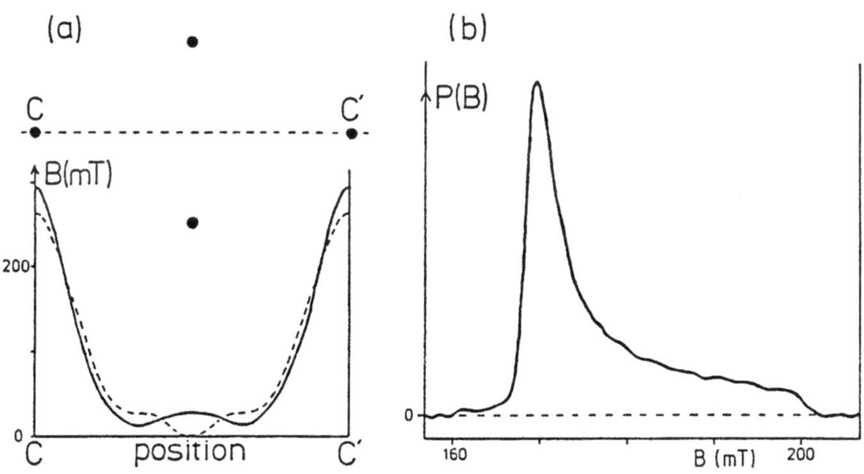

Figure 7 Neutron and muon results from Nb

(a) Spatial variation of field between two flux line cores by neutron diffraction (after [6]).

(b) Probability distribution of field values by muon spin rotation (after [7]).

different distributions of the values of field; thus, μSR results such as those in figure 7(b) can distinguish between the two possibilities. In the case illustrated, the dashed curve is the correct one. The analysis of the μSR results is considerably more indirect: one must first assume a spatial field distribution and calculate from it the distribution of field values via equation (4) in order to compare with experiment. The infinity in the theoretical distribution (figure 2(c)) is removed by a broadening function [7] which represents slight deviations of the flux lines from their ideal lattice positions, due to "pinning" by inhomogeneities in the sample.

In high-T_c materials, the neutron diffraction technique becomes considerably more difficult, because the intensity falls off as $1/\lambda_L^4$. The width of the field distribution obtained by μSR also falls, but only as $1/\lambda_L^2$, so that $<\Delta B^2>^{\frac{1}{2}}$ is easy to measure; however, detailed lineshape information like that in figure 7(b) is difficult to obtain [8]. Despite the low intensity, the neutron diffraction is needed, because it is the only technique which gives direct information about the spatial arrangement of flux lines. However, the distribution of values of B is only obtained indirectly because of the need to calibrate the relationship between intensity and ΔB, the phase problem mentioned earlier and the need to measure higher-order diffracted intensities. On the other hand, μSR only gives indirect information about flux line positions but direct information about P(B). It should be mentioned however that the calculation of the field distribution requires fourier inversion of μSR data which is (i) incomplete, due to the "dead time" near $t = 0$ (ii) subject to Poisson statistics (iii) only available over a time range of a few τ_μ. Thus, the two techniques are complementary and both need to be performed to obtain full information about the mixed state in the new high-T_c materials.

6. Future developments in high T_c materials?

We may hope shortly to observe by neutron diffraction a "single crystal" of a flux lattice in a high-T_c material. This will allow more quantitative analysis and the investigation of the flux lattice structure as a function of field direction in these very anisotropic materials [9][10]. It may also be possible to search for flux lattice "melting", especially in the Bi-based materials which give evidence for this effect in macroscopic measurements [11]. It will be of interest to see whether μSR can also show melting-related phenomena. There are predictions that crystal anisotropy should give a double-peaked P(B) for certain field directions [12] which μSR would be ideal to observe. The very large ratio of λ_L to ξ in high-T_c materials should give a long tail in P(B) to high fields, arising from the fields in the flux-line cores. There has not yet been a convincing demonstration of the existence of the long tail. Failure to observe it in the near future would indicate that we do not yet understand the mixed state and therefore the pairing in these materials. There is clearly some work still to do!

References

1. See e.g. G.J. Dolan, F. Holtzberg, C. Feild and T.R. Dinger, Phys. Rev. lett. **62**, 2184 (1989).
2. H.F. Hess, R.B. Robinson, R.C. Dynes, J.M. Valles, Jr. and J.V. Waszczak, Phys. Rev. Lett. **62**, 214 (1989).
3. E.H. Brandt, Phys. Rev. B **37**, 2349 (1988).
4. E.H. Brandt, Physica C **162–164**, 257 (1989).
5. E.M. Forgan, D.McK. Paul, H.A. Mook, P. A. Timmins, H. Keller, S. Sutton and J.S. Abell, Nature **343**, 735 (1990).
6. J. Schelten, H. Ullmaier, G. Lippmann and W. Schmatz, Proc. 13th Intl. Conf. on Low Temp. Phys. **3**, 54 (North-Holland 1974).
7. D. Herlach et al., Proceedings of μSR90, to be published in Hyper. Inter.
8. B. Pümpin, H. Keller et al., to be published.
9. T. Forgan, Nature **329**, 483 (1987).
10. N. Schopohl and A. Baratoff, Physica C **153–155**, 689 (1988).
11. P. L. Gammel, L. F. Schneemeyer, J. V. Waszczak and D.J. Bishop, Phys. Rev. Lett. **61**, 1666 (1988).
12. S.L. Thiemann, Z. Radović and V.G. Kogan, Phys. Rev. B **39**, 11406 (1989).

Part VI

Conclusions

Exotic Atoms in Condensed Matter: Conclusions*

N.W. Ashcroft

Laboratory of Atomic and Solid State Physics and Materials Science Center, Cornell University, Ithaca, NY 14853, USA

1. Muon Physics; A Commodities Market

I think I should begin with two apologies. First of all I must apologize for interposing between you and lunch. I will try to be brief. Secondly I must apologize because you have already had a lecture from an obvious neophyte in the area of muons in condensed matter. I can only hope that these remarks have some mildly recognizable overlap with what you will find in the proceedings! But, speaking especially as a neophyte, I have enormously enjoyed this workshop, and I have learnt a great deal. What you are about to hear is the distillation of at least some of the things I have learnt. I do not want to imply that you have necessarily focussed on the same things; in fact it is far more likely that you have appreciated much more than I have. But in examining the theme and content of the Workshop, I am struck by the fact that it is divided, roughly speaking, into four components of something vaguely resembling the commodities part of a quasi-economic system, namely:

- The Supply of Mesons
- Their Distribution
- The Demand and Utilization of Mesons
- Future Investment in Mesons

There is an evident supply-side of the system and I will say a little about it because that is the *sine qua non* of the entire operation. There is a distribution side which gets the mesons to where we need them to go; I find some interesting points here as well. On "Demand, etc", I was struck by the remarkable ingenuity and resourcefulness of the experimentalists; there is obvious utilization, and it drives demand. Then finally there are the so-to-speak "venture capitalists", the people who are worrying about paying back through application at least some fraction of the ever declining economic investment that has been made in the basic science (for example, in muon catalysed fusion). I will attempt to give my own observations on these four groups, very briefly, and then to go on to say just a little bit about some aspects of muons in the condensed state (by which I mean solid AND liquid, and I hope that possibly in the future liquids will play more of a role in the physics of the condensed state of matter than they currently do).

1.1 The 'Supply' Side – The Production of Muons

Nagamine: History (namely since 1971) and ideas for new sources;
Ramsauer technique - $\mu+$ versus $\mu-$

Simons: Basic μ-ology: energies, decay modes and lengths;
fluxes ($\sim 10^7$/sec on 100 g), Ramsauer effect

* [Based on a transcription of a tape-recording taken during the summarising lecture of the Workshop]

Now this will be blessedly brief, but on the supply-side, the mildly copious production of muons and their subsequent use has a history, as I took it from remarks of Professor Nagamine, going back to about 1971. I was particularly intrigued by the new and appealing suggestions for sources of muons for the future and it looked to me to be rather optimistic, possibly even overoptimistic, though I am not fully cogniscent of the complexities of the field. From Dr. Simons I was introduced to a good deal of "basic muology" and I gained a particular appreciation of the characteristic energies and decay lengths, the decay modes, the characteristic fluxes that one is dealing with, the hopes for the future, the importance of the Ramsauer technique, and so forth. It is clear that this is a very important part of the future: the ability to develope efficient sources of muons of *both* signs.

1.2 Transportation – Getting Muons into Atoms/Solids

Hartmann: Capture probabilities, elements vs. compounds;
cascade processes; applications to analysis;
medicine, etc.

Kravtsov: Semiclassical (WKB) methods in Coulombic systems;
de-excitation of muonic hydrogen;
Coriolis transitions

Popov: Charge exchange processes;
WKB as applied to light elements

Schneuwly: Fundamentals of μ-capture;
remarkable chemical bond effects;
"Infectious" muons; hidden variables!

Getting muons into atoms and solids is surely the sine qua non of the topic of this conference: Dr. Hartmann informed us about capture probabilities, the complexity in going from elements to compounds, the cascade process and the applications (even today) to analysis – to chemical analysis, to medicine, and so forth. Drs. Kravtsov and Popov acquainted us with the rather successful application of the semiclassical technique in some of the key problems concerning the cascade processes, and the Coulomb de-excitation of muonic hydrogen, the charge exchange processes and their application to the light elements where the semiclassical (quasi-classical) techniques continue to be reasonably valid. I was much fascinated by our co-chairman's description of the remarkable chemical effects which are apparent, especially the infectious muon as it transfers itself between atoms, and the associated possibility that we heard this morning of some interesting new physics involving transfer and decay rates.

1.3 The Consumer State – Inventive and Productive Application

Armstrong: "μ-parents", pions as probes of
organic molecules; transfer vs. capture;
importance of the liquid state;
large mesic molecules

Bianconi: Deep-level physics;
prepared "holes" (via μ-ons)
versus electromagnetic excitations

Bucci: μSR and magnetic order;
effects of spatial and spin correlations

Cox:	Muonium substituted molecules; striking decline of hyperfine constant with band gap (dielectric behavior: $\epsilon \approx 1/\text{band gap}$)
Forgan:	Neutrons and muons: complementarity; probing the spaghetti flux state of superconductors
Gough:	The nature of new superconductors and μSR; inferences of nodeless wavefunctions in high T_c materials; the importance of μSR for new materials
Grenacs:	Asymmetry in nuclear polarization; possible applications to μ-ons?
Keller:	The μSR method; application to diffusion, magnetic microfields; μ-onium in insulators and semiconductors; probing the mixed state in superconductors
Maier:	Negative muons in semiconductors; sensitivity to the bound (d-a) state; "the unselfish muon revealing all before suicide"; the x-ray trigger
Nagamine:	Probing 'just outside the nucleus'; spatial distribution of hyperfine fields; μ-substitutions and exotic bonding

Now I proceed to the consumer state, the third group of people of my list, namely those involved in the inventive and productive application of muons. We've heard from Dr. Armstrong a little about the parents of the muons, (the pions) as probes, and the way we can use them rather successfully in organic molecules; we also heard about the important questions that arise between capture and transfer. I was very gratified to see the possibly emerging importance of the liquid state of matter in deciding these issues. We were reminded by Dr. Bianconi of the commonality of deep-level physics between the muon problem and the problem where one prepares a hole in a strongly bound state, thereby likening the presence of a muon at such a site to that of a hole prepared by an electromagnetic excitation. From Dr. Bucci we heard about muon spin resonance and its immensely successful application to the elucidation of magnetic order. In revealing this magnetic order we see both aspects of the muon coming into play; the fact that the muon resides on a charged particle (to which it is correlated spatially) as well as the fact that the muon carries spin along with its own charge. Interrogation of this spin tells us much!

Dr. Cox showed the (again, notable to me) decline of the hyperfine constant with bandgap. One may wonder whether it is telling an interesting tale about the dielectric behavior, which in some of these narrow gap materials is deeply linked to the bandgap itself, and whether as we pursue this line further, this is going to be quite a fruitful field of investigation, not just of the magnetic aspects of a solid, but also of the electronic and the dielectric aspects. From Dr. Forgan we saw an elegant complementarity of neutrons and muons in probing what he referred to as the "spaghetti states", i.e. the mixed states in high T_c superconductors. One can not help wonder whether in the case of high T_c compounds, where the melting of flux lattices is emerging as a crucially important piece of physics in this area, the "spaghetti" will actually turn into one of the other innumerably interesting but dauntingly complex forms of Italian pasta.

Dr. Gough made for me a very important point. He picked up on the remarkable utility of muon spin resonance, μSR, and its application to the new superconductors, a technique from which we are now getting fundamental information on the nature of the paired state. And he also emphasized the fact that the pair state now appears to be nodeless. But more generally he strongly emphasized the value of muon spin resonance for the study of *new materials.* We only have to look back to the history of condensed matter physics in the last few years to see that our ability to make new discoveries is intimately and increasingly linked to our ability to actually create new materials which bring with them potentially new states of order. Having these analytic probes at hand, which get immediately at microscopic detail (through a technique like muon spin resonance) is immensely useful.

Dr. Grenacs reminded us about an interesting asymmetry in nuclear polarization and the possible extension of that to muons. And (coming to the end of my survey of the consumer state) Dr. Keller also explained to us very beautifully the power and general applicability of the muon spin resonance technique, the possible application to fusion, the way it is utilized in the determination of the magnetic microfields, its application to the behavior of muonium in insulators and semiconductors, and once again to the probing of the mixed state of superconductivity. Dr. Maier made a particularly nice remark, I thought, illustrating the ineffably generous nature of muons. I do not think I quote him quite accurately but it was something like this: "... the unselfish muon, revealing all before it commits suicide", to which I would add "selflessly in the service of physics and its applications". On the way to its end it gives us enormous and valuable information on the nature of the acceptor- and donor-type states that form in a transitory way when the muon finally reaches a nucleus of its liking. And necrology aside, it also it reveals some interesting time structure that one might be able to utilize in the future muon physics in the condensed state.

Dr. Nagamine said quite early on in our Workshop that the ability of the muon to probe the area just outside the nucleus, to survey the spatial distribution of the hyperfine fields there, to look at the interesting chemical and molecular effects that come about when we make muon substitutions, and to elicit inferences on the way the condensed state manifests itself is really what this workshop is about; we have been primarily concerned with the nature of the subsequent exotic bonding. So this brings me to the last group of speakers, the venture capitalists or the "adventure planners" who are worrying about (and who certainly hope to solve) some very important problems for the future.

1.4 Venture Planning

Cerofolini:	Alternatives in fusion; a non-simple H-H potential?
Gerstein:	Muonic molecules and conditions for fusion in these
Scrinzi:	μCF and its possibilities; rotational character; role of condensed state?

The alternatives in fusion were discussed by Dr. Cerofolini and some interesting suggestions were made that the hydrogen-hydrogen interaction, which we thought was known so well may in fact have an interesting twist of "spaghetti," of its own again at short distances. As we heard today from Dr. Gerstein, the conditions under which a muonic molecule are formed obviously play a crucial role in the ability of any muon catalysed fusion process to take place. The role of the condensed phase, and especially of the rotational states which may exist, were also emphasized by Dr. Scrinzi and this, as we have seen in the last year or two, has become a problem which has risen very much to the forefront and has even seen the light of day in the popular press. It may well be that the confusion from cold-fusion will be with us for quite some time.

2. A (Short) Day in the Life of a Muon

In thinking about the second major aspect of this fleeting physics - about the way in which mesons react before they "commit suicide" (to quote again from Dr. Maier) – it seems to me that there are three basic phases. The first phase is where we worry about the source itself. The second phase is where the muons enter condensed matter in a high energy state, (or in a future scenario, will enter in a thermal state, with what we hope will be the new generation of sources). The third phase, which I would call the phase of local physics, in where the muons finally settle down and decide to do something after coming to physical terms with their surroundings.

2.1 Beams for the Future

When we think about beams for the future, and especially so if the emphasis is to be on condensed matter physics (and new physics at that), we ought surely to strive to match as best we can the energies of these beams to the scale of the physics which goes on; there is no question about the fact that condensed matter physics is eV physics, and there is also no question about the richness of physics associated with this energy range. It is difficult to imagine another field that rivals it in terms of pace and scope of discovery. And so one might hope in the future to accommodate the natural scale of condensed matter physics to the sources themselves. One might even hope eventually for thermal and hyper-thermal beams of muons, perhaps focussed, certainly polarized and certainly of both signs. What in general terms would such beams elucidate? There are, of course, some questions that might be answered that are not addressed by other forms of spectroscopy.

In general if we have a beam of particles which couples to condensed matter, the information which we usually get is a dynamic wavenumber- and energy-dependent structure factor or scattering factor; let's call it $S(\vec{q},\omega)$. The interest with respect to muons is that by determining such a function we may be able to probe new ranges of wavevector $\vec{q}$ and new ranges of frequency (i.e., excitations of the system) as compared with more conventional probes, say photons and electrons. One can imagine for the future that the alphabet soup which we have been living with for many years in the electron case, where we have low energy diffraction (LEED), high-energy electron diffraction (HEED) and so on, is merely going to suffer just a simple replacement: we substitute μ for E (or at worst Pμ and Nμ, if we worry about the signs). It may be a long way off, but the key point of course is that once we get well controlled beams, we immediately anticipate a spectroscopy in a new range of wavevectors and energy.

But what about the possibility of bunched muon beams? And also the possibility of bunching with different particles so that once we have relative polarization in the problem, we can imagine optical cooling, the kind that goes on with atomic beams today with such enormous success. We might even imagine the limiting case where we cool enough to actually condense bunches into clusters (that's probably for the very far future, but we should not ignore that possibility either since it is a route to ultradense Coulomb matter).

2.2 Kinetics and Loss

Now I come to the second part of this process: the muons enter the solid. I don't intend to give here any discussion of the loss processes which go on in the thermalization or hyper thermalization of the muons from extremely high energies. I assume, but I should certainly like to hear otherwise, that this part of the problem is non-controversial. Probably that is the case because we know much about the problems associated in bringing charged particles to rest in such a variety of contexts; and probably most of the physics is sorted out at least over reasonable energy ranges. So bringing down a high energy beam from the MeV to the keV level is not, I suspect, a contentious issue. But what about the later stages, when we bring muons down to the eV range?

Now here it is possible that the loss mechanism will begin to involve the collective and particle-hole excitations of the condensed state. Why? Simply because as I've already hinted we are bringing the muon down to the natural energy scale of this physics. As an example of this, we may have a negative muon interacting with all the collective excitations of a metal. For electron spectroscopy, this is a very well known topic. When we probe with an electron microscope, for instance, we're measuring the dynamic scattering factor of the electron gas and once again in the very process of bringing the muon down to the low energy range we should be involving a fairly fundamental physics of the condensed state.

2.3 Local Physics

That brings me to the third part: the local physics. Here we regard the muons as pretty much captured, or about to be captured, or at least very localized. So far as I could tell there seem to be two key parts to the physics here. One part is weak interaction physics; the second is what we are essentially dealing with at this workshop. It is what I might call the "Coulomb physics" and it is fundamentally a many-body problem. If I understand the system correctly, weak interaction physics essentially gets its day in the sun when the length scale of a muon, e.g. in a bound orbit, homes in on the nuclear length scale (1 fm = 10^{13} cm). And we can easily see when that will be about to happen: we know that the Bohr radius for the muon when Z = 1 is just about $a_o/200$ and clearly if we were to choose a value of the atomic number (the Bohr radius goes down inversely with atomic number) to pick up the extra factor of 50 we have a Bohr radius of 5 fm, just the size of the nucleus ($\sim 1.1 A^{1/3}$fm$\sim 1.1 \times (110)^{1/3} \sim 5$ fm)– and we see immediately that weak-interaction physics is expected to be important at atomic numbers greater than 50. So the interesting question this brings into play – and I heard many interesting comments on this – is whether we get a free decay or whether we get capture; and even if the capture does begin to dominate life, there still seems to be a useful branching ratio of competing processes even for heavier elements. But it is an interesting question how all this is modified by strictly atomic and environmental effects.

Setting the weak interaction physics aside for a moment, we come to the main thrust of this meeting – the Coulomb physics part. First: turning to the fundamentals of the capture (in the non-nuclear sense) process; I was much taken by the remarkable oscillatory behavior with atomic number of the capture probabilities. One cannot help but think, when one sees an oscillatory structure like this, that the shell structure of the atom has to play some kind of role even though in terms of sheer electron density this structure is never very strikingly evident. I don't know whether this is true or not, but in any event, the way in which the muon gets down to its final state – this is the link with what Dr. Bianconi was telling us – is going to involve a number of processes which in the electronic context have actually been well studied: I'm referring to the known shake-up and shake-off processes that occur in other spectroscopies. One is asking quite fundamental questions, namely in the process of the downward cascading, where exactly does the energy go in reaching the fundamental state? This is very complex, in at least the photoexcitation case; it is a non-trivial many-body process.

But suppose that we do have muons in the vicinity, say, of an atom. We have been discussing applications in this meeting, and the focus has been on the fact that the muon has a very different mass from the electron, it possesses a spin, and it can possess a charge of either sign. The mass difference is especially interesting: it is intermediate between the electron and the proton. We can see from trivial arguments that if we can localize the muon in a range d, we must suffer (independent of whatever forms of bonding we might suspect there to be) an energy localization that cannot be less than

$$e_{loc} \approx h^2/m_\mu d^2 \approx (h^2/m_e e^2)(m_e/m_\mu)(a_o/d)^2(e^2/a_o)$$
$$= (m_e/m_\mu)(a_o/d)^2 Ry \approx 10^{-3} Ry$$

This is just the uncertainty principle at work. The answer is a pretty large energy. If we use the same argument, for instance, for an atom in a solid, we get a couple of orders down from this. The estimate tells us immediately that we must always expect substantial zero-point energy. A muon must have a lot of kinetic energy, even when we get what we might call a nominally trapped state. And that, I suspect, is why the positive muon is such an efficient quantum diffuser in ordinary metals. But, I think, it is probably worth reminding ourselves from time to time that when the muon has decided to settle at some place in the solid, just before the moment of trapping it is not at all a quiescent object. The dynamics of the muon, and what it might be sampling in the course of its dynamical excursions, look to me quite interesting.

With respect to its spin – well, having not studied it too much in the past, my personal view is that the muon spin resonance work is just magnificent; it is giving us a very detailed look at states of magnetic order. And it seems to me to be such a beautiful technique that I can't help wondering whether it might be extended to states of *near* magnetic order, that's to say to highly fluctuating magnetic states. For instance, consider ^{3}He (it must seem slightly outrageous to suggest ^{3}He, as it must be quite awful to envisage the associated refrigeration problems!); but ^{3}He, as we know, has a superfluid state (it's a Fermi system). It has no intermediary coupling like phonons, and there seem to be pairs because of the importance of spin fluctuations. It becomes an even more interesting material to "re-consider" because if the same mechanism happens to be operating in high T_c compounds (namely if there is pairing without a phonon intermediary) it is a very plausible system to start thinking about more seriously in terms of a model for these other situations. Extensions to near magnetic states of order in other systems are also plausible for the menu; might we possibly be able to look finally for not just the effect of the spin itself but the effects of the spin if it were to couple to, say, an orbit of the muon?

Now, where do we seek such an orbit? Well, it seems to me there has been a slight asymmetry in this meeting; there appeared to be far more of a preference for negative muons than for positive muons. Let me suggest something which may be a little extreme, but its purpose is to provoke at least some consideration of the possibility of muons moving in orbits with more conventional length scales (say electronic scales). The basic energy scale that we see is (a_μ=muon Bohr radius):

$$e^2/2a_\mu = (m_\mu/m_e)(e^2/2a_o) \approx 200 \text{ Ry}.$$

Now, could we propitiously reduce such an energy scale down to the natural scale of condensed matter? Could we by choice of material get this energy into the sub-eV range? For example, what value of the dielectric constant would we need in order to reduce this energy scale to an energy typical of condensed matter physics? As we remember from semiconductors physics this scale goes with the square of the dielectric constant:

$$(1/\epsilon^2)(m_\mu/m_o)(e^2/2a_o).$$

In this argument we need, however, an orbit size large enough that the very concept of the dielectric constant makes sense. We have to average over several atomic sizes to do this. If we took, for instance, a system whose dielectric constant was larger than 20 or so, we begin to get into this range. But of course a very important criticism could and should be raised: Would the muon survive in such an orbit? Wouldn't it decay immediately by scattering, or other quasi-fatal events? Possibly, but suppose that the system was well chosen so that particle-hole scattering, which might probably be the most serious source of scattering, could be diminished. This is a highly material specific question but it may be that there is a window somewhere in the intimidating list of constraints that generally limit the possibilities. The point is simply that muon spin resonance has been a brilliant probe of magnetic order and magnetic fluctuations as I've said earlier. But going from spin to charge it would be rather interesting to see if we would get muons working for us in orbits where they exploit, for example their positive charge (around, say, negative ions). Although for technical reasons these suggestions might eventually turn out to be remote, I think systems that begin to develop fluctuations and order, of say, the ferroelectric type, are worth musing over, at least over a glass or two of Erice Marsala.

Let me now finish up with the three main aspects, I would say, of condensed matter: the fact that we have electronic states of order of various types; the fact that we have magnetic states of order of various types, and finally that probably one of the major industries in condensed matter physics is the study of structural transitions and their associated dynamics. Why it is, for instance, that given such a simple fundamental Hamiltonian, nature chooses a crystal as a ground state, and why does it often appear to get there in a step-wise fashion? When we freeze a liquid we most often go to a body-centered-cubic (bcc) state. I should now parenthetically say that between those two states (the bcc with a Wigner-Seitz polyhedron of almost spherical appearance, and the liquid with a completely spherical cell) we may now find a quasicrystal. But we usually first find the bcc state, and then by a progression we seem to encounter as we go down in temperature, more and more asymmetric structures, (and with them a variety of interesting structural phase-transitions). So let us just think of a couple of these.

The metal-insulator transition is probably the most fundamental of all; it divides the two great classes of condensed matter physics, the conductors and the non-conductors. Can we look at this division, just a little bit more, with the muons that we hope to get in copious numbers and at low energies? We might mention, for instance, muonium in a variety of contexts. But specifically what about muonium in a metallic environment or, let us say, in a near metallic environment such as we may get if we move the system from the insulating to the metallic state by the application of some parameter, or a change of some parameter. Muon spin resonance near metal-insulator transitions (maybe it has been done, I do not know) is a possible probe if the spin orbit coupling notion mentioned above makes sense. The main point I am raising here is that the especially interesting aspect of the μSR is the local information that we get. For instance, might muon probes possess the ability to start distinguishing between percolation-type transition to the metallic state, band overlap types and so forth? It is worth another few hours over marsala.

Naturally we always expect the possibility of a bound state in Coulomb field. But in the presence of screening the main effect is to take the Coulomb potential to something like the Yukawa potential. The Yukawa parameter is connected immediately with the density of electrons. It is well known, once again, that this potential can possess a bound state provided the screening is not too strong. Picking up the bound state in a metallic environment is therefore not ruled out *a priori*. The constraint, of course, is in the overall density and as we move the density lower and lower, we expect localized or insulating states, but we do not achieve them continuously. We go abruptly, and examining the nature of this abrupt transition possibly with the muons is something we may think about. So my last example here is simply raising the possibility of muons as structural probes, both electronic structure and the real space structure it largely determines.

We heard from Dr. Maier about muons in semiconductors, and especially the dramatic effects of temperature. The interesting point about the semiconductors – most of the simplest ones at any rate – is that if we raise the temperature up towards melting we take them immediately from the semiconducting state to the metallic state. Silicon is a very good example. Of course this happens at quite high temperature. But the point here is that the structural state is deeply connected with the electronic properties. They behave in a very interconnected fashion. I can't help wondering whether we can get at this part of the problem using muon probes. For example, an interesting piece of physics which came out recently suggests that the process of melting is now being nucleated around impurities, around grain boundaries, dislocations and defects, and most especially at or near surfaces. From EXAFS, for instance, there is a lot of evidence that the melting process now is one which is nucleated by some dramatic defect, and the most dramatic one we usually have in a solid is its termination (i.e. the surface)! And I wonder again whether, for instance, by carefully choosing a site (maybe we have to implant it) we can now probe even the structural aspects of the solid-solid transitions using muon techniques.

3. Belief!

So, those are just a few thoughts, mainly blown in during the scirocco of the other day. I was also musing on whether there might be a question of belief, that is whether we believe in this muon technique. I have seen ample evidence today and in the last few days, to convince me that the subject has an interesting future. The issue of belief in physics was the subject of some commentary in Physics Today a few months ago. People were debating the early state of the Universe, and of course, they were making some statements, some pretty solid and firm statements, about what was going on at 10^{-32} sec after the Big Bang. We might venture to surmise that certain elements of belief enter here. And when it all ended up – there was some correspondence and a rather nice cartoon in Physics Today. In this cartoon a kind of fundamentalist or revivalist preacher was asking people to just believe in what has been advanced here: "Brothers and sisters, 10^{-33} seconds after the Big Bang the heat was enormous" exhorted the preacher, and then the believers incanted "Verily it was over 10^{32} degrees". And the preacher went on: "Matter and Antimatter arose, and the universe was filled with particles ..." and the believers echoed "Hallelujah, they annihilated each other". Then they chanted "Amen! Quarks and Gluons" and even "Yea, Leptons!".

Well I thought that, as a tribute to our two able organizers, I would take the liberty of just adding "... and later still we had Muons, and not only had we Muons, but they were Muons of both signs!". I take that as a great sign for the future and want to thank again Giorgio and Hubert for organizing this meeting.

Index of Contributors

Springer Proceedings in Physics

Managing Editor: H. K. V. Lotsch

MIX
Papier aus verantwortungsvollen Quellen
Paper from responsible sources
FSC® C105338

If you have any concerns about our products,
you can contact us on
ProductSafety@springernature.com

In case Publisher is established outside the EU,
the EU authorized representative is:
Springer Nature Customer Service Center GmbH
Europaplatz 3, 69115 Heidelberg, Germany

Printed by Libri Plureos GmbH
in Hamburg, Germany